JACKALS, GOLDEN WOLVES, AND HONEY BADGERS

This book explores the fascinating and complex lives of the honey badger, the African jackals (black-backed and side-striped), African golden wolves, and Eurasian golden jackals. In recent years, interest in these creatures has grown exponentially, through wildlife documentaries and media clips showing the aggressive, fearless, and tenacious behaviour of the honey badger, with jackals often presented in a supporting role.

Written by renowned journalist and educator Keith Somerville, this accessible volume includes historical narratives, folklore, and contemporary accounts of human–wildlife relationships and conflicts. It traces the evolution of the species; their foraging and diet; the development of their relationships with humans; and their commensal, kleptocratic, and symbiotic relationships with other carnivores, raptors, and birds. It also charts the recent expansion in European jackal numbers and ranges, now including as far west as the Netherlands and as far north as Finland.

Blending historical observations by non-scientists, colonial officials, administrators, and early conservationists with contemporary scientific accounts, it presents a new multidisciplinary approach that will interest researchers, scientists, and students in wildlife conservation, human–wildlife relations, zoology, biology, and environmental science.

Keith Somerville is a Member of the Durrell Institute of Conservation and Ecology at the University of Kent, UK, where he is a professor at the Centre for Journalism. He is a Senior Research Fellow at the Institute of Commonwealth Studies, a Fellow of the Zoological Society of London, UK, and a Member of the IUCN CEESP/SSC Sustainable Use and Livelihoods Specialist Group, and a fellow of the Royal Historical Society.

Routledge Studies in Conservation and the Environment

This series includes a wide range of inter-disciplinary approaches to conservation and the environment, integrating perspectives from both social and natural sciences. Topics include, but are not limited to, development, environmental policy and politics, ecosystem change, natural resources (including land, water, oceans and forests), security, wildlife, protected areas, tourism, human–wildlife conflict, agriculture, economics, law and climate change.

Threatened Freshwater Animals of Tropical East Asia
Ecology and Conservation in a Rapidly Changing Environment
David Dudgeon

Conservation Effectiveness and Concurrent Green Initiatives
Li An, Conghe Song, Qi Zhang, and Eve Bohnett

Religion and Nature Conservation
Global Case Studies
Edited by Radhika Borde, Alison A Ormsby, Stephen M Awoyemi, and Andrew G Gosler

Jackals, Golden Wolves, and Honey Badgers
Cunning, Courage, and Conflict with Humans
Keith Somerville

Case Studies of Wildlife Ecology and Conservation in India
Edited by Orus Ilyas and Afifullah Khan

For more information about this series, please visit: www.routledge.com/ Routledge-Studies-in-Conservation-and-the-Environment/book-series/ RSICE

JACKALS, GOLDEN WOLVES, AND HONEY BADGERS

Cunning, Courage, and Conflict with Humans

Keith Somerville

Routledge
Taylor & Francis Group

LONDON AND NEW YORK

earthscan
from Routledge

First published 2023
by Routledge
4 Park Square, Milton Park, Abingdon, Oxon OX14 4RN

and by Routledge
605 Third Avenue, New York, NY 10158

Routledge is an imprint of the Taylor & Francis Group, an informa business

© 2023 Keith Somerville

British Library Cataloguing-in-Publication Data
A catalogue record for this book is available from the British Library

Library of Congress Cataloging-in-Publication Data
Names: Somerville, Keith, author.
Title: Jackals, golden wolves, and honey badgers : cunning, courage, and conflict with humans / Keith Somerville.
Description: First edition. | Abingdon, Oxon ; New York, NY : Routledge, 2023. | Series: Routledge studies in conservation and the environment | Includes bibliographical references and index.
Identifiers: LCCN 2022022714 (print) | LCCN 2022022715 (ebook) | ISBN 9781032059082 (hbk) | ISBN 9781032059075 (pbk) | ISBN 9781003199793 (ebk)
Subjects: LCSH: Jackals. | Wolves. | Honey badger. | Human-animal relationships.
Classification: LCC QL737.C2 S575 2023 (print) | LCC QL737.C2 (ebook) | DDC 599.77—dc23/eng/20220624
LC record available at https://lccn.loc.gov/2022022714
LC ebook record available at https://lccn.loc.gov/2022022715

ISBN: 978-1-032-05908-2 (hbk)
ISBN: 978-1-032-05907-5 (pbk)
ISBN: 978-1-003-19979-3 (ebk)

DOI: 10.4324/9781003199793

Typeset in Bembo
by codeMantra

CONTENTS

FIGURES

ABBREVIATIONS

BCE	Before Common Era (formerly BC)
BP	Years before present time. The date for "present" being 1950
CE	Common Era (formerly AD)
CITES	Convention on International Trade in Endangered Species of Wild Fauna and Flora
CKGR	Central Kalahari Game Reserve, Botswana
DWNP	Department of Wildlife and National Parks, Botswana
EWT	Endangered Wildlife Trust (South Africa)
GKE	Greater Kafue Ecosystem
GLTFCA	Greater Limpopo Transfrontier Conservation Area, South Africa, Mozambique, and Zimbabwe
GLTP	Greater Limpopo Transfrontier Park, South Africa, Mozambique, and Zimbabwe
GMA	Game Management Area
IUCN	International Union for the Conservation of Nature
ka	thousand years ago
KAZA	Kavango-Zambezi Transfrontier Conservation Area
KNP	Kruger National Park, South Africa
KTP	Kgalagadi Transfrontier Park, Botswana and South Africa
KZN	KwaZulu-Natal province, South Africa
ma	million years ago
NCA	Ngorongoro Conservation Area, Tanzania
NP	National Park
SNP	Serengeti National Park, Tanzania
TFCA	Transfrontier Conservation Area
UNDP	United Nations Development Programme
VOC	Dutch East India Company

WAPC	Wild Animal Poisoning Club (South Africa)
WildCRU	Wildlife Conservation Research Unit, University of Oxford
WWF	World Wide Fund for Nature
ZAWA	Zambian Wildlife Authority
ZCP	Zambian Carnivore Programme
ZimParks	Zimbabwe National Parks
ZSL	Zoological Society of London

GLOSSARY

Eocene   56 to 33.9 million years ago, henceforth ma.

Pliocene   from 5.33 to 2.6ma.

Oligocene   about 33.7–23ma. The climate cooled and many modern mammals evolved.

Miocene   23–5.3ma, preceding the Pliocene.

Quaternary period   the last 2.6 million years.

Pleistocene   part of the Quaternary, stretches from 2.6ma to 11,700 years ago.

Early Pleistocene   from 2.6ma to c781,000 years ago.

Middle Pleistocene   from c781,000 to 126,000 years ago.

Late Pleistocene   from 126,000 to 11,700 years ago.

Holocene   from 11,700 to present.

GLOSSARY OF OTHER TERMS

Anthropocene   a term often used to apply to the current phase of the Holocene.

Anthropogenic   effects related to human activity.

Apex predators   the predatory species at the top of a food chain.

Boma   an enclosure, usually surrounded by a fence of thorn bush or other fencing materials, set up to protect a homestead which would include livestock.

Commensal   a biological interaction in which members of one species gain benefits, while those of the other species are neither benefitted nor harmed.[1]

Hominid/Hominidae   member of the group consisting of modern and extinct humans, great apes, and their ancestors.

Hominin/Homininae   a member of the group of species (genus) that are human or human ancestors. This includes the *Homo* species and all of the Australopithecines and other ancient forms like Paranthropus and Ardipithecus.

Mesopredators   a mid-ranking predator, which typically preys on smaller animals. Jackals, golden wolves, and honey badgers fall into this category, except in the absence of larger carnivores, where they assume the roles of apex predators.

Mesopredator release   when populations of medium-sized predators rapidly increase in ecosystems after the removal of larger, apex carnivores.

Neolithic   relating to the Later Stone Age. This period, also sometimes called the New Stone Age, was characterised by use of stone tools produced by polishing or grinding.

Oldowan   the stone-tool making culture of hominid/hominin ancestors, from 2.5 to 1.7/1.5mya. First defined by Louis and Mary Leakey, following tool and fossil discoveries at Olduvai Gorge, Tanzania.[2]

Palaeolithic period   the age in which early humans made chipped-stone tools, sometimes called the Old Stone Age. The period started around 2.5mya.

Plasticity   the ability to change and adapt. This may refer to diet, foraging and hunting techniques, and use of habitat or timing of activity to avoid threats.

Scats   faeces, particularly of carnivorous animals.

Symbiosis   any type of a close and long-term biological interaction between different species than be mutually beneficial, commensal, or even parasitic, not excluding predation or competition.[3]

Taxonomic   relating to systems for naming and organising things, especially plants and animals into groups with similar qualities.

Human Evolution Chronology[4]

55ma

First primates evolve.

15ma

Hominids (great apes) split off from ancestors of the gibbon.

8mya

Chimpanzee and human lineages diverge from gorillas, with the evolution of hominids put between 8 and 5ma.

7ma

Around 10–7ma, chimpanzees and hominins diverge, with the last common ancestor of both chimpanzees and humans living in this period. *Sahelanthropus tchadensis*, found in Chad, may be the first of the evolving hominin ancestors to diverge from hominids.

4.4ma

Ardipithecus appears: an early "proto-human" with grasping feet.

4ma

Australopithecines appeared, with brain the size of a chimpanzee's.

2.3ma

Homo habilis first appeared in Africa.

2yma

Homo erectus appears. Fossil evidence shows the species has dispersed across the globe.

1.85ma

First "modern" human hand evolves, according to fossil evidence.

1.6ma

Hand axes appear in fossil record – they are a major technological innovation.

800,000 years ago

Evidence of use of fire by *Homo* species.

350,000–200,000 years ago

Fossils believed to be among the first *Homo sapiens*, called Omo I, discovered in south-western Ethiopia. New research suggests that the emergence of *Homo sapiens* could be as early as 350,000 years ago.[5]

300,000 years ago

Evidence of early *Homo sapiens* in Morocco.

Neanderthals (*Homo neanderthalensis, Homo sapiens neanderthalensis*) evolve and by 200,000 years ago spread across Europe and Asia.

60,000 years ago

Modern human migration from Africa that led to modern-day non-African populations.

45,000–40,000 years ago

Homo sapiens move into Europe, followed by the disappearance of Neanderthals.

Notes

1 E.O. Wilson (1975) *Sociobiology: The New Synthesis*, Cambridge, MA: Harvard University Press, p. 354.
2 Ibid.
3 A. Douglas (2010) *The Symbiotic Habit*, Princeton, NJ: Princeton University Press, pp. 5–12.
4 H. Devlin, Tracing the tangled tracks of humankind's evolutionary journey *Guardian* 12 February 2018, https://www.theguardian.com/news/2018/feb/12/tracing-the-tangled-tracks-of-humankinds-evolutionary-journey accessed 13 February 2018.
5 Celine M. Vidal et al. (2022) Age of the Oldest known *Homo Sapiens* from Eastern Africa, *Nature, Nature, Age of the Oldest known Homo Sapiens from Eastern Africa | Nature*, accessed 13 January 2022.

INTRODUCTION

Why honey badgers and jackals

The origins of this book lie in sightings of black-backed jackals and honey badgers in the Central Kalahari Game Reserve (CKGR) of Botswana and regular encounters with jackals (black-backed, side-striped, and golden wolves) in southern, central, and east South Africa and honey badgers in Botswana, Namibia, South Africa, and Zimbabwe over more than 40 years.

I was fascinated by the relationship between the badger and the jackal in the CKGR, with pale chanting goshawks thrown in for good measure. This led me to research what was known about their interesting interactions – be they symbiotic, commensal, or kleptoparasitic – and then on to delving into the nature of the jackal species (plus golden wolves) and the honey badger, their histories, interactions with humans, the persecution of them, and the prospects for them in a human-dominated world increasingly hostile to wildlife.

Although I had seen honey badgers in Zimbabwe, Namibia, and Botswana before and jackals/golden wolves on every wildlife-related trip to Africa from Malawi in 1981 to the present, I date the origins of this book to 4.39pm on 25 June 2015 in Botswana's CKGR. Driving back to Tau Pan Lodge from a trip to Deception Valley, we saw clouds of dust being kicked up in the air, with the slowly setting sun shining orange through the dust. As we got closer, we could see it was a honey badger digging vigorously. Behind it, watching its every move, was a black-backed jackal, and, in a tree above, a pale chanting goshawk. We edged closer and settled to watch. The badger was not the least bothered by the proximity of the jackal and simply ignored the raptor. It just kept digging; then it suddenly grabbed at something, while a small animal – possibly a gerbil, rat, or one of the larger species of mouse – squeezed past the badger only to be snapped up by the jackal. The badger emerged, chewing on something, and proceeded to its next digging spot. The jackal came a little too close, sniffing at the badger, which whirled round and threw itself at the jackal grunting aggressively. The

DOI: 10.4324/9781003199793-1

jackal retreated and then moved back towards the badger, eliciting another lunge and snarling grunt. The jackal then kept its distance, while the badger started digging again, and the goshawk hovered overhead. This was repeated several times, though with no signs of the badger, jackal, or goshawk getting anything more to eat. Two years later, again in the CKGR, quite near to Tau Pan, we saw a similar jackal-badger bout, with similar results (Figure 1.0).

This led me to write a short piece for a University of London School of Advanced Study website on the political economy of the honey badger and its relationships with the jackal, the goshawk, and other animals with which it interacted.[1] I then started studying the honey badger's interactions with a variety of carnivores, its much-discussed relationship with the honeyguide bird, and its reputation for fearlessness, aggression, and taking on carnivores or attacking ungulates many times its size, as well as venomous snakes and scorpions. Honey badgers have a reputation for aggressively driving off predators and potential threats and even successfully challenging lions and other carnivores, such as hyenas, for their kills. This toughness led the South African Defence Force under apartheid to adopt the name "ratel" for one of their armoured fighting vehicles – the honey badger is also called a ratel, particularly in South Africa and India.

I was hooked, and this book was the inevitable next step – not just about the badger, but it would also include jackals. I had become very interested in the jackal family and its close relative, the golden wolf. I had seen them far more often than badgers, and the more I read about them, the more they seemed worthy of close attention and inclusion. They had admirable-seeming family relationships – abiding ties between mated pairs and the altruistic but self-improving nature of those offspring who remained with the adults to help bring up the next generation of pups. The ability of the black-backed jackal to survive about 350 years

FIGURE 1.0 Jackal and honey badger in the Central Kalahari Game Reserve, Botswana. Copyright: Keith Somerville.

of concerted persecution and killing by guns, dogs, traps, and poison was more than worthy of respect and investigation. The history of the golden wolf – once a jackal, now a wolf – at the centre of Egyptian rituals regarding death was similarly fascinating as was its recategorising as wolf, while the rise, recovery, and range expansion of the Eurasian golden jackal is a story in itself.

Courage, cunning and aggression – myths and realities of honey badgers in the human imagination

In recent years, the legendary fearlessness and intelligence of the honey badger have become the subject of documentaries, clips on YouTube, and the adoption of the nickname "honey badger" by a rugby player known for his toughness and aggression. Carter et al. referenced this in their comment that

> [t]he honey badger is an intriguing animal with very particular behavioural characteristics … [often called] the most fearless animal in the world … an interesting animal as there are many myths surrounding its nature and behaviour. It is well known as a fearless creature that can be quite aggressive when threatened.[2]

It has also been called the "meanest animal in the world," "pound for pound, the most powerful creature in Africa," and the "bravest animal on earth."[3]

Most of the legends of the honey badger centre on its supposed invincibility, its fearsome aggression that will deter lions, leopards, hyenas, wild dogs, and cheetahs. The San Diego Zoological Society examined many of the stories and judged as myths the alleged invincibility, noting that lions do occasionally kill honey badgers and both leopards and hyenas hunt them for food and that they do become sick or even die as a result of substantial numbers of bee stings and can succumb to snake venom – though they do have an amazing survival rate from cobra, mamba, and puff adder bites.[4] They listed a number of TV and film documentaries that have contributed to the badger's indestructible reputation:

- *Snake Killers: Honey Badgers of the Kalahari* – the 2001 National Geographic film detailing the lives of wild honey badgers; three years' worth of footage shot in conjunction with the first intensive study of these amazing carnivores
- *Ultimate Honey Badger* – the 2014 National Geographic film follows the life of Badgie as she learns how to survive on her own in Africa
- *Honey Badgers: Masters of Mayhem* – the 2014 PBS *Nature* program follows a team of South African researchers in search for the truth behind the highly intelligent animal.[5]

In 2019, Oosthuizen and Felmore wrote in *Africa Geographic* about the courage and tenacity of the honey badger, starting with the Afrikaans expression, *so taai soos 'n ratel* – which translates to "as tough as a honey badger."[6] Their appearance, stocky with powerful claws and a confident, bouncing gait, adds to the reputation, which is then reinforced in the popular view by video clips on YouTube shot by tourists in southern Africa, showing a mother badger rescuing her cub from an adult leopard, another showing two badgers fighting off a group of six young lions, and a third which shows a badger being crushed by a large python, only for the badger to extricate itself from the python's coils and drag it off into the bush, after a tug of war with two black-backed jackals.[7] The honey badger's gurking, grunting, booming roars when threatened or attacked add to its tough, aggressive, indestructible image.[8] In addition, the honey badger has, what one writer described as, "a secret weapon to defend itself."[9] Situated at the base of its tail are two anal glands that can squirt out a foul-smelling liquid, whose odour can be detected 40m away.[10] The animals generally expel the substance to mark their territory, but they will also release a "stink bomb" when threatened or frightened. Some naturalists have speculated that the smell can be used to pacify bees when the badgers raid nests.[11]

Honey badgers are credited with being highly intelligent as, unlike most non-primate mammals, are having the ability to use tools.[12] Captive honey badgers have been known to work together to unlock gates and use rocks, a rake, mud, and sticks to escape from their enclosure, according to a documentary produced by BBC, called *Honey Badgers: Masters of Mayhem*[13] – watching the documentary, the discerning viewer might be somewhat suspicious that rocks, implements, and logs have been placed in such a way to encourage the honey badger to use them. This documentary and others, such as Nat Geo Wild's *Grit: Honey Badger Tough*,[14] have played an interesting role (akin to that of natural history documentaries about meerkats[15] – leading to series such as *Meerkat Manor*[16] and the irritating *Compare the Market* ads with a Russian-accented meerkat)[17] – in bringing honey badgers to a wider public. This has begun to establish honey badgers as wildlife superheroes and loveable rogues – with one Australian rugby union player, Nick Cummins, being nicknamed the "honey badger" because of his aggression and general toughness.[18]

One popular almost comic book celebrates this fearsome reputation and is titled *Randall's Guide to Crazy Nastyass Animals: Honey Badger Don't Care*. It includes the passage, with apologies for the language, "This honey badger's vicious as fuck and *will* bite your balls off. You think I'm kidding? Well, I'm not! Honey badger don't give a shit."[19]

In a similar vein, there is a comic-style book called *Who Would win? Hyena vs Honey Badger*.[20] This picture book with a jokey narrative asks who would win in a fight between a honey badger and a spotted hyena. After giving some basic and reasonably accurate simple facts about each animal, the book sets out a fictional fight between them that sees the badger effectively triumph. It is another example of the popularisation of the fearlessness and somewhat exaggerated fighting skills of the honey badger. They are fearless and tough, but they do get killed by larger predators. A hyena could kill a honey badger, but most of the time, a show of aggression and some bites from a badger seem to show why larger carnivores think fighting and killing them is not worth the effort.

Jackals in myth, religion, and popular representations

One of the leading researchers of jackals and their behaviour noted that, unlike the almost superhero representation of honey badgers, jackals of all species (and here I include golden wolves) were long seen as "as skulking scavengers with base and reprehensible behaviour."[21] This low-life image as an eater of carrion, a thief who steals brazenly from lion or cheetah kills, and a sneaky killer of lambs and goats, is belied by scientists' accounts, based on long observation, of a clever, courageous and family-oriented mammals where "males and females form long-term pair-bonds, often lasting a life time. They hunt together, share food, groom each other, jointly defend their territory, and provision and defend their pups together. Some of the pups stay with their parents and at the age of one year help raise the pups."[22] They undoubtedly do kill domestic stock – and are fiercely persecuted for it, but they are much more than scavengers and stock-killers, often performing ecological cleansing roles by cleaning up diseased, rotting carcasses and may also help some stock farmers by killing grain-eating or grass-mowing rodents.

In historical terms, jackals, as they were called, though we now know them as golden wolves, were worshipped. Their disposal of the dead – animals and even people's bodies – being celebrated in the way they were characterised by the Egyptian dynasties as integral to human transition from life to the afterlife. Anubis, the jackal-headed deity, was the god of the afterlife and of mummification rituals – one of the oldest deities venerated by Egyptians, appearing on tombs of the First Dynasty (c. 3150–2890BCE) onwards, and believed to be a replacement for the earlier pre-Dynastic jackal God, Wepawet.[23] There was a cult around first Wepawet and then Anubis, making the jackal sacred as an incarnation of the deity. He is depicted mainly as a black canine with jackal/dog-like pointed ears and muzzle, usually with a muscular man's body.[24] Anubis was the son of the god of the sun, sky, and of kings, Ra, and was central to human death, mummification, judgement of the soul, and entry into the afterlife – which was an incredibly important role in a cultural system obsessed with death.

In India, jackals appeared as servants of major deities and are mentioned in the religious works that emerged from and shaped Indian culture – the *Brahmanas*, *Mahabharata*, and *Ramayana*.[25] Hindu deities have incarnations or avatars in the form of animals, from fish to elephants.[26] The goddess Durga (aka Devi or Shakti) is represented as a protective mother and goddess of war, strength, and protection.[27] She is usually depicted riding a lion or a tiger and has eight or ten arms, each holding a special weapon of one of the gods.[28] Wilkins notes that in some of the accounts of Durga, a jackal is depicted as her.[29] Other early accounts of animals in Indian oral history and cultural systems come from the *Pañcatantra* (Sanskrit: पञ्चतन्त्र, also called the *Five Treatises*), a collection of interrelated animal fables in Sanskrit verse and prose, dating from about 200–300CE but based on older oral traditions.[30] Jackals occur regularly in a number of fables and tales in the *Pañcatantra*. The names of the two jackals who are referred to most often, Karataka and Damanaka, are described by the translator Chandra Rajan in his foreword as "Wary and Wily" – with virtues or vices of intelligence and guile ascribed to them. This representation in oral history and literature establishes in Indian culture the reputation of the jackal for cunning and a perceived idea of it adapting behaviour to the presence of humans.[31] Jackals are often depicted negatively but at other times as clever, with a certain courage and resourcefulness, unlike Rudyard Kipling's much later depiction of the jackal Tabaqui in *The Jungle Book*, as a low, snivelling, treacherous wretch – "Tabaqui, the Dish-licker ... the wolves of India despise Tabaqui because of the truth he runs about making mischief ... and consuming rags and pieces of leather from the village rubbish-heaps."[32]

In Western culture and religion, jackals generally get a poor press – as in the Bible, where they represent desolation and despair (see Chapter 5). The depiction of cunning, scavenging jackals is common, even finding its way into Charles Dickens's *A Tale of Two Cities* as a description of the young but careless lawyer Sydney Carton, with Chapter 5 of the book titled "The Jackal."[33] The jackal motif of evil or cunning was resurrected in the title of Frederick Forsyth's novel about the attempt to assassinate President de Gaulle of France, *Day of the Jackal*.[34] A more encouraging development has been the publishing of a children's book depicting a jackal family in a positive and accurate manner, Betty Dinneen's *The Family Howl*.[35] This more even-handed approach was replicated in a number of TV films, such as *The Year of the Jackal* and *Jackals Out of Africa – The Secrets of Nature*, giving the characterisation of the jackals hunting seals on the Namibian coast.[36]

Chapters and acknowledgements

The aim of the book is to set out a factual and scientifically supported narrative of the lives of honey badgers, jackals, and golden wolves, with illustrations from folklore, religion, and the accounts of hunters, travellers, amateur naturalists, and game wardens – often standing on the shoulders of the scientists who have invested so much time and effort into studying the animals.

Chapter 1 is about jackals and golden wolves – their distribution, social behaviour, breeding, feeding, and relations with other carnivores and with humans; Chapter 2 looks at their evolution; Chapters 3 and 4 at the early history of human interactions and images of jackals and golden wolves in Africa; Chapter 5 looks at the history and current status of the Eurasian golden jackal, particularly its expansion into central, western, and northern Europe; Chapters 6 and 7 bring the history of Africa's jackals up to date; Chapter 8 is about the distribution, social behaviour, breeding, feeding, and relations of honey badgers with other carnivores and birds and with humans; Chapter 9 deals with the evolution and history of the honey badgers as seen by humans; and, finally, Chapter 10 covers the honey badger in today's world, especially its conflict with beekeepers.

I would like to thank a number of conservation scientists, palaeontologists, and experts on traditional medicine who have generously given their time and expertise to help me:

> Dr Colleen Begg of the Niassa Carnivore Project for her guidance and advice and for her invaluable honey badger studies and her willingness to read over the honey badger chapters; Dr Andrew Loveridge for reading and improving Chapter 1; Dr Ross Barnett for casting an expert eye over Chapter 2; Nick Huisman, European Wilderness Society, for his assistance with Chapter 5; Professor Nicoli Nattrass, Institute for Communities and Wildlife in Africa (iCWild), University of Cape Town, for her advice and reading Chapters 6 and 7; Liz Campbell, WildCRU (University of Oxford), for her help on the issue of what is and is not a jackal or wolf; Rob Davies of Nottingham Trent University, based in Kasungu National Park in Malawi; Dr Amy Dickman, WildCRU and Ruaha Carnivore Project; Professor Graham Kerley, Director, Centre for African Conservation Ecology & Department of Zoology, Nelson Mandela University, for his provision of a number of invaluable papers; Dr Claudio Sillero-Zubiri, WildCRU; and, Dr Viv Williams, University of Witwatersrand, for her invaluable assistance in tracing data on traditional medical uses of jackals and honey badgers in South Africa; and last but not the least, Alaitetei Laltaika, Ngorongoro Conservation Area.

At Routledge, Matthew Shobbrook and my editor have been of huge help turning my scrawl into a proper book.

Most of all, I must thank my wife, Liz, for her patience and her help with German translations and grammar, and my son, Tom, for keeping my feet on the ground and just being himself.

Notes

1 K. Somerville (2018) The political economy of the honey badger – just don't 'ratel' its cage, *Talking Humanities*, https://talkinghumanities.blogs.sas.ac.uk/2018/05/24/political-economy-honey-badger-just-dont-ratel-cage/ accessed 21 June 2021.
2 S. Carter et al. (2017) The honey badger in South Africa: Biology and conservation, *International Journal Avian & Wildlife Biology*, 2(2), https://medcraveonline.com/IJAWB/the-honey-badger-in-south-africa-biology-and-conservation.html accessed 14 October 2021, no page numbers.
3 Cited by Colleen Margaret Begg (2001) Feeding ecology and social organisation of honey badgers (*Mellivora capensis*) in the southern Kalahari, University of Pretoria doctoral thesis November 2001, p. 1.
4 San Diego Zoo (no date) *Ratel/Honey Badger (Mellivora capensis) Fact Sheet: Population & Conservation Status*, https://ielc.libguides.com/sdzg/factsheets/ratel/population, accessed 19 October 2021.
5 Ibid.
6 N. Oosthuizen & T. Felmore (2019) Honey badger – Africa's most fearless and tenacious carnivore, *Africa Geographic*, 29 March 2019.

7 https://www.youtube.com/watch?v=hLG_Q8FJda0; https://www.youtube.com/watch?v=NvlalDNxccw; https://www.youtube.com/watch?v=JgKN3BuvC3E – all accessed 20 April 2022.

8 Oosthuizen & Felmore, 2019.

9 Ibid.

10 SANBI (no date) The honey badger, South African National Biodiversity Institute, https://www.sanbi.org/animal-of-the-week/honey-badger/ accessed 4 November 2020.

11 J. Kingdon (1979) *East African Mammals: v. 3B: An Atlas of Evolution in Africa*, San Diego, CA: Academic Press.

12 B. Panesar (2020) Honey badgers: Adorable but fierce little mammals, *Live Science*, 22 August 2020.

13 BBC Natural World (2014) *Honey badgers: Masters of Mayhem. Natural World*, https://www.bbc.co.uk/programmes/b0418x7x accessed 4 November 2020.

14 NatGeo Wild (no date) *Grit: Honey Badger Tough*, https://www.sky.com/watch/title/programme/9b951513-8112-40cf-9381-0f055aa06888 accessed 4 November 2020.

15 BBC, *Meerkats United*, IMDB, https://www.imdb.com/title/tt1860282/ accessed 4 November 2020.

16 *Maarkat Manor*, https://www.bbc.co.uk/iplayer/episodes/b006v05d/meerkat-manor accessed 4 November 2020.

17 Compare the Market, https://www.comparethemarket.com/ accessed 4 November 2020.

18 Wikipedia (no date) Nick Cummins, https://en.wikipedia.org/wiki/Nick_Cummins#:~:text=Cummins%20is%20also%20known%20by,season%20of%20The%20Bachelor%20Australia accessed 4 November 2020.

19 Randall (2011) *Randall's Guide to Crazy Nastyass Animals. Honey badger don't care*, Kansas City, MO: Andrew McNeel Publishing, p. 1.

20 J. Pallotta (2018) *Who Would Win? Hyena vs Honey Badger*, New York: Scholastic Inc.

21 P.D. Moehlman (1987) Social Organization in Jackals: The complex social system of jackals allows the successful rearing of very dependent young, *American Scientist*, 75(4), 366–75, p. 366.

22 Ibid.

23 J.J. Mark (2016) Anubis, I world history encyclopedia, https://www.ancient.eu/Anubis/ accessed 30 November 2020.

24 Ibid.

25 J. Keay (2010) *India a History from the Earliest Civilisations to the Boom of the Twenty-First Century*, London: Harper Collins, Updated edition, p. 2.

26 Ibid.

27 D.R. Kinsley (1989) *The Goddesses' Mirror: Visions of the Divine from East and West*. New York: State University of New York Press, pp. 3–4.

28 A. Gaur (2018) Durga, https://britannica.co./topic/Durga accessed 28 December 2020.

29 W.J. Wilkins (1900) *Hindu Mythology*, New Delhi: Rupa Publication (reprint in 2013 of 2nd edition), p. 423.

30 S. Visnu (2006) *The Pancatantra*, London: Penguin Books Ltd. Kindle Edition.

31 Ibid., loc 262.

32 R. Kipling (no date) *The Jungle Book*, Kindle Edition, loc 70.

33 C. Dickens (no date) *A Tale of Two Cities*. Pandora's Box. Kindle Edition.

34 F. Forsyth (1971) *Day of the Jackal*, London: Arrow Books.

35 B. Dinneen (1981) *The Family Howl*, New York: Macmillan.

36 The Nature Conservancy (1991) *The Year of the Jackal*, Partridge Films/PBS, Nature – Year of the Jackal Nature Documentary – YouTube, accessed 6 January 2022; W. Paschinger (2007) *Jackals Out of Africa – The Secrets of Nature*, Jackals Out of Africa – The Secrets of Nature – YouTube, accessed 6 January 2022; Nat Geo Wild (2021) *Gangster Jackals, Jackal on the Hunt, National Geographic*, accessed 6 January 2022 – short film showing jackals hunting seal pups on Namibian coast – main interest in this short film is the Gangster sub-title.

1

JACKALS AND GOLDEN WOLVES

Dramatis personae

Dramatis personae

1. Black-backed jackal (also known as silver-backed jackals)
Order: Carnivora
Family: Canidae
Genus: Lupulella (previously classified as Canis)
Species: Lupulella mesomelas (previously rendered as Canis mesomelas)
Sub-species: Lupulella mesomelas mesomelas (Cape black-backed jackal)
and Lupulella mesomelas schmidti (East African black-backed jackal)

Related jackal species

The Sundevall side-striped jackal (*Lupulella adustus adustus*) in central and northern areas of southern Africa; Equatorial side-striped jackal (*Lupulella adustus lateralis*) in western, central, and parts of East Africa; Kaffa side-striped jackal (*Lupulella adustus kaffensis*) in the Horn of Africa, South Sudan, and north-western Kenya (being absent from arid areas of northern Kenya) and Uganda.[1]

The African golden wolf (*Canis lupaster* – with sub-species in North Africa, *Canis lupaster lupaster*; West Africa, *Canis lupaster anthus*; and East Africa, *Canis lupaster bea*) is less closely related. It was originally counted as a member of the golden jackal species of Eurasia (*Canis aureus* – with six sub-species spanning the geographical range from Europe (south-eastern, central, northern, and now parts of western Europe – see Chapter 5 for details of their expanding range) east through Palestine, the Levant, Central Asia, West Asia, and into South and Southeast Asia), but genetic research "suggests that the African golden wolf diverged from the Eurasian golden jackal more than a million years ago and is deserving of its own species."[2] The European and Asian golden jackal (*Canis aureus* – with the six sub-species) is related but, as Figure 1.1 shows, comes from

DOI: 10.4324/9781003199793-2

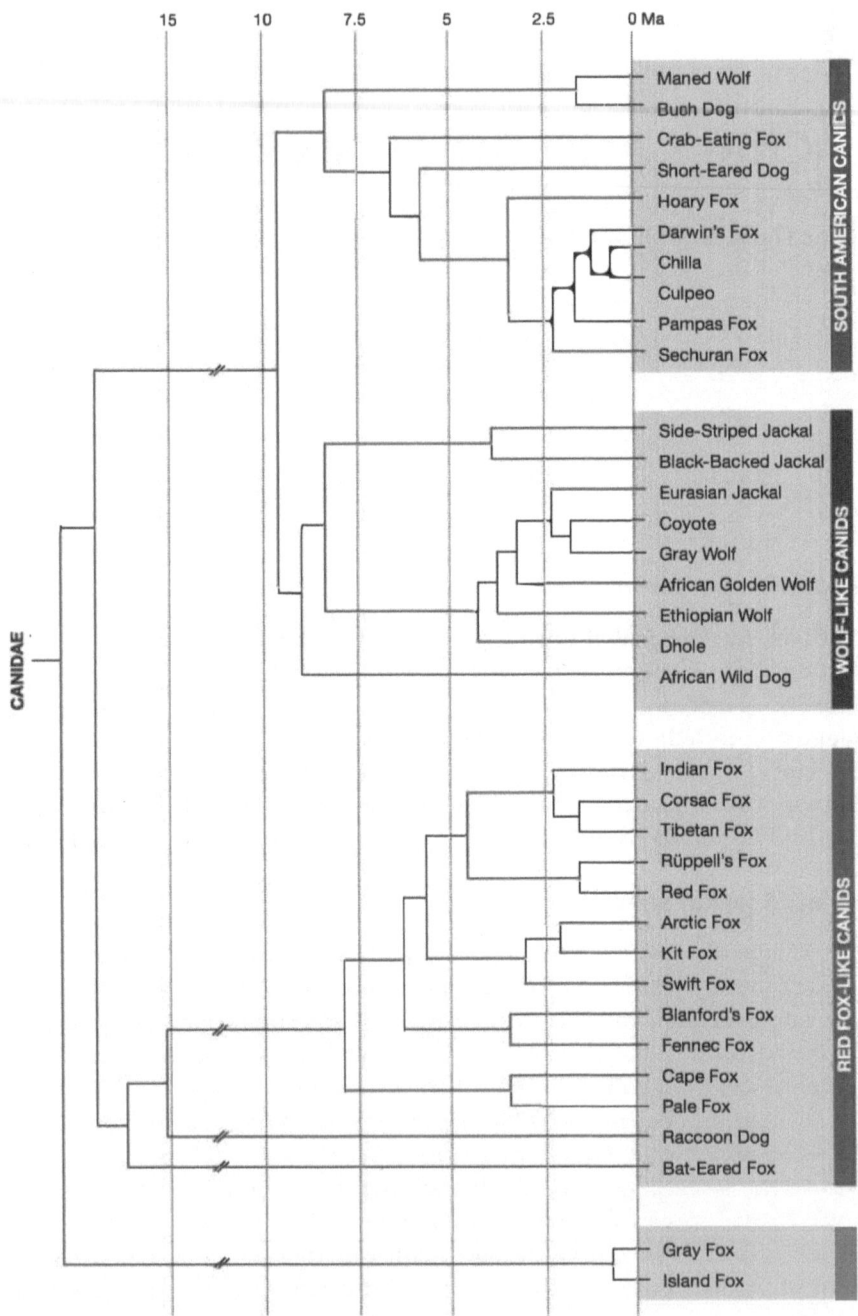

FIGURE 1.1 Phylogenetic tree of the canidae. José E. Castelló (2018) *Canids of the World*, Princeton, New Jersey: Princeton University Press/Princeton Field Guides, p. 15.

a separate branch on the phylogenetic tree of the *canidae* to the black-backed and side-striped jackals, which are the most closely related.

In their 1990 Canid Action Plan report for the International Union for the Conservation of Nature (IUCN), Ginsberg and Macdonald recorded that all the species of jackals and the African golden wolf could be labelled as "Requiring No Immediate Protection,"[3] as there was no overall threat, even though numbers and densities are low in some areas and they are threatened by habitat loss and human persecution in some parts of their range.

Size, weight, and physical characteristics

Head and body: female 66–85cm, male 69–90cm; tail: 27–38cm; height at shoulder: 38–48cm; weight: female 5.9–10kg; male 6.4–11.1kg (Figure 1.2).[4]

The black-backed jackal is reddish brown, with a distinctive black or silvery-black saddle of fur running from the neck to the tail. The bushy tail is reddish-brown and black, usually with a black tip.[5] It is a medium-sized *canid* – females are a little smaller than males and less richly coloured.[6] It has slender, long legs for its size. The jackal has a pointed, fox-like muzzle and large pointed ears.[7] The East African sub-species is slightly heavier than the southern African one while being less rufous/russet in colouring.[8]

FIGURE 1.2 Black-backed jackal. Copyright: Keith Somerville.

Distribution and conservation status

The distribution of the southern African or Cape black-backed jackal extends from southern Angola, Namibia, Botswana, and South Africa to Lesotho, Mozambique, eSwatini (Swaziland), and Zimbabwe, occurring as far north as the Zambesi escarpment. The species is rare in the Zambezi valley and absent north of the Zambezi, where the Sundevall side-striped jackal is found instead. The black-backed and side-striped jackals occur together or with overlapping ranges over much of south and central Zimbabwe, Mozambique, Angola, north-eastern South Africa, eSwatini, and northern Botswana (Figure 1.3).[9]

In East Africa and the Horn of Africa, the black-backed jackal occurs in Djibouti, Eritrea, Ethiopia, Kenya, Somalia, South Sudan, Sudan, Tanzania, and Uganda. In parts of the range, notably Kenya, Somalia, South Sudan, and Ethiopia, it is sympatric with the Kaffa side-striped jackal, and in Kenya, Sudan, Tanzania, and Uganda with the Equatorial African side-striped jackal. In Eritrea, Ethiopia, South Sudan, and Sudan, the black-backed jackal is also found alongside the North African golden wolf, and in Kenya and Tanzania with the East African golden wolf.

Social behaviour and breeding

The family is the basic social unit of the black-backed jackal. The mating pair raise their pups and will tolerate the continued presence in the core territory of the previous year's progeny, especially if they assist in feeding and guarding the new generation of pups.[10] Territories are marked with urine and scats, and are defended against intruding jackals. Young jackals disperse from their birth territory when the parents no longer tolerate them, such as when they cease to bring back food for the group or guard pups at the den. Some sub-adults disperse when new pups are produced, as not all help raise the next generation. Within the family, play is an important part of bonding, and the pups play vigorously to develop hunting and social skills, the latter including the use of threats or violence to establish dominance. When young jackals disperse, they initially become nomadic and may follow migratory herds of ungulates. They have been observed in groups of five or six nomads. Those that don't establish pair bonds and breed may remain effectively nomadic.[11]

When male and female black-backed jackals pair up, they become monogamous and usually mate for life. The pair will breed and often hunt together.[12] A mated pair, and their pups, will groom each other, share food, hunt cooperatively, and, according to Moehlman, "care for their sick or injured partners."[13] They share equally in most activities, such as marking and defending their territory and foraging. They are most active at dawn and dusk but may also forage and move around during the day, and may adopt a nocturnal activity pattern around areas of human habitation and on farmland, where they are very likely to be persecuted as livestock killers. They will have a core territory around a den

Distribution Map

Canis mesomelas

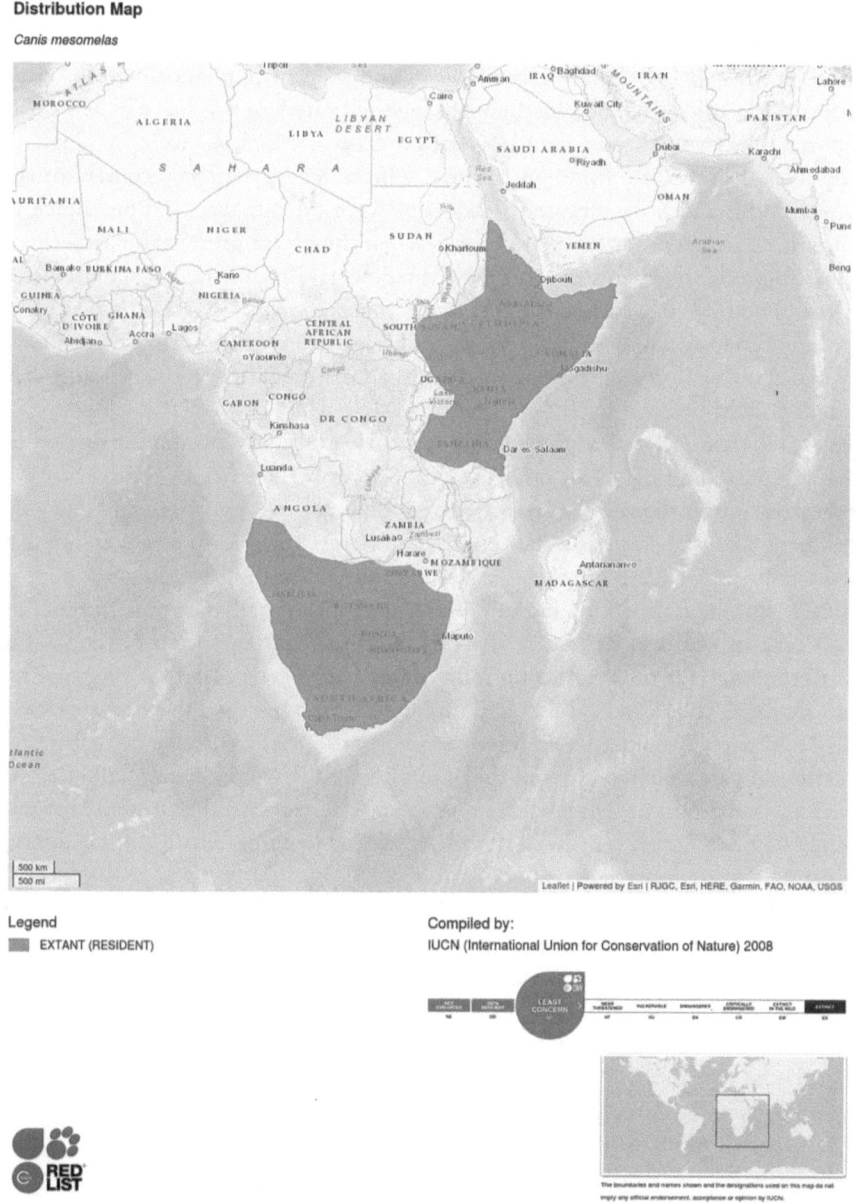

Legend

▪ EXTANT (RESIDENT)

Compiled by:

IUCN (International Union for Conservation of Nature) 2008

FIGURE 1.3 IUCN map of the distribution of black-backed jackals.

site when they have pups and forage over a large home range which can vary, according to food availability and the presence of other jackals or larger carnivores, from 2 to 33km². Unmated/nomadic adults range over larger areas, nearer the top end of the range size, and may overlap with the outer limits of the territories of mated pairs.[14]

A study of black-backed jackal territory marking in South Africa's Addo Elephant National Park (NP) examined whether marking with scats was carried
out on prominent sites like rocky outcrops and piles of elephant or rhino dung.
Rhino dung middens and, in their absence, elevated rocky surfaces were the
favoured sites.[15] The study found 211 jackal scats of which 47 were on elephant
dung piles, 12 on rhino middens (rhino are more scarce, which accounts for the
lower figure), and 6 on elevated rocks; 124 were on dirt tracks. The choice of
dung, rhino middens, and elevated rocks when they were present in jackal territories "suggests these sites may amplify the detectability of their scats via visual
and/or olfactory cues to make them more conspicuous."[16] Markings would be
regularly renewed, especially on rhino middens, where rhino dung would have
covered the scats, or where elephant dung had disintegrated and new dung had
been deposited.[17]

A mated pair will defend their home territory, especially around the den when
there are pups, and drive off other jackals. Successful defence of the territory is
heavily reliant on the resident pair working together to scent mark and drive off
intruders. Research suggests that conflict over territorial disputes is frequently
"gender specific with competing males and females failing to engage opposite
genders in combat, possibly due to … males being generally larger than females.
The death or removal of one of the territory-holding pair often results in the
loss of territory for the remaining individual."[18] But when there is large source
of food, such as the carcass of a large ungulate which has died of disease or been
killed and partially eaten by a larger carnivore, too many jackals may gather for
a territorial pair to drive off, and they are forced to share the food,[19] albeit amid
snarling, snapping, and shows of aggression. The territorial pair will try, and
usually succeed, in preventing subordinate jackals from breeding and producing
offspring in their territory.[20]

Defence of territory, particularly the den site, involves physical displays (raising hackles, facial expressions, and postures) and vocalisations. Vocalisations are
a sharp shrill howl at night, yapping to communicate with each other and a
cackling noise when fighting.[21] Often intruders will try to avoid physical conflict while attempting to stay in the territory by adopting submissive behaviour
towards the territory holders.[22] Often, the intruder is trying to use a waterhole
or feed from a large carcass. Nattrass et al. suggested that while black-backed
jackals are not pack animals like wolves or wild dogs, they

> can access a set of social conventions usually found in packs to recognise
> and reinforce hierarchies. This also allows them to hunt co-operatively in
> quickly and loosely formed packs if the opportunity arises … studies paint
> a rich picture of a highly adaptable animal whose behaviour, diet and even
> breeding varies across time and space.[23]

An example of this behavioural adaptability was when a researcher observed a
black-backed jackal attacking a springbok that was trying to get out of a water

hole in Etosha NP in Namibia. The attack attracted five other jackals, who joined in the hunt. Having killed the springbok, the jackals then ate from it according to their level of dominance, with subordinate jackals displaced from the carcass until the dominant jackals had finished, providing "another example of how this normally solitary hunter can access wider hierarchical and social conventions/ behaviours when necessary to help co-ordinate collective efforts and to provide ordered access to resources."[24]

Dominance and challenges to that dominance involve highly ritualised posturing or actions, which are intended to be threatening and the exercise of power without resort to actual violence. Most aggressive displays to establish or reinforce dominance don't involve physical contact and so don't cause harm to the subordinate jackal, with the result being flight or a submissive display placating the dominant animal.[25] Ferguson noted that threatening behaviour was exhibited in 44 of 116 interactions between two or more jackals he observed.[26] Dominance signals include baring teeth, erection of hackles along the back, and an elevated tail. If this doesn't get the necessary submissive reaction (flattening down of ears, a submissive facial expression like a grin, or the subordinate jackal rolling onto its back), dominance displays can be followed by body-slamming – with the dominant animal swinging its hindquarters into the forelegs and torso of the submissive one.[27]

Van Lawick observed a range of behaviours when studying black-backed jackals, which followed the wildebeest, gazelle, and zebra migrations in the Serengeti. He said that non-territorial jackals would become nomadic at times following the herds and that, on occasions, he witnessed six or more nomadic jackals moving together as a loose pack; territorial pairs did not leave their territories during the migration.[28] The nomadic jackals coalescing in groups would exhibit dominance displays and sometimes physically fight as they sorted out their social hierarchy, including body-slamming. Often, these interactions would be followed by a period of what seemed to be play, as the jackals reinforced their bonds and their status, something which he said was not replicated when he observed golden wolves in the Serengeti.[29] Friendly behaviour, such as grooming, is used to reinforce bonds between animals and is a common type of social behaviour. This latter behaviour is typical of much of the interaction between mated pairs and their family group.

In a study in the early 1980s of black-backed jackals in the Kwang Pan area of the Kalahari Gemsbok NP (now Kgalagadi Transfrontier Park – KTP), the Transvaal Highveld near Bloemhof (now North West Province), and the Suikerbosrand district of Transvaal (now Gauteng), 46 were given ear tags or collars and their social organisation and movement patterns monitored by Ferguson et al.[30] They found that pups were generally born in August and September. Pups stayed close to their dens until about 12 weeks of age, after which they would start accompanying adults to forage but rarely going more than 2km from the den. They would disperse around 7 months old, unless they stayed to help raise the next litter. When as sub-adults they moved out of the core area of the adults' home range, they usually stayed in areas that overlapped the periphery of the parental territory.[31] Adult home ranges rarely overlapped with the territories of other mated adults, though in the Kalahari, there was about a 10% overlap between two paired adult territories.[32] There was more overlap between adult ranges in the farming area of Suikerbosrand. Large variations have been found in the size of home ranges (from 2.1 to $91.5km^2$), with the size of a range for an adult pair, according to Ferguson et al., "probably the result of a combination of factors such as prey availability, population density and metabolic rate of the predator. The large variation ... indicates that Black-backed jackals can adapt to widely divergent ecological circumstances."[33] The ranges in the Kalahari were generally smaller because of the abundance of prey such as springhares, hares, rats, and other rodents. As was found in the Lake Ndutu research, the South African jackals mated for life, with the bond only seeming to be broken by the death of one adult.

James found that across their East and southern African ranges, black-backed jackal territory size was highly variable but that the average South African jackal territory was $18.2km^2$, while those in Zimbabwe and Kenya's Rift Valley were much smaller, ranging between 0.6 and $3.0km^2$. It follows that the density of jackal populations varies with as few as one jackal per $2km^2$ up to 22 per km^2,[34] the primary factor in territory size and density being prey and scavenging availability. A good example of the importance of prey availability on territory size and the extent of defence and resulting conflict is provided by the black-backed jackal populations resident at Cape fur seal colonies along the Namibian coast. There, territories have been recorded as ranging from 7.1 to $24.9km^2$, with aggressive interactions between neighbouring jackals being rare because they have a substantial and easily obtained source of food[35] (this will be described in greater detail in Chapter 7). Anthropogenic food sources like waste dumps and vulture restaurants (areas set aside on reserves or game farms where carcasses are left out to aid the recovery of vulture numbers) can lead to higher jackal densities because of the abundance of food sources.[36] This has certainly been the case at the vulture feeding station, using carcasses of dead livestock from neighbouring farms, at the Mankwe Wildlife Reserve in South Africa's North West Province. Black-backed jackal numbers increased after the introduction of the vulture feeding programme there and declined when it was stopped. But jackal numbers remained stable in nearby Pilanesberg NP, where no vulture restaurants were provided.[37] Given that dispersal of young from the birth territory is believed to be driven by competition with the adults for food, Nattrass et al. pointed out,

> This hypothesis is supported by evidence showing that an increase in local food availability, such as the opening of a vulture restaurant, results in genetically distinct clusters of black-backed jackals as the benefits of dispersal fall relative to staying ... black-backed jackals are thus facultative cooperative breeders, capable of breeding as lone pairs and forming extended family groups when ecological conditions (abundant food and limited vacant territories) favour philopatry (staying in a particular, usually natal area) over dispersal.[38]

Minnie et al. found that in the KTP, adults had an average home range of $11km^2$ (in a range from 3 to $22km^2$) compared to $85km^2$ (range: $2-575km^2$) in sub-adults, while in farming areas in South Africa's North West Province, adults had an average home range of $28km^2$ (range: $3-92km^2$).[39] Population density also varies in both the southern African and East African ranges of the black-backed jackal, with 35–40 jackals/$100km^2$ in the Giants Castle Nature Reserve, KwaZulu-Natal Province of South Africa; 50 jackals/$100km^2$ in the Serengeti NP, Tanzania; and, 400–700 jackals/$100km^2$ in the Tuli Game Reserve, Botswana (McKenzie, 1990). Macdonald et al. found that in Hwange NP, Zimbabwe, black-backed jackal density was 53.9–79.1 per $100km^2$, going up to 69.3–97.1 in the breeding season.[40]

In another study, at Benfontein Game Farm in South Africa's Northern Cape, Kamler et al. found that a typical jackal family group consisted of the mated pair, up to three non-breeding adults (pups from previous litters) and young pups. They occupied ranges of $9.4 \pm 1.2km^2$, while subordinate jackals on the farm had slightly larger ranges of $9.8 \pm 0.7km^2$.[41] Dispersal was limited, and there were a number of subordinate jackals who did not move from close proximity to adult territories, chiefly due to the high density of their main prey, springbok. The proximity of subordinate adults to the ranges of breeding adults was matched by quite regular extraterritorial forays by adults into the ranges of neighbouring jackal pairs – three out of five alpha adults made periodic forays into the territories of other, and many of these were during the summer months when jackals left the game farm and visited livestock farms, with the motivation appearing to be predation of sheep. The researchers believed that high social status, seasonal availability of springbok and other wild prey, breeding cycles, and the availability of sheep to substitute for wild prey were the reasons for regular movements out of their own territories.[42] Loveridge and Macdonald studied the dispersal of black-backed jackals and concluded that territorial forays and incursions can be used by territorial pairs to extend their territory, explore for vacant areas, or gain access to abundant food sources; they noted that incursions by young jackals did not lead to aggressive reactions by the holders of the territory, possibly through tolerance of the movements of young jackals or perhaps because of avoidance strategies by the young animals.[43] In farming areas, both domestic livestock and game farms, the high density of potential prey combined with farmers culling resident adult jackals can destabilise territorial occupation by mated pairs and lead to a higher turnover of jackals within territories.[44]

Within a mated pair's territory, only the dominant pair will mate – any adult helpers who have stayed to help raise the young will disperse before pairing up and mating. Mating seasons vary greatly according to geographical location and, it is thought, according to food availability.[45] In some areas of southern Africa, mating may take place between May and August, but in the Western Cape, Gauteng, Mpumalanga, and North West Provinces of South Africa, pups are usually born between July and September.[46] In the Lake Ndutu area, referred to earlier, mated black-backed jackal females usually give birth to pups during the dry season, between June and November. While these jackals are highly varied in their diet, the grass rat (*Arvicanthis niloticus*) is an important part of the diet, and the rat is at the peak of its yearly cycle during the jackal birthing period and so is available as relatively easily caught prey that can be taken back to the den for the pups once they start eating solid food.[47] Dens are frequently located in old aardvark or warthog burrows or are dug out under rocks or tree roots.

Pups are born in the den and stay there for the first weeks of life. Kingdon put the average litter size at four, but numbers range from 1 to 9.[48] In her 1989 study, Moehlman said black-backed jackals produced large litters in an average of 5.7, with an average of 1.3 pups surviving to adulthood. Of these, 24% of surviving pups stayed to help in raising/guarding the following litters, with no difference in between male and female in staying.[49] Pups are blind at birth and open their eyes after about ten days. By 3 weeks old, they will emerge from the den. The mother will have spent most of those 3 weeks in the den.[50] The pups will continue to use the den, though with more and more time spent outside, from 7 to 14 weeks of age. While still largely dependent on their mother's milk, pups will start to eat meat regurgitated by the adults following hunting or scavenging trips and by 14 weeks are strong enough to start accompanying the adults on foraging trips.

The first 3 months of life are when pups are most vulnerable to predation or simply getting lost and then starving. As Moehlman observed,

> All observed mortality has occurred within this initial 14-week period. This is the time within which the pups are most vulnerable to food deprivation and predation, and the time period in which major contributions are made by helpers and parents. The presence of helpers correlates positively with pup survivorship.[51]

The role of an older pup from the previous litter enables the mated pair to forage while having a guard at the den. Without a helper, one adult would have to forego foraging to guard the den or, as happens for about nearly a fifth of the time, leave the pups in the den unguarded. Guarding would involve giving vocal warnings to the pups to retreat into the den if a potential threat was seen and then threatening any potential predator. Single jackals can often deter a single hyena even to the extent of biting its rump,[52] though it would have trouble with several hyenas or a leopard, the latter periodically taking young and adult jackals.

The presence of helpers not only assists in guarding the pups but also feeding, with helpers providing up to a third of the food for the pups.[53] The helper or helpers will hunt/forage and bring back food for the pups.[54] In her study at Lake Ndutu, Moehlman noted that in 28 months observing the jackals, 11 out of 15 litters of jackal pups had helpers, and these had a much higher pup survival rate. The helpers were all offspring of the parental pair and full siblings of the pups being raised.[55] She recorded that four families of jackals she observed in 1976–77 had 12 surviving offspring, of which four (one per family) stayed as helpers for the adults with the next litter of pups, with the others dispersing to look for territories and mates. A fifth family observed had four pups, none of which stayed to help raise the next generation.[56] Observations by many researchers and on films (from the documentaries referenced in the Introduction) also show the role of helpers in developing social skills and basic hunting techniques among the pups. Moehlman calculated that, on average, each helper added 1.5 surviving pups to the litter and that

> an adult helper gains more (yield 1 pup per adult) by being a helper of its parents than by finding a mate and raising its own pups aided only by its mate … Helpers may also derive benefits from extended experience on the home territory in terms of increased survivorship, increased long-term reproductive success and acquisition of a portion of the home territory.[57]

This suggests that there is a clear survival and breeding advantage for jackal families where older pups that become helpers, and that the helpers themselves gain greater experience prior to dispersal from the natal territory.

Black-backed jackals reach sexual maturity at 10–11 months, by which time the next litter is likely to have been born. As noted, just under a quarter of surviving pups stay as helpers, with the majority dispersing. In most observed family groups, dispersal of the pups occurs between 1 and 2 years of age in jackals on both farmland and protected areas, though some leave as early as 6 months to a year old. As Minnie et al. concluded,

> It is unclear what drives dispersal, but it may be due to intraspecific competition with dominant individuals, and the need to establish a territory, find food and a mate and … dispersing black-backed jackals may have one of four options … (1) stay in their natal territory as a helper; (2) move into vacant territories; (3) move into nearby territories to be incorporated into those territories' resident groups; or (4) float between their natal territory and adjacent territories.[58]

Once they reach sexual maturity, black-backed jackals have a predicted lifespan of between 3 and 12 years in the wild,[59] the lifespan seemingly unaffected by whether or not they stay with their parents as helpers.

The Owens observed a black-backed jackal family over a long period in the Central Kalahari Game Reserve in Botswana. They noted that as the pups matured, there was often fighting between males and their father over food, with the offspring body-slamming the dominant male to establish a right to feed.[60] Such conflict may well be a spark for dispersal. There is no clear distance limit on dispersal and may involve a migration of as much as 100km.[61] Dispersing jackals can move quickly, with one monitored individual in north-western South Africa moving 87km in four nights and then later moving on another 30km. When it died 13 months later, it had moved 126km from its natal territory.[62]

The mated territorial pair will not usually tolerate breeding by other jackals in their territory. Nattrass et al. reported that helpers don't normally breed at their natal den and don't display sexual behaviour, but there have been exceptions recorded, with more than one female breeding at a den.[63] They note two cases where black-backed jackal pups, differing by a few weeks in age, were pulled from the same natal den, suggesting that the adult breeding male may have mated with the helper, who then gave birth in the same den as the adult female, though it is possible that a female helper may have been mated by an unrelated male. They also cite a report that a professional hunter had said that he had killed a breeding pair and six pups in a den on a farm and had gone back the next night and killed another female emerging from the same den with swollen teats. "In his assessment, this was a helper with her own litter because the food supply on that particular farm in the South African Karoo could support a dual litter."[64]

The main obstacles for dispersing jackals are not only defended jackal territories they have to cross in areas where food sources are not abundant but also well-maintained, predator-proof fences in areas where livestock farmers wish to totally exclude jackals and other potential livestock killers.[65] Often burrows dug under fences by animals such as aardvarks and warthogs create openings, and any gap means jackals are often able to cover large distances when dispersing through farmland and livestock areas. This has often rendered ineffective culling or other measures aimed at limiting or completely excluding jackal presence and so has not been a bar to widespread dispersal and repopulation of areas where numbers had been reduced; in addition, removal of established family groups from a territory on farmland creates a vacuum that can be easily filled by dispersing adults or sub-adults.[66]

The periods of activity of black-backed jackals will vary somewhat according to the abundance or scarcity of prey, the types of prey or other sources of food (fruit, for example), the presence of larger carnivores, and whether jackals are close to human habitations or farms. Ferguson et al. radio-tracked jackals, in the KTP and in two farming areas in what was then Transvaal, the latter in stock farming areas. Generally, jackals have peaks of activity between 1900 and 2400 and then 0500 and 0900, with far less activity between 0900 and 1900.[67] In the KTP, the times varied somewhat, with much activity between 1700 and 2100 and some but far less between 0700 and 0800.[68] Their research suggested strongly that activity was governed by the main activity patterns of their prey – chiefly rodents such as the striped grass mouse (*Rhabdomys pumilio*) and the multimammate mouse (*Praomys natalensis*).[69]

Diet, foraging, and hunting

Black-backed jackals are carnivorous but can be classed as omnivorous because of the diversity of their diet, including vegetable matter and fruits.[70] Prey species include beetles, termites, grasshoppers, crickets, winged ants, spiders, scorpions, crabs, lizards and snakes, ground-dwelling birds, their eggs and nestlings, shrews, mice, rats, springhares, porcupines and other rodents, mongoose, young bat-eared foxes, small ungulates and the young of wildebeest, kongoni, impala and lesser kudu, young and adult gazelles, steenbok, dik-dik, duiker, springbok, young Cape fur seals, lambs, and young goats and calves; they also feed from the carcasses of dead wildlife or livestock and also fallen fruits, nuts, berries, roots, other vegetable matter, and even grass.[71] Moehlman described black-backed jackals as "highly opportunistic omnivores,"[72] something which enables them to survive in a variety of habitats and adapt to loss of some food sources, as long as other opportunities are available. She noted that in northern Tanzania, a major source of food is fruit that is obtained by extensive foraging – the date-like fruit of the *Balanites aegyptiaca* tree, which may involve jackals travelling 6–8km to find sufficient fruit while hunting for rodents along the way. When ripe fruit was available, black-backed jackals systematically visited Balanites trees to forage. They also searched for dung buried by dung beetles to get the beetle larvae developing within dung balls underground and tried to catch adult beetles, sometimes grabbing them in the air.[73] Cape hares were also hunted, but unless an unwary one could be ambushed, the hares were often fast and agile enough to elude the jackals in eight out of nine instances witnessed by Lamprecht, and they made up a relatively small proportion of their intake.[74]

A study at Mokolodi Nature Reserve, near Gaborone, Botswana, found that diet varied and included a wide variety of sources, including human refuse. Researchers discovered the diversity of foods through analysis of 237 scats between November 1995 and February 1997. They reported the following results:

> Seasonality of prey occurrence in scats was pronounced for small mammals, miscellaneous fruits and invertebrates ... mammals were the most common food resource (32.4, $n = 168$), followed by anthropogenic items (14.8), fruits (12.9), invertebrates (1.8), birds (8.5), unidentified items (3.5) and reptiles (1.4).The presence of domestic mammals and poultry remains in scats reveals their importance in the diet of jackals and the tendency of jackals to frequent human settlements in search of food.[75]

Larger mammal remains were dominated by kudu, impala, and warthog – while jackals might take impala fawns and very young warthog, most of these remains will have been scavenged from carcasses from leopard or hyena kills or animals that died of disease or other natural causes.

Temu et al. from their study of black-backed jackals, Sundevall side-striped jackals, and golden wolves in the Ngorongoro Crater, northern Tanzania, also referred to them as "omnivorous opportunists, feeding on variety of foods including invertebrates, reptiles, birds, small to medium sized mammals, plant materials, and carrion" but noted that their size and nutritional requirements indicated that the jackals could survive on an entirely invertebrate diet.[76] The jackals and golden wolves shared the crater with a very substantial number of predators, with very high densities of spotted hyenas, large lion prides, leopards, and cheetahs.[77] The black-backed jackals spent most of their time in grassland areas with medium to long grass and so more cover but also more rodent prey than in the short grass areas.

During the wet season, both the black-backed jackals and the golden wolves hunted and ate substantial numbers of Thomson's gazelle fawns, with mated pairs of jackals and wolves cooperatively hunting the fawns, with one of the pair distracting or warding off the mother and the other catching and killing the fawns.[78] Lamprecht collated statistics for hares and gazelle hunts by black-backed jackals and found that despite the relative lack of success in hunting hares, the jackals had a 60% success rate over 25 witnessed hunts of hares or gazelles. He compared the rate with the overall hunting success rates of lions (26%), spotted hyaenas (35%), cheetahs (70%), and wild dogs (70–86%), concluding that the jackals were successful hunters, especially if more than one jackal was involved in a hunt.[79]

Black-backed jackals in Ngorongoro searched for the carcasses of ungulates killed by larger predators and tried to steal scraps while the predators (mainly lions and hyenas) were feeding, a practice with the danger of injury or death, or waited for the hyenas or lions to move away and then at times competed with golden wolves and vultures for what remained.[80] When there is a large carcass that jackals have access to, the territorial boundaries of pairs do not prevent large numbers of jackals gathering to feed, albeit with some aggressive behaviour and even fighting. In the Ngorongoro Crater, scavenging provided black-backed jackals 38.8% of their food intake during the dry season and 27.5% in the wet season.[81] At Nxai Pan NP in Botswana on 27 April 2018, I witnessed more than 15 jackals circling and almost mobbing a pride of lions (three female, a large male, and four cubs) feeding from a buffalo kill. With a certain amount of squabbling, the jackals milled around, occasionally darting in to grab a scrap and then retreating to avoid the half-hearted charges of the lionesses. The jackals were able to feed on some scraps, clotted blood, and other detritus of the kill at the site where the buffalo appeared to have been killed, before the male lion had dragged the remains into the shade of a bush 20 metres away.

Jackals are also known for scavenging around human habitations looking for remains from slaughtered animals and other edible refuse across their East and southern African ranges,[82] something I witnessed in tourist camps in Namibia's Etosha NP and adjoining private reserves in 2008 and 2017, and at Okonjima in Namibia in 2008, 2017, and 2019. Kaunda and Skinner, who studied black-backed jackal diet at Mokolodi Nature Reserve in Botswana, said that 14.8% of the jackals' diet was of human origin in the form of waste. The waste material was broken down as follows:

> First, remains from domestic animals typified by white chalky faeces containing abundant poultry claws, feathers and bone chips, and fish remains (e.g. fish scales and bones) resulting from the activities of anglers who operate in the nearby rivers and dams. The second type, derived from feeding on exploitable scraps from refuse disposal sites, typified by dark brittle

faeces containing plastic bags, bottle tops and fragments of glass, together with some commercial fruit pips ... Small quantities of sheep, donkey and cattle remains were found in scats.[83]

In her examination of the diet of the jackals in North West Province of South Africa, Van der Merwe found a wide variety of foods in an analysis of 43 scats obtained from a mixture of habitats – protected areas, game farms, and livestock farms. It indicated that jackals consumed significant numbers of small mammals, although evidence of consumption of large mammals was found in 50% of scats, with impala being the most frequently present at 34%, with no other large mammal making up more than 5% of the jackal diet.[84] The breakdown of the presence of particular remains in scats was large mammals occurring in 59.1%, small mammals 77.3%, birds 9.1%, invertebrates 27.3%, seeds/fruits 13.6%, anthropogenic 4.6%, and reptiles 4.6%.[85] Jackals will take lizards and snakes, including highly venomous ones. The Owens recorded seeing two jackals cooperating to kill a nine-foot black mamba in the Central Kalahari[86] – one bite would kill a jackal.

Across most of their range, according to Tambling et al., black-backed jackals prey on smaller ungulates that hide their young but tend to avoid larger ungulates that hide their young and those smaller ungulates whose young follow the parents.[87] Jackal diets are influenced both by the presence or absence of apex predators and the abundance or otherwise of prey species of a suitable size. The presence of larger carnivores can both limit hunting choices for jackals and provide scavenging opportunities. Where larger predators are absent or only present in small numbers, black-backed jackal predation can have the effect of influencing numbers of the prey species, such as springboks in the Northern Cape of South Africa, where apex predators have been exterminated in farming areas or are scarce. Where apex predators are missing from farmland, predation on livestock by jackals is common, and jackals can be successful livestock killers despite deterrence measures and retaliatory killing of jackals. But where there are apex predators and "consequential carrion provisioning, peaks in the availability of juvenile ungulates appear to be less important for foraging black-backed jackals."[88] Tambling concludes that

> the pattern of consumption of livestock by black-backed jackal seems to mimic the patterns exhibited when black-backed jackals consume ungulates in the absence of apex predators ... [but] it remains unclear whether jackals select wild prey more than domestic prey ... consumption of wild versus domestic prey may however also be dependent on the composition and catchability of wild prey available to black-backed jackal.[89]

When it comes to hunting wild ungulates, in most cases jackals will go for newborn or juvenile gazelles, springboks, and impalas, but they have been witnessed killing adults. Kamler et al. reported that there were recorded cases of black-backed jackals killing adult gazelles, impalas, and springboks – though killing adults almost always involves more than one jackal cooperating; however, a single jackal was observed and photographed killing a healthy adult impala near Little Mombo Camp in the Moremi Game Reserve in Botswana in April 2009.[90] The jackal pursued an adult impala ewe, which injured its leg in the chase and was caught and throttled to death by the jackal. The jackal was seen to feed from the impala before the kill was taken by a spotted hyena. Black-backed jackals in East Africa are considered to be major predators of gazelles, both fawns and adults, and, "due to heavy predation on fawns, jackals were thought to have a significant negative impact on gazelle populations."[91] Klare et al. found from their research in South Africa's Northern Cape, near Kimberley, that in the springbok lambing seasons in spring and autumn, springboks contributed substantially to food intake, 76.5% and 76.0%, respectively. Other small or medium-sized ungulates such as common duiker, blesbok, or domestic sheep were detected in scats only sporadically.[92] Outside the springbok lambing seasons, a much wider variety of small prey was detected in scats, and fruit was particularly evident (just over half of food intake) during summer months when prey was scarcer. The jackals' hunting abilities prompted Klare et al. to insist that black-backed jackals should be considered members of the large carnivore guild and not treated as scavengers and foragers that may hunt.[93]

In general, jackals will opportunistically scavenge for carcasses or hunt prey – whether wild or domesticated – that offer a good chance of success in that moment, so they will consume a preponderance of insects in southern Africa deserts, seals, dead fish, and penguins on the Namibian coast, springboks in the Karoo, ostrich eggs in Addo Elephant NP (using one egg as an anvil against which to crack another),[94] rodents in other areas of southern Africa, and fruit in many areas according to the season.[95] Jackals will also kill bat-eared foxes, mongooses, and Cape foxes if other prey is unavailable or opportunity presents itself. In areas where livestock is plentiful, dead stock will be consumed and young, sick, or old livestock killed. The availability of livestock, especially goats and sheep, clearly influences black-backed jackal diets. As Minnie et al. observed, domestication of sheep and goats has deprived them of much of their anti-predator defences, making it easier for jackals to successfully hunt them.[96] Some studies, referenced by Minnie et al., of black-backed jackal diets on livestock farms indicate that livestock may contribute to a large proportion of the diet (25–48%), but others show that this is not the case (e.g. 16% of diet).[97] Another indication that black-backed jackals can vary their diet to include livestock, where it is easily available, eating less wild prey as a result, but that "this shift in diet is context-dependent, as several studies show that black-backed jackals on farms consume more small mammals and small ungulates than on nature reserves."[98]

As Nattrass et al. concluded in 2020, "The scientific record is replete with examples of jackals' inventiveness in finding food."[99] A good example of invention was given in Fourie's study of black-backed jackal diet in the Karoo NP before and after the introduction of additional springbok. The augmentation of springbok numbers led to increased jackal predation on them. But after lions were reintroduced and their hunting produced increased scavenging opportunities on large ungulates killed by the lions, jackals began to eat more large ungulate carrion and consumption of springboks declined to pre-augmentation levels.[100] Nattrass et al. concluded that this demonstrated how jackals, like other similar-sized predators such as North American coyotes, can "cope with rapid and substantial shifts in diet as the resource-base shifts."[101]

Black-backed jackals often foraging alone or in pairs but on occasions, where ungulates like springboks, gazelles, and impalas are among the prey hunted, they may come together as a family group or even involve non-family members in hunting as a pack. Minnie et al. wrote that sometimes in Botswana, they have been observed to form "temporary 'packs' of six to 12 individuals to attack and kill an adult impala; and in Namibia they displayed similar co-operative hunting to kill an adult springbok."[102] Jackals will, if necessary due to scarcity of prey or carrion, cover large distances while searching for prey or other food. A South African Department of Agriculture study of jackals in the early 1920s showed that jackals could easily cover 45km in a night's foraging and that dispersing young adults might move 55–100km from their natal territory.[103]

When hunting small ungulates, gazelle fawns and adults, as well as small livestock such as sheep and goats, black-backed jackals usually kill their prey by strangulation, biting, and gripping the neck to seal the trachea, rather than tearing at the prey before death.[104] Similarly, they will grab young Cape fur seals by the throat; Kolar witnessed 83 jackal kills and numerous unsuccessful attempts, of which 80 kills were of newborn pups and 3 were yearling seals taken before the new pups were born.

> Several times two jackals were observed cooperating in killing a seal. The kills lasted from five to 20 minutes. The jackal would grab the pup on the neck or throat and hold it until the pup either suffocated or died of exhaustion. The pups could escape if they managed to pull their opponent into the water … or if the jackal was chased away by an adult seal.[105]

Jackals only attacked adult seals if they were sick or badly injured. Kolar recorded that

> [o]nce a jackal had killed a young seal it had to open the tough skin. This procedure could take several minutes, and the jackal always started from beneath one fore flipper, although other observers in this area have seen jackals would also open the skin at the anus. As it was difficult for the jackals to open the skin they would turn the carcass inside out to reach the flesh. This resulted in the characteristic "sleeping bag" skin by which one can recognise a seal consumed by jackals.[106]

Along their Atlantic coast range, jackals take seals, cormorants, penguins, and other sea birds and forage for whale, fish, and other carcasses along the shore and scavenge for scraps of meat from the seal processing plants, where culled seals are butchered for meat, fat, and skins.[107]

While black-backed jackals will scavenge human waste sites, such as meat processing facilities and middens, they are generally thought to be more wary of foraging around human habitations than side-striped jackals or golden jackals; observations from Botswana show that human waste and animal remains at rubbish dumps, fishing camps, poultry or livestock farms, and at lodges or campsites in safari areas do provide a source of food even in close proximity to human presence.[108] I can certainly confirm this through direct experience at the busy Okaukuejo rest camp and park HQ in Etosha NP, Namibia. Hundreds of tourists and large numbers of staff, with constant coming and going of vehicles, do not stop black-backed jackals from entering the rest camp under or through gaps in fences and very actively foraging, including approaching people eating meals to wait patiently for any scraps they can grab, as in Figure 1.4. This jackal stood within five yards of us in Etosha NP, waiting for us to finish our braai (aka BBQ) to try grab any discarded food remains or rifle through the bins by the chalets.

Whether scavenging carrion from a carcass or hunting, jackals have been observed to cache surplus food that can't be eaten then or taken back to the den for cubs. In Botswana, they have been monitored on five occasions caching meat

FIGURE 1.4 Black-backed jackal, Okaukuejo rest camp, Etosha NP, Namibia. Copyright: Keith Somerville.

from freshly caught prey and then retrieving it on two occasions.[109] On the Namibian coast, Hiscock and Perrin recorded jackals caching carrion or meat from seal pups.[110] Cached items are generally hidden some distance from where the kill took place or where meat was scavenged from a carcass. Nattrass et al. reported that farmers told them that lambs would disappear completely, with no trace of their carcasses, leading to a belief that they had been partially eaten and then cached.[111]

One interesting aspect of the fruit consumption by jackals is that they play a role in the dispersal of seeds. Kamler, Klare, and Macdonald researched the potential for jackals, Cape foxes (*Vulpes chama*), and bat-eared foxes (*Otocyon megalotis*) to disperse the seeds of fruits and other plants they ate in semi-arid areas of South Africa.[112] All three species ate the fruit of the bushveld bluebush (*Diospyros lycioides*), while jackals also ate fruit of the mesquite (*Prosopis*), the buffalo thorn (*Ziziphus mucronate*), and the brandy bush/raisin tree (*Grewia flava*). Jackals were particularly efficient at wide seed dispersal as they ate large quantities of fruit when it was available and deposited their scats around their territories and boundaries as scent markers.[113] At the Benfontein Reserve in the Northern Cape, jackals were estimated to eat 5.5kg/km^2 of fruit annually crossing all seasons, with a peak of bluebush fruit consumption in the autumn.[114] Kaunda and Skinner found that jackals' consumption of brandy bush fruits occurred across the seasons and that jackal defecation around their territories assisted greatly in the dispersal of seeds and growth of new plants.[115] Kamler, Klare, and Macdonald found that jackals played an important role in seed dispersal and were one cause of the spread of the invasive mesquite.[116]

Interactions with other carnivores

Interactions with other carnivores – whether conflictual, commensal, parasitical, or symbiotic – are important aspects of the distribution, habitat selection, hunting or foraging success, and infant and adult mortality of all carnivores, black-backed jackals not the least.[117] Jackals hunt, scavenge, forage, and compete for food, benefitting from large carnivores killing ungulates that are too large for jackals to hunt and competing or stealing scraps at carcasses. They are vulnerable to attack by large carnivores. They also have, as detailed in the Introduction, an interesting, certainly commensal, and probably parasitic relationship with the honey badger. The effects of interactions with other carnivores can be positive, negative, or a mixture of both – they are rarely neutral.

The presence of large carnivores can be positive through the provision of carcasses from which to scavenge, or they can deter jackals from using particular areas and so affect their distribution and ranges.[118] While lions may occasionally kill jackals when the latter are trying to steal from a kill, generally jackals are able to assess well how close they can approach a kill and when they can dart in with minimal risk to seize scraps. As I witnessed at Nxai Pan, a dozen or more jackals may gather at a lion kill and rush in and out to grab scraps. Once they have fed, the lions often move off for a short distance enabling the jackals to feed. Schaller and Lowther noted lions, hyenas, and leopards would all kill jackals.[119] Black-backed jackals are preyed on by leopards quite often. Ray et al. recorded that jackals are occasionally killed by lions and wild dogs, mostly in conflicts over carcasses and the killing of pups or sub-adults by spotted hyenas, which eat them.[120]

The relationship between black-backed jackals and cheetahs is less straightforward. Jackals will harass and try to mob cheetahs on kills and often succeed in stealing part of the kill,[121] especially if one jackal can get the cheetah to chase it, while the other rushes in to steal part of the kill. Jackals will kill and consume cheetah cubs. The carnivore expert George Schaller said he witnessed two jackals pursuing a gazelle fawn and four adult gazelles. Before they had caught one, a cheetah sprinted after them and killed the fawn. The jackals moved towards the kill, presumably hoping to steal some of it, but they then saw Schaller and ran off.[122]

Black-backed jackals (as well as the other jackal species, golden wolves, and honey badgers) are often referred to as mesopredators, being predators that are ranked lower in the food chain than apex predators like lions, tigers, wild dogs, spotted hyenas, and leopards. They are ranked as such because of body size, subordination to or threats to them from larger carnivores. They may thrive and expand in numbers and range in the absence of larger predators and become the apex predators. Prugh wrote that the extermination or other absence of large predators can lead to "mesopredator release ... the expansion in density or distribution, or the change in behavior of a middle-rank predator, resulting from a decline in the density or distribution of an apex predator."[123] But the absence of apex predators can cut both ways. While it may remove threats to jackals and reduce the competition they face for access either to prey species or carrion, it can also remove major sources of food such as carcasses of large ungulates killed by large carnivores.

Black-backed jackals may share habitat with side-striped jackals in southern and East Africa, and with golden wolves in East Africa and the Horn of Africa, as well as Cape and bat-eared foxes. They periodically kill bat-eared foxes and Cape foxes for food.[124] Kamler et al., in their study at Benfontein in the Northern Cape, noted that larger carnivores had been exterminated and black-backed jackals were the apex predators and were recorded predating five bat-eared foxes and a Cape Fox during the 3-year study period.[125] Jackal diet in the area was dominated by springboks, hares, and springhares, with the foxes presumably opportunistically killed to be eaten. It was also likely that jackals killed foxes to reduce competition for food, even though their diets did not overlap to a substantial extent.[126] The study also showed that Cape foxes were found in low densities on or avoided jackal territories, as did bat-eared foxes, and Cape foxes were largely nocturnal to avoid the crepuscular or diurnal jackals.[127] Jackals appear to suppress Cape Fox numbers in their territories and foraging areas, while bat-eared foxes coexisted with jackals despite periodic predation by the jackals.[128]

Black-backed jackals coexist across part of their range with side-striped jackals – they occur together over most of south and central Zimbabwe and northern

Botswana. The side-striped jackal is present across suitable woodland and grassland habitats in Central Africa as far south as Zimbabwe and northern South Africa, while the black-backed is distributed across South Africa, Namibia, and Botswana and into Zimbabwe, as far north as the Zambesi escarpment. Loveridge and Macdonald radio-tracked 11 jackals of each species near Hwange NP in Zimbabwe, finding that there was some spatial partitioning by the two species, with black-backed jackals favouring grassland and side-striped, thicker bush or woodland, and seeming to avoid grassland, even though where it does not overlap with the black-backed jackal, it regularly uses grassland as part of its habitat.[129] There was also some partitioning in terms of periods of activity, with the black-backed jackal more active during the day. When they did use the same areas at the same time, Loveridge and Macdonald observed that black-backed jackals chased the side-striped away or that the side-striped retreated and would avoid the presence of the other species, even when there was no chasing.[130] Not all the encounters and chases involved competition over food and chases covered 15–50m. Where food was involved in a contact, the side-striped often retreated or took cover in the bush and waited for the black-backed to leave.[131] Nattrass et al. also recorded that black-backed dominated side-striped, even though they were smaller than them.[132] They also noted that black-backed jackals were more willing to take the risk of feeding at or waiting near carcasses being fed on by lions or spotted hyenas than side-striped jackals.[133]

In the Serengeti, black-backed jackals and golden wolves occupy different habitats and breed at different times, which may lessen competition and conflict. The black-backed usually give birth during the dry season when grass rat numbers are at their peak, and fruit and berries are widely available. The golden wolf produces pups in the wet season, when the Thomson's gazelles are giving birth, providing the wolves with plentiful fawns to hunt and placentas and stillbirths to scavenge.[134] Moehlman observed that the golden wolves had smaller territories than the black-backed jackals and appeared to have weaker pair bonding than them.[135]

Black-backed jackals are dominated by spotted hyenas and generally by brown hyenas.[136] They will try to steal from carcasses being fed on by spotted hyenas, and I have seen many examples of this from kills I've seen in the Maasai Mara, Serengeti, Ngorongoro Conservation Area, and Kruger; they are easily seen off and usually only get scraps until the hyenas finish. Jackals have a more complicated relationship with brown hyenas and will scavenge alongside them on Namibian beaches with seal colonies but almost compete with them for carcasses in arid, inland habitats such as Damaraland in Namibia and the Kalahari in Botswana and South Africa. Brown hyenas are capable of killing or chasing off jackals, but often the jackals will be more daring and even nip at the rumps of brown hyenas in a way they would be far less likely to do with spotted hyenas. In some areas of South Africa's North West Province and in arid regions of Namibia outside the range of lions and wild dogs and where leopards and cheetahs have been exterminated or severely depleted, brown hyenas and black-backed jackals are apex predators. Van der Merwe found in parts of North West Province that there was considerable dietary overlap between jackals and brown hyenas, but that jackals killed more of their own food and relied less on scavenging than brown hyenas.[137] Both relied, to some extent, on scavenging from large carcasses, and there was a slightly variable dominance hierarchy at carcasses. Through size, weight, and jaw strength, a single brown hyena could dominate one jackal, but if several jackals gathered, they could intimidate a single hyena.[138]

Jackals also suffer from the depredations of both spotted and brown hyenas. In Serengeti, Lamprecht found the spotted hyena to be a regular robber of jackals which had killed gazelle fawns and even hares. He witnessed hares and gazelles being killed by black-backed jackals 17 times and on four occasions lost the entire animal to spotted hyenas and twice lost about half of the kill. He never saw a jackal successfully defend its kill from a hyena.[139] Jackals are more successful in defending their dens and cubs from marauding hyenas, and this may be because of the investment made in rearing cubs that makes the jackals more aggressive towards the hyenas, and the addition of a helper to defend the den should be taken into account. Hyenas appear to hunt jackal pups opportunistically and do not put in a great deal of effort.

As noted at the start in the Introduction, black-backed jackals have an interesting relationship with honey badgers. The jury is still out on the nature of that relationship, though the author veers towards a commensal one, with the jackal benefitting from it but with no evident value to the badger but no great loss either. It is possible that it could be interpreted more harshly as kleptoparasitism by the jackal, with the badger gaining nothing and potentially losing prey items to the jackal. The relationship may also involve pale chanting goshawks benefitting from honey badger foraging, notably digging rodents from their burrows, without conferring any benefit on the badger.[140] This relationship between the badger, black-backed jackal, and goshawk, which I witnessed on two separate occasions in the Central Kgalagadi Game Reserve, involves the badger foraging for rodents, reptiles, insects, and birds' eggs. When badgers detect prey underground, they use their strong claws and powerful front paws to dig out its prey and snap it up. But when the prey is fast or numerous, some may escape presenting the waiting goshawk and jackal with an opportunity to benefit from the badger's foraging. It is hard to see what the badger gains from this regular association, apart from perhaps an early warning system when its safety is compromised while head-down digging furiously. The gains for the jackal and goshawk are clear to see. But the badger's reputation for ferocity in defending itself makes the early warning explanation tenuous. Gorta supports this view, observing that jackals benefit from hunting in close proximity with badgers as "only jackals obtain a net benefit by exploiting prey flushed from honey badger diggings," concluding that at best this is a commensal relationship or perhaps a "producer-scrounger" one, with the jackal reducing the eneregy expended to get food by following badgers and benefitting from their strenuous digging.[141] He added that on one occasion at Okaukuejo in Etosha NP, a jackal grabbed a yellow mongoose flushed out by a badger, but that the badger then attacked the jackal and retrieved the mongoose.[142]

A photographer, Willem Kruger, witnessed an interesting black-backed jackal–honey badger interaction over the carcass of a giraffe killed by the lion pride, and one which indicated that the jackals had a healthy respect for the aggression and strength of the honey badger:

> When we first arrived ... a male lion was still feeding on the carcass on the grassy bank next to the road. The black-backed jackals kept a safe distance away from the carcass and the lion ... The next morning we drove north along the Nossob River back to the sighting. On arrival we saw that the lion had abandoned the carcass and the black-backed jackals had clearly made use of this opportunity – there were 15 of them around the kill! ... While watching the jackals, we heard movement in the grass bank next to our vehicle. To our surprise, a honey badger was moving through the grass very close to our vehicle. The honey badger passed right underneath our vehicle and approached the carcass. Some of the jackals were disturbed by the approaching badger, while others just kept on eating. The badger approached the carcass very cautiously, and surprisingly the jackals did not back away at all. He took his time to sniff around but did not try to eat anything. A few of the jackals followed the badger all around the carcass and this seemed to make him nervous. Eventually, the jackals persuaded the badger to leave the carcass without a fight and he made his way over the sandbank on the opposite side of the road. ... We followed him as he made his way towards the Kwang waterhole, followed by a jackal or two. He reached the waterhole and started to drink in the early morning sunlight under the watchful eye of a nearby jackal ... [later in the day] we once again passed the carcass and to our surprise found the same male honey badger now feeding on the carcass! As is expected, the jackals were not impressed with the presence of the badger. Some of them tried to intimidate the badger but unlike that morning, the badger was unperturbed and even showed aggression towards those that bothered him. However, most of the jackals kept a safe distance from him. Every now and again one of the jackals would approach the badger from behind, smelling him and one even tried to bite him! The badger's only reaction to this was to turn around and show his teeth. The jackals would then make hasty retreats to a safe distance.[143]

Jackals have a rather perilous relationship with domestic and feral dogs, even though they may kill them on occasions. There might be fatalities or serious injuries to jackals in conflict with packs of domestic dogs –and large dog populations appear to be linked with low-density jackal presence.[144] Jackals themselves will attack domestic dogs on occasions. Black-backed jackals are vulnerable to diseases found in domestic dogs, especially where veterinary care and vaccinations of dogs are low or non-existent. Studies have shown that jackals test positive to most transmissible canid diseases wherever the two species overlap and are "significant vectors of rabies in southern Africa."[145] Most of Namibia's reported rabies cases involving wildlife between 1986 and 1996 were in black-blacked jackals and were thought to be in populations near human settlements, as jackals in protected areas away from dogs had low rates of rabies.[146] The large jackal population in Namibia and its ability to adapt to the presence of humans and their dogs are believed to have a role in jackals spreading canine diseases between domestic dog populations which are geographically separated. A canine distemper virus (CDV) outbreak among domestic dogs in widely separated coastal areas was blamed on jackals, who passed the virus through their dispersal and large ranges in some areas, enabling the virus to enter new populations of domestic dogs.[147]

In Zimbabwe, statistics gathered by Bingham et al. showed that between 1950 and 1996, 78.8% of rabies cases in black-backed jackals occurred on commercial farms, 11.3% in the communal sector, 9.6% in urban areas, and just 0.3% in protected areas.[148] They also indicated that rabies spread from black-backed to side-striped jackals.[149] Rabies has also spread from domestic dogs, especially free-roaming ones in communal farming areas, causing epidemics of rabies in both black-backed and side-striped jackals in Zimbabwe. Jackals can also contract parvovirus through contact with dogs.[150] Jackals also suffer from and can play host to babesiosis, which can be transmitted between jackal species and to and from jackals and domestic dogs.[151]

Research in Kenya has shown that black-backed jackals can be infected by *Ehrlichia canis*, a tick-transmitted disease, which can be transmitted between jackals and domestic dogs.[152] In the Serengeti-Maasai Mara ecosystem of East Africa, jackals are very vulnerable to CDV spread from domestic dogs kept by pastoral communities to guard their livestock. In 1978, a severe outbreak of CDV killed a large number of jackals in this ecosystem when the disease grew to epidemic proportions.[153] Alexander et al. reported that a CDV outbreak in 1994 led to a large but temporary decline in black-backed jackal numbers in the Maasai Mara Reserve.[154] In 1994, lions in the Serengeti became infected and many died from CDV, with jackals and hyenas blamed for transmitting the virus to lions through contact with domestic dogs.[155] The growth in human populations and growing encroachment of people into wildlife areas, as well as the ability of jackals to adapt to human presence and opportunistically feed on waste produced by human activity, could have long-term implications for jackal numbers because of epidemics resulting from jackal–dog interaction.[156] With an estimated 500 million dogs living alongside humans globally and a significant number in sub-Saharan Africa free-roaming or totally feral, the chances of disease transmission to numerous canids like jackals are substantial and increasing.[157]

Interactions with humans

Black-backed jackals are not popular with humans in farming communities in East and southern Africa, the latter in particular. In Chapters 4–7, there will be extensive details of the war fought against jackals in South Africa by farmers following

European occupation and the development of large sheep farms – a war that continues today with huge numbers of jackals, probably numbering in the millions, trapped, poisoned, or shot over more than 300 years. In this section, I will give the broad outlines of human–jackal interactions to set the scene for the later historical narrative and description of the current state of coexistence and conflict.

Much of the conflict is in areas that are not formally protected and on rangelands that are too arid for intensive crop production. The result is that livestock husbandry or game farms are crucial to human livelihoods and bring humans and their stock into conflict, often fatal to the jackals and sometimes to the livestock.[158] In many of these areas, the prey base of the jackals has been depleted – notably through the destruction of the massive herds of springboks and the numerous other antelopes and wild ungulates that inhabited southern Africa. The larger carnivores – especially lions, wild dogs, and cheetahs – were exterminated in the 18th and 19th centuries, with some leopards remaining but black-backed jackals and smaller cats like caracals becoming the apex predators and turning to killing sheep, goats, and even young cattle to supplement the reduced prey base resulting from human hunting and changes in land use from unfenced rangelands to livestock farms with greatly reduced wild prey. On game farms, which have become more and more common in the last 50–60 years, jackals are not welcome, as they will predate the young of wild ungulates being bred to be hunted or to be sold on in the extensive game animal sector, and adults of smaller game species like springbok, duiker, steenbok, and dik-dik. Livestock and some game farmers "kill predators in retaliation for livestock predation or as a pre-emptive measure to reduce predator numbers and the probability of predation."[159] Predator-proof fences are used, not always with great success, to keep jackals from livestock farms.[160] Minnie noted the adaptability of jackals and their survival in the face of widespread and centuries-long persecution, adding that they "are the dominant cause of livestock predation in South Africa, causing financial losses reputed to be over US$90m per year."[161]

Despite centuries of culling and the development of more efficient predator-proof fences, the adaptability of the black-backed jackal to different diets, habitats, and ways of avoiding humans, plus what James calls "remarkable population plasticity," has enabled them to survive and breed to replace those culled.[162] This has constantly raised questions about the efficacy of predator control measures in the face of a highly resilient species. Nattrass et al. have suggested that culling doesn't seem to work, nor does taking no action, so "farmers might be better off having a dominant territorial pair on their land, rather than killing them, thereby creating a 'sink' attracting (perhaps several) dispersing jackals."[163] The drawback there is that if the livestock is not in some way secured, jackal pairs might tolerate other intruding jackals because of the abundant food supply. Nattrass et al. cite a South African hunter on the difficulties of controlling numbers, even with a dominant breeding pair in place:

> The story that the good jackals keep others away is not entirely the truth. I sat on a particular farm … within two hours I had shot 11 adults, without

> moving from my spot. It was June and there were five pairs and a really old male whose mate had almost certainly died of old age. How come the dominant jackal pair had not done their work?[164]

Minnie found that in areas with extensive hunting of jackals, increased breeding filled the gaps created. Neighbouring areas with no or low hunting provided reservoirs of jackals from which animals dispersed into the empty areas where jackals had been exterminated or reduced in numbers.[165] With an adaptable species like black-backed jackals, there is clearly no easy, single answer to0 human–jackal conflict over livestock predation. The same also applies, as Sillero-Zubiri, Reynolds, and Novaro found, with golden wolves.[166]

The other human effects on black-backed jackal populations, though not ones that seem to threaten survival in any particular region or reduce numbers substantially, are jackal mortality on roads, with road kills being one of the major causes of adult mortality in areas of high human activity, and hunting for skins (perhaps a by-product of culling for other reasons) with the use of skins and furs to make karosses (traditionally used as capes or rugs and widely sold to local people and tourists).[167]

Side-striped jackals
Order: Carnivora
Family: Canidae
Genus: Lupulella (previously Canis)
Species: Lupulella adustus (previously rendered as Canis adustus)

With three sub-species – *Lupulella adustus adustus* (Sundevall side-striped jackal of Central and southern Africa from Angola, Gabon, and Congo in the west, through DR Congo and into Zambia, Zimbabwe to Mozambique, northeastern South Africa, and eSwatini in the east), *Lupulella adustus lateralis* (Equatorial Africa side-striped jackal of West, Central, and East Africa from Senegal and Mauritania in the west across the Sahel belt and northern Central Africa to Kenya, Tanzania, and Mozambique in the east), and *Lupulella adustus kaffensis* (Kaffa side-striped jackal of Ethiopia, Somalia, and probably Kenya, Uganda, and South Sudan).[168]

Size, weight, and physical characteristics

Although the background colours of the coat and the size and prominence of the stripe along both flanks may differ between the sub-species and, to some extent, on a geographical basis within each sub-species, the side-striped jackal generally has a tan, buff, or greyish-yellow coat with a white stripe from elbow to hip, often with a white tip on the tail. Stripes are not always very obvious. Head-and-body length: 65–90cm. Tail length: 27–41cm. Shoulder height: 38–50cm. Weight: 5.9–14kg, males somewhat larger than females; males mean of 9.4kg,

FIGURE 1.5 Side-striped jackal. Copyright: iStock.

females 8.3kg.[169] There is some variation in recorded body sizes with Castelló recording the three sub-species as having body lengths of between 62 and 77.5cm for males and between 60 and 77cm for females, with body weights of 7.3–12.1kg for males and 7.2–10kg for females (Figure 1.5).[170]

Distribution and conservation status

The side-striped jackal "has the second widest distribution of the 12 species of canid occurring on the [African] continent, distributed in cc33% of the land area. Only the golden jackal, *Canis aureus* [now called the African golden wolf], a generalist feeder, has a wider distribution, at c40%."[171] The ranges of the three sub-species are given above, and all are listed as of Least Concern, though with localised variations. The Sundevall side-striped jackal is locally Near Threatened in South Africa.[172] They are not as numerous as black-backed jackals within their ranges and are described by Ginsberg and Macdonald as "are rare, but not threatened, throughout their range."[173] They are not seen, unlike black-backed jackals, as major threats to stock, so persecution is lower (except where they may be treated as vermin, along with black-backed, where ranges overlap), and they are not killed for their skins, although in Uganda, the Buganda people use their hearts as a traditional cure for epilepsy (Figure 1.6).[174]

The side-striped jackal is the least studied of the jackals. Side-striped prefer moist savannas and denser bushes or woodlands.[175] The West African ranges of the Equatorial side-striped jackal are less researched than the Central, East,

Distribution Map

Canis adustus

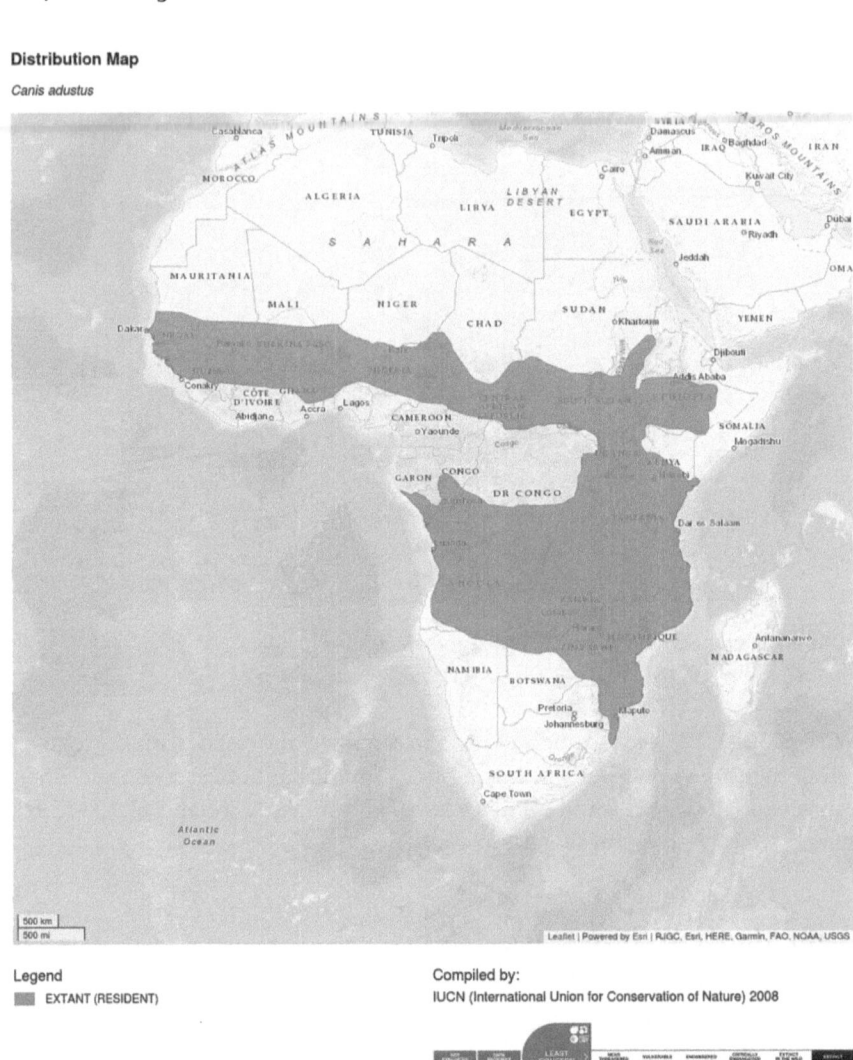

Legend

▮ EXTANT (RESIDENT)

Compiled by:

IUCN (International Union for Conservation of Nature) 2008

FIGURE 1.6 IUCN map of the distribution of the three species of side-striped jackal. Copyright: IUCN. *Canis adustus* (Side-striped jackal) (iucnredlist.org).

and southern African ones. It is believed to include Benin, Burkina Faso, Côte d'Ivoire, Gambia, Ghana, Guinea, Guinea-Bissau, Mauritania, northern Nigeria (Gombe, north of Benue, and Borgu Game Reserve in the west),[176] Senegal, Sierra Leone, and Togo.

Social behaviour and breeding

Like black-backed jackals, side-striped live in mated pairs, which can develop into small family groups, or as solitary, nomadic individuals that have yet to pair up.[177] Home ranges are quite small and may be as low as from 0.2 to 1.2km^2 in some areas[178] but can be as large as 12–20km^2.[179] Densities vary seasonally according to food availability – though is less likely on farmland or around human settlements, where food availability seems more even, with side-striped jackals being avid foragers of human waste. They will even venture into towns at night in search of food.[180] In the Zimbabwean highveld, towards the southern edge of the Sundevall range, there were 20–30 side-striped jackals per 100km^2, with numbers increasing to 80–120 in the breeding season.[181]

The mating seasons vary across the ranges of the three sub-species – June–July/September–October in East Africa; June–November in South Africa; and June–July, but with some mating throughout the year, for the Equatorial sub-species range.[182] Gestation period is 57–70 days. Litter sizes average 3–6, with a mean of 5.4. Pups are weaned at 8–10 weeks. They reach sexual maturity at 6–9 months and may disperse at 11–12 months, though some sub-adults stay with the parents to help raise the next litter, leading to family groups of 7–9. Their life expectancy is around 10–12 years.[183] Most of the data here is based on the Sundevall and Equatorial sub-species with little research data on the Kaffa side-striped jackal.[184]

Diet, foraging, and hunting

Side-striped jackals, possibly even more so than black-backed jackals, are opportunistic omnivores with a wide prey base. Prey ranges from termites, through lizards, snakes, ground-based birds, and their eggs, to gazelles and other small antelopes and occasionally old, injured, or sick adults.[185] Atkinson, Macdonald, and Kamizola studied the diet in the highveld region of Zimbabwe, examining 2,752 scats, finding that "half of the fresh-weight biomass of the diet of side-striped jackals *Canis adustus* comprised small and medium-sized mammals, and a further third consisted of fruit."[186] Multimammate mice, bushveld gerbils, and scrub hare were most common prey species, while mobola plum, chocolate berry, wild fig, and waterberry were the most commonly eaten fruits. They noted that captive jackals always preferred meat to fruit. The authors concluded that prey consumption did not increase when there was a high density of prey species and that "the feeding of jackals is largely geared by searches for fruit, and that mammals are taken opportunistically when they are encountered."[187] Fruit made up the largest single weight of food digested (30%) and, combined with small mammal remains, made up half the weight of food digested. Small amounts of remains of domestic cattle were found in the scats examined, and a separate study suggested there was less evidence than for black-backed jackals of regular predation of livestock.[188] The research also found that side-striped jackal foraging strategies appear to be "driven by cost-effective searches for food. Jackals spend longest in those habitats which contain the widest variety and abundance of easily obtainable food."

Interactions with carnivores

There is little published research on the relationships of side-striped jackals and other carnivores. As noted above in the black-backed jackal section, jackals are subordinate to lions, leopards, and spotted and brown hyenas, though the latter are far less common in the regions populated by side-striped jackals. They would be as vulnerable as black-backed to predation by leopards and opportunistic or competition-based killing by lions and hyenas. In the author's limited experience of observing them in Hwange NP in Zimbabwe, the Serengeti NP in Tanzania, and Kasungu NP in Malawi, they did not exhibit the same behaviour of congregating in numbers beyond the usual territorial family group around lion or other predator kills as black-backed jackals did.

As detailed above, side-striped jackals behave in a generally subordinate fashion to black-backed jackals, avoiding contact or being easily chased away from food sources or territorial cores by the black-backed. Like black-backed jackals, side-striped are infected by the canine diseases such as rabies, parvovirus, and CDV. Foraging around human settlements brings them into contact with domestic or feral dogs, and they have suffered epidemics of rabies in Zimbabwe, especially in communal farming areas.[189]

Interactions with humans

There is far less evidence of side-striped jackals preying extensively on livestock; they are still blamed and persecuted. Often fatalities occur in areas of overlap with black-backed jackals as a result of traps or poison used against the black-backed. Side-striped jackals, like the other species, are frequently blamed for stock-raiding, leading to persecution, though this seems to have had little serious effect on side-striped jackal numbers. Farmers shoot, snare, and poison jackals, rarely discriminating between species. In the 1950s, side-striped jackals were trapped and poisoned in Uganda to try to limit the spread of rabies.[190] Guilt by association is amplified by the side-striped's regular foraging for livestock remains and human waste at night.[191]

Eurasian golden jackals
Order: Carnivora
Family: Canidae
Genus: Canis
Species: Canis aureus

There are six sub-species – the Persian jackal (*Canis aureus aureus*), the European jackal (*Canis aureus moreoticus*), the Syrian jackal (*Canis aureus syriacus*), the

Indian jackal (*Canis aureus indicus*), the Sri Lankan jackal (*Canis aureus naria*), and the Indochinese jackal (*Canis aureus cruesemanni*). There is some dispute over whether the Indochinese jackal is a separate sub-species "as its classification is based mainly on observation of captive animals."[192]

Size, weight, and physical characteristics

Giannatos described the jackal as "a medium-sized canid usually the size of a cocker spaniel dog." The animals in northern, central eastern, and south-eastern Europe, Asia Minor, and Caucasus (and now spreading into western Europe) belong to the sub-species *Canis aureus moreoticus*, which seems to be one of the largest jackals (Figure 1.7).[193]

Across the six sub-species, the colour and length of coat vary with season and region but are usually a pale gold-brown or brown tipped with yellow and grey. Fur is coarse and not usually very long.[194] Along the back, the fur is often black and grey, while the head, ears, and sides can be rufous. Underside is frequently ginger or light brown with a white area under the throat and on the chest. The tip of the tail is black. There is sexual dimorphism of approximately 15% in body weight, with males larger and heavier than females. Head-and-body length: 60–90cm. Tail length: 20–30cm. Shoulder height: 38–50cm. Weight: 7–13kg – as above, the European jackals (*Canis aureus moreoticus*) are slightly larger than the other sub-species.[195]

FIGURE 1.7 Eurasian golden jackal. Copyright: iStock.

Distribution and conservation status

The Eurasian golden jackal is the most widely distributed of the jackals. Its sub-species were widely distributed but until recently only in south-east Europe, West Asia, South Asia, to Burma and Thailand,[196] but over the last 30–40 years, the jackals of south-east Europe and the Balkans have increased in numbers and dispersed north and west into Central and eastern Europe, reaching as far north as Estonia and Finland and into Germany, France, Belgium, the Netherlands, and Denmark.[197] Its European range is expanding very quickly and widely, with ever-increasing reports of jackals moving into the areas of northern Europe and western Europe where they have not been present before.[198] Bulgaria has long housed the largest European golden jackal population and appears to have been the focal point for population increase and dispersal over several decades.[199] The expansion of the golden jackal in Europe will be covered in greater detail in Chapter 5 (Figure 1.8).

Like its African cousins, the Eurasian golden jackal sub-species are very adaptable in terms of habitat, ability to survive in close proximity to humans, and omnivorous diet.[200] They are locally common throughout their range. They can survive in semi-arid desert, grassland, scrub, foothills of mountain ranges, forests, and jungles (the latter in South and Southeast Asia) and have expanded from tropical and temperate regions into the colder forested areas of Central and North-eastern Europe.[201]

The start of the movement of Eurasian jackals into Europe is put at the early Neolithic period in human development, between 7000BCE and about 1900BCE. The dispersal from south-eastern Europe and West Asia was through the Balkan Peninsula, and this has been particularly evident in the late 20th and early 21st centuries.[202] Researches into the area of origin of the golden jackals have involved molecular testing, which has revealed demographic expansion in the Indian subcontinent, which can be traced back ~37,000 years ago during the late Pleistocene, and the "results suggest that golden jackals have had a potentially longer evolutionary history in India than in other parts of the world, although further sampling from Africa, the Middle East and south-east Asia is needed to test this hypothesis."[203] In India, the jackals are found in protected areas, semi-urban settings, and rural farming landscapes, only being absent from the higher regions of the Himalayas, being particularly common in grassland and farming areas of western and northern India with substantial livestock husbandry, providing carcasses for jackals to scavenge.[204]

Social behaviour and breeding

Golden jackals have social systems broadly based on a mated pair with pups and perhaps sub-adults as helpers.[205] But they are flexible and are sometimes found in loose packs of related individuals,[206] which may coalesce and break up according to local conditions and food availability. In areas where food is abundant, they have been recorded in groups of up to 20 members.[207] Areas with a very

Distribution Map

Canis aureus

Legend

■ EXTANT (RESIDENT)

Compiled by:

IUCN (International Union for Conservation of Nature) 2018

FIGURE 1.8   IUCN map of the distribution of the Eurasian golden jackal species. Copyright: IUCN. *Canis aureus* (Golden jackal) (iucnredlist.org).

concentrated food supply will support a high density of jackals; the presence of human food sources, such as waste dumps, has been shown to increase the numbers of golden jackals, notably in the Israeli-occupied Golan Heights.[208] Vocalisations are used to communicate within pairs, family groups, and packs – using a mixture of yapping, howls, and screaming yells when excited or scared.

Scent marking using scats and urine will be used to mark territory boundaries to deter intrusions. Macdonald and Sillero-Zubiri found that "patterns of scent marking behaviour that had never been seen at low densities became conspicuous in a population living in large groups on small territories, and they also displayed such wolfish behaviour as mustering for territorial patrols."[209]

The size of a pair's or a pack's territory varies from 1km² in high-quality habitats to 12km² in low-quality ones.[210] In food-abundant regions, golden jackals can live in high densities, with the result that large populations that are viable in the long term can exist in quite small areas.[211] Giannatos said that in a fertile area on an alluvial plain at Mornos in Greece, 30 jackals inhabited an area of 18km² (0.6 per km²), while in less fertile areas, density could be one jackal per 2.5km².[212] He also reported that populations can expand quite fast in favourable conditions, citing the Greek island of Samos, off the Turkish west coast, where in a period of 3 years, a rapid increase from 13–14 to 20 distinct groups was recorded.

Macdonald's extensive study of the social behaviour of golden jackals at a site with very abundant sources of human-produced food in Israel found that jackals obtained 92% of their food from a single feeding site created by people and that the reliable supply of food meant that jackals had formed into larger groups – one of 10 and one of 20, with a stable composition and neighbouring territories close to the feeding site.[213] The site was on a nature reserve 10km south of the Palestinian town of Jericho in the Israeli-occupied West Bank. Jackals had survived there despite the 1964 Israeli jackal poisoning programme to contain rabies, because Israel had only occupied the area during the June 1967 war. On the reserve, there were two feeding sites at which jackals could regularly obtain food, in addition to prey species. One was a specific carnivore feeding site at which dead poultry, even entire carcasses of cattle and donkeys, and other off would be deposited to feed carnivores. The other site was a large waste pit into which refuse from visiting tourists and a cafe were dumped.[214] There were about 70 jackals living on the reserve, and 35 at a time could be seen at feeding sites.

Macdonald's study provides fascinating detail of the social behaviour within and between the two groups. He said that just before dusk, jackals would start to move towards a feeding site:

> They would meander slowly into the open, select a slight ridge or mound, and there sit or lie until dusk. Frequently several individuals sat together … Most of this time was passed in stretching, yawning and sleep; there were also bouts of play … consisting of feints and chases, shoulder-barging contests and attempts to grab each other's scruff (often ending in both partners clashing their teeth together).[215]

He also described the friendly greetings when jackals met or prepared to join up on what he called the "dusk patrol," with perhaps two or three jackals starting out and being joined by others along the way, practically always following the same route. When groups of jackals met at the feeding site, there might be short disputes over food between individuals or several jackals, which occasionally went beyond aggressive postures and vocalisations and resulted in brief fights, which seemed fierce but didn't seem to result in serious injury. Strangers, who did not seem to be part of the two main groups, would often be chased away by several jackals from the local groups.[216] Macdonald concluded, quite understandably, that the group size, behaviour at food sources, and the nature of conflicts were an adaptation to this unusual abundance and regularity of feeding that enabled groups to form and stay together and competition not to escalate into fights that would result in death or serious injury.[217]

Mating varies according to geographical region, climate, and peaks of food availability. In Romania, mating usually occurs in October or a little later and with a gestation period of about 61–63 days, pups are born from late January onwards. In other regions mating would take place in different months – February–March in southern Russia and former Soviet Central Asia and October–February in Israel.[218] Litters can be up to six pups, but two to four is more usual.[219] The pups are weaned within 50–90 days, and sexual maturity is reached around 11–12 months, though may be later in sub-adults that stay to act as helpers with the next litter.[220]

Diet, foraging, and hunting

Golden jackals exhibit opportunistic feeding behaviour, with an extremely varied diet and are combined hunters/foragers. They prey on small mammals, birds and their eggs, amphibians, reptiles, and even invertebrates; they take carrion and refuse from waste dumps when available and feed on vegetables or fruits.[221] As part of its foraging behaviour, the golden jackal has become tolerant of human disturbance allowing it to become established in farmland and around populated areas to take advantage of a variety of forms of waste produced by human activity, thereby providing "ecosystem services by removing discarded animal remains and facilitating detritus food chains."[222] Despite evidence of regular feeding on human-produced waste, Tsunoda and Masayuki's research into the diet of Eurasian jackals found that small mammals were the staple food across their range but that jackals switched intake according to availability of different foods available on a local or seasonal basis.[223] The reduced availability of small mammals in peri-urban or urban areas means that consumption of waste and plant material increases there, compared with farmland. In higher latitudes in Europe, into which jackals have been dispersing for a number of decades, evidence suggests a greater intake of small mammals, meaning they are in competition with the smaller red fox,[224] and their size and ability to hunt in mated pairs or small family groups would give them an advantage over foxes as predators.

The Tsunoda and Masayuki study of published sources on jackal diet recorded 259 different food species: 81 mammals, 31 birds, 8 reptiles, 3 amphibians, 5 fish, 50 invertebrates, and 81 plants. The main foods in terms of proportion of scats containing them were small mammals (mainly rodents) 41.8%, plants (mainly fruit) 35.1%, and domestic animals 31.5% – the latter mainly from scavenging remains rather than predation.[225] Invertebrates were also eaten, particularly in the warmer months. In Israel, there is evidence of jackal predation on newborn calves, in areas where cattle are left to graze unattended all year round, while in the Greek Peloponnese, some predation of sheep and goats has been recorded.[226] A study across jackal habitats in south-east Europe found that small mammals made up 54% of the biomass of food intake and that where domestic animal remains were present, they were primarily consumed as carcasses, with "scavenging of improperly discarded (e.g. illegally dumped at the side of the road or left in the fields or woods) carcasses of livestock" an important part of the overall food intake, with no documented cases of jackals hunting domestic livestock.[227] Where apex predators such as Eurasian lynx are present in the jackal range, jackals can feed on leftovers from kills, but they are less likely to feed from wolf kills, "as wolves actively drive golden jackals from their territories."[228] In India, examination of scats in the Kanha NP revealed occurrence rates in jackal scats of 64% rats or mice (common around villages), 11% reptiles, 9% birds, 8% fish, 2% hares, 2% chital deer, 17% fruit, 6.5% garbage, and 3% grass or seeds.[229]

Interactions with carnivores

In much of its Eurasian range, the golden jackal is sympatric with the wolf, the brown bear, the lynx, and smaller carnivores such as foxes, otters, weasels, stoats, and badgers. In India, it is found alongside the wild dog or dhole, the tiger, and leopard and, in the Gir region of Gujarat, the Asiatic lion, as well as the Asiatic brown bear, the sloth bear, the wolf, and smaller *felids* and *mustelids*, including the honey badger. In Sri Lanka, it is sympatric with the leopard and smaller cats and in Indochina, with small numbers of tigers, dholes, leopards, and smaller carnivores.

The jackal's relationship with an apex predator such as the tiger has not been studied in detail, but the jackal has long had the reputation, totally misplaced, that it is the tiger's provider, leading it to prey. This idea dates back to folklore and to British hunters' sightings of jackals in close proximity to tigers with no sign of aggression or conflict. A 19th-century naturalist, F.J. Hill, wrote that "the name of lion or tiger provider has stuck to jackals, even though only the 'most credulous' would believe the jackal actually leaders lions or tigers to prey."[230] Rather than being providers who lead tigers to kills, jackals will follow hunting tigers or detect their kills and scavenge from them. Henry Shakespear was a prolific hunter who wrote of his experiences hunting big game in India in the 1850s; described jackals as scavengers of tiger kills, including bodies of people killed by man-eaters; and noted the habit of jackals following tigers in the hopes of benefitting from any uneaten kill.[231] As in Africa with black-backed and side-striped jackals, leopards may kill them for food.

There is a lack of research or even anecdotal accounts of the nature of the jackal's relationship with dholes or Indian wolves, even though in Europe and parts of West Asia, the wolf is often thought to exclude jackals from their territories by direct attack or the threat thereof (see Chapter 5 for more detail on interactions in Eurasia) because of dietary overlap and so competition. In Greece, for example, wolves are mostly found in semi-mountainous areas, while jackals are still scarce in comparison with neighbouring countries and are generally found in areas with few or no wolves.[232] In Hungary, the jackal is believed to avoid areas settled by wolves, but both are found in remote mountainous areas.[233] In Iran, there have been documented cases of wolves killing jackals – notably a jackal being killed by a collared wolf that was part of a study.[234] As will be shown in Chapter 5, there are indications that jackals avoid area with wolves, though there is no conclusive evidence, and the recent expansion of jackals across Europe has yet to be investigated in terms of jackal dispersal into areas with significant wolf populations.[235] It is worth noting that wolves inhabit mountainous and heavily wooded areas, which are not ideal jackal habitat.[236] Shakarashvili et al. did note, however, that in parts of Anatolia, Southern Caucasus, Iran, Afghanistan, Pakistan, northern India, and some other parts of Asia, the ranges of the grey wolves and the golden jackals overlap without signs of excessive conflict or wolf exclusion of jackals.[237]

In India, domestic and feral dogs are reported to be a cause of major wildlife loss, with 80 species reportedly attacked, particularly wild ungulates but also carnivores, with 45% of attacks leading to the death of the animal attacked.[238] India has an estimated dog population of 60m, with a large proportion outside heavily populated urban area free-running and a large urban and peri-urban feral dog population. Fox species were the most common target for dogs among carnivores, followed by jackals and wolves.[239] In Greece, there is also evidence of a high number of stray dogs in rural and peri-urban areas and regular interactions between domestic or feral dogs and jackals near livestock enclosures, villages, and isolated human settlements. But there is no conclusive evidence of the influence of dogs on the presence or absence of jackals.[240] Giannatos did find, though, that the presence of stray dogs around waste dumps did appear to keep jackals away.[241]

Interactions with humans

Like their black-backed and side-striped relatives, golden jackals come into conflict with humans on a number of levels – suspected or actual predation of livestock, conflict with domestic dogs, human hunting of jackals, in particular, or generalised culling and persecution of carnivores and loss of fruit and other crops that jackals feed on. The latter can involve low-level and low-value crop-raiding or regular crop-raiding that can severely reduce harvests.[242] In the case of golden jackals in parts of Europe and Asia, crops such as figs, grapes, watermelons, and sugarcane are among the most affected by jackals, which will take significant quantities of fruits and other plant matter as part of their diet.

In Hungary, jackals have dispersed in reasonable numbers in recent decades, having previously been hunted out or excluded by agricultural developments from most of the country in the middle of the 20th century. Since their expansion, hunters have complained of jackal predation of deer, wild boar, and game birds, while farmers have alleged livestock predation by jackals, leading to controversy over their return and strongly divided opinions about how to deal with it.[243] Diós, Ulicsni, and Molnár carried out a survey among hunters, who believed that numbers of roe and fallow deer had declined since the return of the jackal, leading to lower annual offtake by hunters. In some areas, hunters said that there had been a 53% decline in fallow deer.[244]

Ray, Hunter, and Zigouris observed in their study of jackals that while golden jackals on the whole create fewer problems for humans than other species, they were still blamed for livestock and game predation and damage to food crops, especially fruits, leading to conflict with hunters and farmers, resulting in culling and predator control programmes that often involved the fairly indiscriminate use of poison on carcasses of dead livestock.[245] Despite this, golden jackals across their range are found in close proximity to humans. In Bulgaria, thought to be one of the focal points for jackal expansion and dispersal in south-eastern Europe, jackals have increased in numbers despite being hunted for sport and as a pest.[246] There was no hard evidence of jackals predating livestock. While more study over the long term is required to ascertain just how much crop and livestock food are taken by jackals and the extent of damage caused to farmers, the jackal is viewed as a serious pest in Bulgaria and hunted as a result.[247] Between 2000 and 2008, between 5,000 and 29,000 jackals are believed to have been killed annually in Bulgaria.[248]

African golden wolf
Order: Carnivora
Family: Canidae
Genus: Canis
Species: Canis lupaster
Sub-species: Canis lupaster lupaster (North African golden wolf), Canis lupaster anthus (West African golden wolf), and Canis lupaster bea (East African golden wolf).

The African golden wolf has been included here and will be referred to frequently in the narrative sections of the book because of its long categorisation as part of the jackal family, only recently changed (see below and in Chapter 2). It has also had a central role in early human beliefs, myths, and representations of jackals, something which has affected human understanding (and, of course, misunderstanding) of the nature, ecological roles, and interactions with humans of all the jackal species, not to mention its deification in Egyptian religious beliefs and long association with death and the afterlife.

Size, weight, and physical characteristics

Head and body length: 72–93cm; tail: 29–34cm; height at shoulder: 38–40cm; weight: 10–15kg. There are no major differences in size between male and female, though the former might be slightly heavier.[249] The East African sub-species, *Canis lupaster bea*, is smaller at 64–85cm for head-and-body length; tail: 20–27.5cm; weight: 6–10kg, with females slightly smaller and lighter than males (Figure 1.9).[250]

FIGURE 1.9 African golden wolf. Copyright: iStock.

Distribution and conservation status

The North African sub-species is found in Western Sahara, Morocco, Algeria, Tunisia, Libya, Egypt, Sudan, South Sudan, Chad, central African Republic, Niger, Nigeria, Burkina Faso, Mali, Mauritania, Eritrea, Ethiopia, and Somalia. It inhabits desert margins and arid areas, farmland, scrub, rocky hills, and mountainous regions, such as the Bale Mountains of Ethiopia. The West African golden wolf is found only in Senegal, though it is possible that it is also present in Gambia. It inhabits semi-arid areas, grassland, and scrub. The East African

golden wolf is found in Kenya and northern Tanzania and has been seen occasionally in Uganda and Rwanda. They are found mostly in dry grasslands,
open bushlands, and semi-deserts but often frequent the outskirts of villages and
towns. None of the three species has a conservation status given by Convention
on International Trade in Endangered Species of Wild Fauna and Flora as they
are believed to be widespread and fairly numerous.[251] The main threats are human persecution, feral dogs, and diseases (Figure 1.10).

Wolf or jackal?

The African golden wolf, especially the North African sub-species, has, over
time, been variously referred to as a sub-species of the Eurasian golden jackal
and as the Egyptian wolf, sometimes also as the Egyptian jackal. The West
African sub-species has been called the Senegalese jackal, and the East African sub-species the golden jackal, as in Hugo and Jane van Lawick-Goodall's
Innocent Killers.[252] As will be noted in the historical chapters, early writers
on the natural world and then amateur naturalists often described the North
African sub-species, which was best known to Greek and Roman writers, as
being like wolves. Over time, naturalists placed golden wolves in the jackal
family, seeing them as sub-species of Eurasian golden jackals. As Campbell
noted, taxonomy was very imprecise, and "[e]arly literature as far back as
Aristotle described a wolf-like animal in north and west Africa and scientific
descriptions of the species were first published in 1820 (*Canis anthus*) as the
Senegalese jackal and in 1832 (*Canis lupaster*) as the Egyptian wolf."[253] Aristotle had referred to wolves from Egypt and noted that they were smaller than
the wolves from Greece.[254] Herodotus also recorded that Egyptian wolves
were smaller than their Eurasian counterparts – given that he would have been
familiar with wolves and jackals, he clearly believed the Egyptian animals to
be closer to wolves.[255]

By the 20th century, all three sub-species were placed as part of the *Canis
aureus* species, the Eurasian golden jackal, in a 1926 classification, which was
entrenched in a 1939 publication *African Mammals*.[256] In her study, Campbell
noted that the golden jackal misnomer continued for 70 years until questions
began to be asked by canine specialists about wolf-like jackals observed in Eritrea in 2002. These had larger ears than most jackals and a longer tail, and local
people called them wolves. While examples of that particular animal have not
been photographed, captured, or more closely identified, this opened up the
question of the classification of the three jackal/wolf species of the northern half
of Africa. Genetic research using samples taken from Israeli golden jackals and
Egyptian ones revealed that the Egyptian ones were more closely related to grey
wolves. Rueness et al. looked at the question of what they called the "Cryptic
African Wolf" in 2011 and came to the conclusion that the DNA analysis of
what had been called the Egyptian jackal (*Canis aureus lupaster*) indicated that it
should actually be placed "within the grey wolf species complex, together with

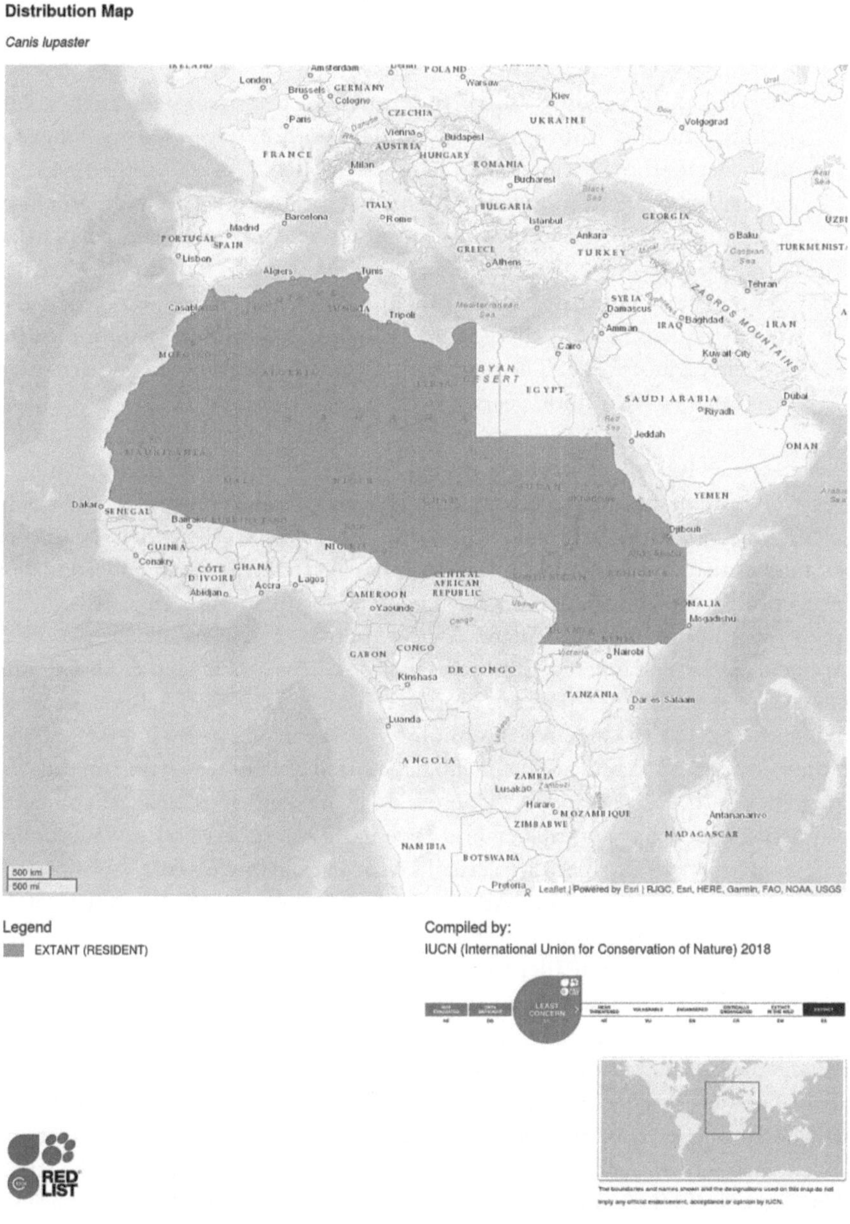

Distribution Map

Canis lupaster

Legend

▓ EXTANT (RESIDENT)

Compiled by:

IUCN (International Union for Conservation of Nature) 2018

FIGURE 1.10 IUCN map of the distribution of the African golden wolf. Copyright: IUCN. *Canis lupaster* (African wolf) (iucnredlist.org).

the Holarctic wolf, the Indian wolf and the Himalayan wolf … [and it] seems to represent an ancient wolf lineage which most likely colonized Africa prior to the northern hemisphere radiation."[257] They noted that the same was true of golden jackals from Ethiopia.

In later studies, DNA samples from "jackals" in Ethiopia, Senegal, Algeria, and Mali were compared. The results

> revealed that not only are these animals distinct from Eurasian golden jackals, but also from grey wolves. The authors proposed to name this new species the African golden wolf, *Canis anthus*, based on the first name believed to have been given to them. Furthermore, rather than an elusive wolf hiding among jackals, they found no evidence of true golden jackals in Africa; all 'golden jackals' sampled across North, East and West Africa were actually golden wolves. [which] wolves split from their closest relatives, grey wolves and coyotes, an estimated 1.3 million years ago during the early Pleistocene. The ancestor of golden jackals diverged earlier, approximately 1.9 million years ago.[258]

The golden wolves which evolved from grey wolves moved into Africa, while the grey wolves remained at the other side of Egypt's Sinai Peninsula, as did golden jackals, with golden jackal remains found in Sinai near the Egyptian–Israeli border, but no further west across Sinai, which has led to unanswered questions about the possibility of hybridisation between golden wolves and golden jackals.[259] The grey wolf is not found in mainland Africa, and the closest extant relative on the continent is the endangered Ethiopian wolf.[260] From their studies in Ethiopia, Rueness et al. concluded that the North African golden wolf "is present in the highlands of Ethiopia, effectively expanding the taxon's range by at least 2,500 km to the southeast."[261] Liz Campbell, who has carried extensive research on the African golden wolf, has concluded, "There is no evidence that golden jackals made it into Africa aside from a possible hybrid zone in Egypt, so all of what were thought to be golden jackals in Africa are actually African golden wolves."[262]

Recent fossil finds in Saudi Arabia's Nefud Desert include a mixture of African and Eurasian species, including *canid* remains that are closest to the African golden wolf, being smaller than Asian wolf species but larger than Eurasian golden jackals. This suggests a possible route to Africa of the golden wolf ancestors that evolved from grey wolves.[263] The North African golden wolves are in size and appearance nearer to their wolf ancestors than the slightly smaller East African golden wolf, which appears to have evolved further from the wolf original, becoming smaller and closer in size to the black-backed and side-striped jackals with which they coexist in Kenya and Tanzania.[264] Viranta et al.'s studies of African wolf DNA and physical characteristics confirmed that "there are significant differences in size between populations of *C. lupaster*, with East African individuals being smaller than North and West African ones."[265] It has also been suggested that the African golden wolf could have developed from "a hybridization event between the grey wolf and the Ethiopian wolf, *C. simensis*" and that different degrees of gene flow between the two species could account for the physical differences between the North and West African golden wolves and the East African ones.[266]

Social behaviour and breeding

Golden wolves are usually found in mated pairs, small family groups, or lone wolves which have dispersed from the family group on reaching sexual maturity. The territories of mated pairs vary according to density of prey and other food sources and the presence of other golden wolves or competing predators. Generally territory sizes are in the range of 1.1–20 per km^2.[267] Most research on golden wolf's social behaviour relates to golden jackals and predates the redesignation as golden wolves. Like the jackal species, they are usually in small family groups based on the mated pair, sometimes with sub-adults staying with the parents to raise the next litter of pups. As with golden jackals, they may form larger packs on a loose basis when there is an abundance of food available.[268]

The North African golden wolf usually produces a litter of pups (4–5 on average per litter) between March and May, having mated in early spring, with gestation of about 60 days. Weaning of the pups takes 56–70 days, and sexual maturity is around 12 months. The West African sub-species give birth between December and March, with a gestation period of 63 days. The birth period coincides with the wet season in regions like the Serengeti, when there is an abundance of prey items – such as rodents and the young of gazelles. Litters vary greatly from 1 to 9 pups, with weaning the same as for the North African wolves. Sexual maturity is thought to be about 12 months, but there has been far less research into this sub-species. The East African golden wolf has a very similar breeding profile to the West African.[269] At the time of sexual maturity, most young wolves disperse, and only a small proportion stay to help with new pups, but Moehlman found that they leave during the dry season, but as many as 70% may return in the wet season and help raise the pups.[270] She looked at records of 14 years of monitoring golden wolves in the Serengeti and concluded that the wolves mated for at least 6–8 years and often for life and that while helpers will bring food back and regurgitate to feed the pups, defend the den and the territory, and guard the pups, but they do not scent mark.[271]

Diet, foraging, and hunting

As with the African jackal species and the Eurasian golden jackal, golden wolves are omnivorous and opportunistic hunters/foragers. Eddine et al. calculated from their observations that over 60% of the golden wolf diet consisted of rodents, lizards, birds, snakes, hares, and both young and adult Thomson's gazelles, and they also ate invertebrates and fruits.[272] In common with the jackal species, they frequently scavenge from kills made by larger carnivores. They will also forage in dung of large ungulates and rhino middens searching for insects, particularly dung beetles. Lamprecht recorded seeing a golden wolf eating 37 beetles in 30 minutes.[273] He also noted that golden wolves would often hunt in pairs when chasing hares and larger prey and that greater success was achieved than by a golden wolf hunting alone. He found they were less successful, even hunting in pairs, than black-backed jackals, with only a 32% success rate compared with 60%.[274]

Foraging and opportunistic hunting take up a large amount of golden wolf time. Temu et al. watched golden wolves in the Serengeti/Ngorongoro ecosystem for a total of 125.1 hours in the dry season and 227.8 hours in the wet season. Foraging took up 68.90 hours and 97.3 hours, respectively.[275] In the wet season, gazelle fawns were born, and they made up a large proportion of food intake and hunting time, while the afterbirth from wildebeest births was another source of food. During his research in the Serengeti, Lamprecht frequently observed golden wolves foraging in the refuse pits at safari camps or another at the research HQ and even between the houses there.[276] Their intake of carrion from kills by large predators or from carcasses of animals that died of disease or natural causes was an important part of their diet, and they are often to be seen, as are black-backed jackals, near lion or hyena kills. Again mirroring black-backed behaviour, they will wait until the predator has finished feeding and moved away before trying to feed, but if a chance offers itself, they will dash in and grab a piece of meat or other edible item.[277] At large carcasses, golden wolves and black-backed jackals might gather together with vultures and compete aggressively for access. When a large carcass was available, territorial boundaries generally broke down, even though threats and some fighting would still take place.

Interactions with carnivores

As with black-backed and side-striped jackals, as described in the preceding sections, golden wolves are subordinate to the larger predators such as lions, leopards, spotted hyenas, and wild dogs. Leopards will predate on golden wolves as they do on the jackal species, and wolves may be killed by the other large predators on a competitive and seemingly opportunistic basis. Golden wolves are, to an extent, dependent on the hunting of larger carnivores for the supply of carcasses of large ungulates which they are not capable of killing themselves. In the highlands of Ethiopia, there is competition for prey (especially rodents) between African golden wolves and Ethiopian wolves (*Canis simensis*). In a 12-month research period, Gutema et al. observed 82 interactions between the two species, 58 of which occurred in what they termed the buffer zone between protected areas and human-inhabited areas, while 24 were in core of the protected area. In the buffer zone, 55 interactions were aggressive, and 52 were deemed to have been won by the golden wolves and only 3 by the Ethiopian wolves, but in the core areas, Ethiopian wolves won 23 out of 25.[278] A major reason appears to be that the Ethiopian wolves avoid human contact if possible and protect the core area, while golden wolves will forage around human habitations.[279]

Interactions with humans

The nature of golden wolf interactions with humans mirrors those of the three jackal species, with conflict arising chiefly from the belief across the golden wolf ranges that they are a major killer of livestock, that they attack domestic dogs, spread diseases like rabies, and raid fruit crops. Further details of wolf–human interactions will be provided in the narrative sections.

Notes

1 J.E. Castelló (2018) *Canids of the World*, Princeton, NJ: Princeton University Press/ Princeton Field Guides, pp. 160–5.
2 N. Nattrass et al. (2017) Understanding the black-backed jackal, *CSSR Working Paper*, No. 399, Institute for Communities and Wildlife in Africa, p. 2.
3 J.R. Ginsberg & D.W. Macdonald (1990) An Action Plan for the Conservation of Canids, IUCN/SSC Canid Specialist Group Foxes, Wolves, Jackals, and Dogs, Gland, Switzerland: IUCN, https://portals.iucn.org/library/sites/library/files/documents/1990-008.pdf accessed 7 September 2021.
4 L. Hunter (2011) *Carnivores of the World*, Princeton, NJ: Princeton University Press, p. 104.
5 Ginsberg & Macdonald, 1990, p. 13.
6 Nattrass et al., 2017, p. 1.
7 Castelló, 2018, p. 156.
8 Ibid., p. 158.
9 J. Loveridge & D.W. Macdonald (2002) Habitat ecology of two sympatric species of jackals in Zimbabwe, *Journal of Mammalogy*, 83(2), 599–607, p. 599.
10 Jonathan Kingdon (1979) *East African Mammals an Atlas of Evolution in Africa, V IIIA*, London: Academic Press, p. 33.
11 Ibid.
12 Ginsburg & Macdonald, 1990, p. 14.
13 P.D. Moehlman (1987) Social organization in Jackals: The complex social system of jackals allows the successful rearing of very dependent young, *American Scientist*, 75(4), 366–75, p. 368.
14 Ginsburg & Macdonald, 1990, p. 14.
15 M.W. Hayward & G.J. Hayward (2010) Potential amplification of territorial advertisement markings by black-backed jackals (Canis mesomelas), *Behaviour*, 147(8), 979–92, p. 979.
16 Ibid., p. 987.
17 Ibid., p. 988.
18 R. James (2014) *The population dynamics of the black-backed jackal (Canis mesomelas) in game farm ecosystems of South Africa*, Thesis for Doctorate of Philosophy, University of Brighton, November 2014, p. 6.
19 Moehlman, 1987, p. 368.
20 L. Minnie et al. (2018) Biology and ecology of the black-backed jackal and the caracal, in G.I.H. Kerley, S.L. Wilson & D. Balfour (eds) *Livestock Predation and its Management in South Africa: A Scientific Assessment*, Port Elizabeth: Centre for African Conservation Ecology, Nelson Mandela University, 178–204, p. 184.
21 Kingdon, 1979, p. 32.
22 N. Nattrass, M. Drouilly & M. Justin O'Riain (2020) Learning from science and history about black-backed jackals Canis mesomelas and their conflict with sheep farmers in South Africa, *Mammal Review*, 50, 101–11, p. 105.
23 Ibid.
24 Nattrass, 2017, pp. 14–5.
25 J.W.H. Ferguson (1978) Social interactions of Black-Backed Jackals Canis Mesomelas in the Kalahari Gemsbok National Park, *Koedoe*, 21, 151–62, p. 152.
26 Ibid., p. 153.
27 Ibid., pp. 154–7.
28 H. van Lawick & J. van Lawick-Goodall (1970) *Innocent Killers*, London: Collins, pp. 135–6.

29 Ibid., p. 136.
30 J.W. Ferguson, J.A. Nels & M.L. de Wet (1983) Social organization and movement patterns of Black backed jackals *Canis mesomelas* in South Africa, *Journal of Zoology*, 199, 497–502, p. 488.
31 Ibid.
32 Ibid., p. 495.
33 Ibid., pp. 497–8.
34 James, 2014, p. 6.
35 Ibid., p. 11.
36 Nattrass et al., 2017, p. 24.
37 Ibid., 2017, pp. 4–5.
38 Ibid., p. 9.
39 Minnie et al., 2018, p. 190.
40 D.W. Macdonald, A. Loveridge & R.P.D. Atkinson (2004) A comparative study of side-striped jackals in Zimbabwe: the influence of habitat and congeners, in David W. Macdonald and Claudio Sillero-Zubiri (eds) *Biology and Conservation of Wild Canids*, Oxford: Oxford University Press, 255–70, p. 258.
41 J.F. Kamler et al. (2019) Social organization, home ranges, and extraterritorial forays of black-backed jackals, *Journal of Wildlife Management*, 83(8), 1800–08, p. 1800.
42 Ibid., pp. 1802–3.
43 A.J. Loveridge & D.W. Macdonald (2001) Seasonality in spatial organization and dispersal of sympatric jackals (*Canis mesomelas* and *C. adustus*): implications for rabies management, *Journal of Zoology*, 253, 101–11, pp. 105–6, p. 108.
44 James, 2014, p. 12.
45 Nattrass et al., 2017, p. 9.
46 Ibid.
47 Moehlman, 1987, pp. 366–7.
48 Kingdon, 1977, p. 35.
49 P.D. Moehlman (1989) Intraspecific variation in canid social systems, in Gittelman, J.L. (ed) *Carnivore Behavior, Ecology, and Evolution*, London: Chapman and Hall, 143–63, p. 150.
50 Moehlman, 1987, p. 370.
51 Ibid., p. 383.
52 Ibid., p. 370.
53 Ibid.
54 P.D. Moehlman (1979) Jackal helpers and pup survival, *Nature*, 75(4), 382–3, p. 382.
55 Ibid.
56 Ibid.
57 Ibid.
58 Minnie et al., 2018, p. 194.
59 James, 2014, p. 6.
60 Mark & Delia Owens (1986) *The Cry of the Kalahari*, London: Fontana, pp. 66–7.
61 Minnie et al., 2018, p. 194.
62 Nattrass et al., 2017, p. 10.
63 Ibid., p. 12.
64 Ibid.
65 N.A. Littlewood et al. (2020) *Terrestrial Mammal Conservation: Global Evidence for the Effects of Interventions for Terrestrial Mammals Excluding Bats and Primates*, Cambridge: Open Book Publishing, University of Cambridge, p. 123.
66 Minnie et al., 2018, p. 194.
67 J.W. Ferguson, J.S. Galpin & M.J. de Wet (1988) Factors affecting the activity patterns of black-backed jackals *Canis mesomelas*, *Journal of Zoology*, 214, 55–69, p. 59.
68 Ibid.
69 Ibid., p. 63.

70 C.J.1 Tambling et al. (2018) The role of mesopredators in ecosystems: potential effects of managing their populations on ecosystem processes and biodiversity, in G.I.H. Kerley, S.L. Wilson and D. Balfour (eds) *Livestock Predation and its Management in South Africa: A Scientific Assessment*, Port Elizabeth: Centre for African Conservation Ecology, Nelson Mandela University, 205–27, p. 210.

71 Kingdon, 1979, p. 31.

72 Moehlman, 1987, pp. 366–7.

73 J. Lamprecht (1978) On diet, foraging behaviour and interspecific food competition of jackals in the Serengeti National Park, East Africa, *Zeitschrift-Saeugetierkunde*, 43, 210–23, https://www.zobodat.at/pdf/Zeitschrift-Saeugetierkunde_43_0210-0223. pdf accessed 20 October 2021, p. 214.

74 Ibid.

75 S.K.K. Kaunda & J.D. Skinner (2003) Black-backed Jackal Diet at Mokolodi Nature Reserve, Botswana, *African Journal of Ecology*, 41, 39–46, p. 39.

76 S.E. Temu, 1 C.L. Nahonyo, 1 & P.D. Moehlman (2016) Comparative foraging efficiency of two sympatric Jackals, Silver-Backed Jackals (Canis mesomelas) and Golden Jackals (Canis aureus), in the Ngorongoro Crater, Tanzania, *International Journal of Ecology*, 1–5, p. 2.

77 See, K. Somerville (2021) *Humans and Hyenas. Monster or Misunderstood*, Abingdon, Oxon: Routledge/Earthascan, p. 174; and, Keith Somerville (2020) *Humans and Lions. Conflict, Conservation and Coexistence*, Abingdon, Oxon: Routledge/Earthscan, pp. 144–8.

78 Personal observations October 1986, October 1988, and September 2000, and in the wider Ngorongoro Conservation Area in May 2015. Also reported from their more substantial periods of observation by Temu et al., 2016, p. 3, and Lamprecht, 1978, pp. 214–5.

79 Lamprecht, 1978, p. 221.

80 Temu et al., 2016, p. 3.

81 Ibid.

82 Lamprecht, 1978, p. 218.

83 Kaunda & Skinner, 2003, p. 43.

84 I. van der Merwe et al. (2009) An assessment of diet overlap of two mesocarnivores in the North West Province, South Africa, *African Zoology*, 44, 2, 288–91, p. 289.

85 Ibid., p. 291.

86 Owens, 1986, pp. 62–3.

87 Tambling et al., 2018, pp. 210–11.

88 Ibid., p. 211.

89 Ibid.

90 J. F. Kamler, J.L. Foght & K. Collins (2009) Single black-backed jackal (Canis mesomelas) kills adult impala (Aepyceros melampus), *African Journal of Ecology*, 48, 847–48, p. 847.

91 U. Klare, J.F. Kamler & D.W. Macdonald (2010) Diet, prey selection, and predation impact of black-backed jackals in South Africa, *Journal of Wildlife Management*, 74, 5, 1030–42, p. 1030.

92 Ibid., p. 1034.

93 Ibid., p. 1040.

94 Nattrass, Drouilly & O'Riain, 2020, p. 104.

95 Ibid.

96 Minnie et al., 2018, p. 181.

97 Ibid.

98 Ibid.

99 Nattrass, Drouilly & O'Riain, 2020, p. 104.

100 R.M. Fourie (2011) The Effect of Management Perturbations on the Diet of the Black-Backed Jackal (Canis mesomelas) in the Karoo National Park (Unpublished

Honours thesis). Department of Zoology, Nelson Mandela Metropolitan University, Port Elizabeth, South Africa, p. 27; see, also, Nattrass et al., 2017, pp. 6–7.

101 Nattrass et al., 2017, pp. 6–7.
102 Minnie et al., 2018, p. 180.
103 W. Beinart (2003) *The Rise of Conservation in South Africa. Settlers, Livestock, and the Environment 1770-1950*, Oxford: Oxford University Press, p. 204.
104 Nattrass et al., 2017, pp. 4–5.
105 B. Kolar (2005) Black-backed jackals hunt seals on the Diamond Coast, Namibia, *Canid News*, 8(2), http://www.canids.org/canidnews/8/Black_backed_jackals_hunt_seals.pdf. accessed 14 December 2021, p. 2.
106 Ibid., p. 3.
107 M.R. Hiscocks & K. Perrin (1988) Home range and movements of black-backed jackals at Cape Cross Seal Reserve, Namibia. *South African Journal of Wildlife Research*, 18, 97–100, pp. 97–8.
108 Nattrass et al., 2017, p. 3.
109 Nattras et al., 2017, pp. 18–9. See also, Kaunda, 2001.
110 M.R. Hiscocks & K. Perrin (1987) Feeding observations and diet of black-backed jackals in an arid coastal environment, *South African Journal of Wildlife Research*, 17, 55–8, p. 57.
111 Nattras et al., 2017, pp. 18–9.
112 J.F. Kamler, U. Klare & D.W. Macdonald (2020) Seed dispersal potential of jackals and foxes in semi-arid habitats of South Africa, *Journal of Arid Environments*, 183, 1–8.
113 Ibid., p. 1.
114 Ibid., p. 4.
115 Kaunda & Skinner, 2003, p. 43.
116 Kamler, Klare & Macdonald, 202, p. 6.
117 S. Creel, G. Spong & N. Creel (2001) Interspecific competition and the population biology of extinction-prone carnivores, in J.L. Gittelman et al. (ed) *Carnivore Conservation*, Cambridge: Cambridge University Press, 37–60, p. 57.
118 Ibid.
119 G.B. Schaller & G.R. Lowther (1969) The relevance of carnivore behavior to the study of early Hominids, Southwestern, *Journal of Anthropology*, 25(4) (Winter, 1969), 307–41, p. 324.
120 J.C. Ray, L. Hunter & J. Zigouris (2005) *Setting Conservation and Research Priorities for Larger African Carnivores*, New York: WCS Working Paper No. 24. Wildlife Conservation Society, p. 99.
121 Ibid., p. 75.
122 G. Schaller (1970) *Golden Shadows, Flying Hooves*, Chicago, IL: University of Chicago Press, p. 213.
123 L.R. Prugh (2009) The rise of the mesopredator, *Bioscience*, 59, 9, 779–90, p. 780.
124 D.W. Macdonald & C. Sillero-Zubiri (2004) Wild canids – an introduction and dramatis personae, in David W. Macdonald & Claudio Sillero-Zubiri (eds) *The Biology and Conservation of Wild Canids*, Oxford: Oxford University Press, 3–38, p. 10.
125 J.F. Kamler et al. (2012) Resource partitioning among cape foxes, bat-eared foxes, and Black-Backed Jackals in South Africa, *Journal of Wildlife Management*, 76(6), 1241–53, p. 1246.
126 Ibid., p. 1247.
127 Ibid., p. 1249.
128 J.F. Kamler, U. Stenkewitz & D.W. Macdonald (2013) Lethal and sublethal effects of black-backed jackals on cape foxes and bat-eared foxes, *Journal of Mammalogy*, 94, 2, 295–306, p. 295.
129 A.J. Loveridge & D.W. Macdonald (2002) Habitat ecology of sympatric species of jackals in Zimbabwe, *Journal of Mammalogy*, 83, 2, 599–607, p. 599.
130 Ibid., p. 603.
131 Ibid.

132 Nattrass et al., pp. 3–4.

133 Loveridge & Macdonald, 2002, p. 605.

134 Nattrass et al., pp. 3–4.

135 Moehlman, 1987, pp. 366 and 369.

136 J. Comley et al. (2020) Do spotted hyenas outcompete the big cats in a small, enclosed system in South Africa? *Journal of Zoology*, 311, 145–53, p. 147.

137 Van der Merwe et al., 2009, p. 288.

138 Ibid., pp. 288–9.

139 Lamprecht, 1978, p. 219.

140 K. Somerville (2018) The political economy of the honey badger – just don't 'ratel' its cage, *Talking Humanities*, https://talkinghumanities.blogs.sas.ac.uk/2018/05/24/political-economy-honey-badger-just-dont-ratel-cage/ accessed 21 June 2021.

141 S.B.Z. Gorta (2020) What goes around comes around: complex competitive interactions between two widespread southern Africa mesopredators, *Canid Biology & Conservation*, 22, 2, 8–10, p. 8.

142 Ibid., pp. 8–9.

143 Honey badger versus jackals, Africa Geographic Photographer of the Year 2018, *Africa Geographic*, 6 March 2018, https://africageographic.com/stories/fantastic-sighting-honey-badger-versus-jackals/, *Written, and photographs, by Willem Kruger* accessed 1 September 2021.

144 Ray, Hunter & Zigouris, 2005, p. 99.

145 Ibid., pp. 99–100.

146 Ibid.

147 S.E. Bellan et al. (2012) Black-backed jackal exposure to rabies virus, canine distemper virus, and bacillus anthracis in Etosha National park, Namibia, *Journal of Wildlife Diseases*, 48, 2, 371–81, p. 372.

148 J. Bingham, C.M. Foggin, A.I. Wandeler & F.W.G. Hill (1999a) The epidemiology of rabies in Zimbabwe. 2. Rabies in jackals (Canis adustus and Canis mesomelas). Onderstepoort, *Journal of Veterinary Research*, 66, 11–23, p. 19.

149 Ibid., p. 20.

150 J.R.A. Butler, J.T. du Toit & J. Bingham (2004) Free-ranging domestic dogs (Canis familiaris) as predators and prey in rural Zimbabwe: threats of competition and disease to large wild carnivores, *Biological Conservation*, 115, 369–78, p. 370.

151 B.L. Penzhorn et al. (2017) Black-backed jackals (*Canis mesomelas*) are natural hosts of Babesia rossi, the virulent causative agent of canine babesiosis in sub-Saharan Africa, *Parasites and Vectors*, 10, 124, 1–6, p. 1.

152 J.E. Price & L.H. Karstad (1980) Free-living jackals (*Canis mesomelas*) – potential reservoir hosts for *Ehrlichia canis* in Kenya, *Journal of Wildlife Diseases*, 16(4), 469–73, p. 469.

153 Ray, Hunter & Zigouris, 2005, p. 101.

154 K.A. Alexander et al. (1994) Serologic survey of selected canine pathogens among free-ranging jackals in Kenya, *Journal of Wildlife Diseases*, 30(4), 486–91, p. 488.

155 Bellan et al., 2012, p. 372.

156 Ibid.

157 J.K. Young (2011) Is wildlife going to the dogs? Impacts of feral and free-roaming dogs on wildlife populations, *BioScience*, 61, 2, 125–32, pp. 125–6.

158 M. Drouilly & M. Justin O'Riain (2019) Wildlife winners and losers of extensive smalllivestock farming: a case study in the South African Karoo *Biodiversity and Conservation*, 28, 1493–511, p. 1494.

159 Ibid.

160 Kamler, Stenkewitz & Macdonald, 2013, p. 295.

161 L. Minnie (2018) Compensatory life-history responses of black-backed jackals undermine population reduction efforts, in G. Giannatos et al. (eds.) *Proceedings of the 2nd International Jackal Symposium, Marathon Bay, Attiki Greece 2018*. Hellenic Zoological Archives. 9, November 2018, pp. 99–100.

162 James, 2014, 79, pp. 84–5.
163 Nattrass et al., 2017, pp. 15–6.
164 Ibid.
165 Minnie, 2018, pp. 99–100.
166 Claudio Sillero-Zubiri, Jonathan Reynolds & Andres Novaro (2004) Management and control of wild canids alongside people, in D.W. Macdonald & C. Sillero-Zubiri (eds), *Biology and Conservation of Wild Canids*, Oxford: Oxford University Press, 255–70, 107–22, p. 107.
167 Ray, Hunter & Zigouris, 2005, p. 100.
168 Castelló, 2018, pp. 160–5.
169 Ginsberg & Macdonald, 1990, p. 11.
170 Castelló, 2018, pp. 160–4.
171 R.P.D. Atkinson, D.W. Macdonald & R. Kamizola (2002) Dietary opportunism in side-striped jackals Canis adustus Sundevall, *Journal of Zoology*, 257, 129–39, p. 129.
172 Castelló, 2018, pp. 161.
173 Ginsberg & Macdonald, 1990, p. 1,
174 Ibid., p. 12.
175 Macdonald, Loveridge & Atkinson, 2004, p. 255.
176 Ibid., p. 12.
177 Macdonald & Sillero-Zubiri, 2004, p. 14.
178 Ginsberg & Macdonald, 1990, p. 11.
179 Castelló, 2018, p. 163.
180 Ginsberg & Macdonald, 1990, p. 11.
181 Macdonald et al., 2004, p. 258.
182 Castelló, 2018, p. 163.
183 Ibid.
184 Ibid.
185 Castelló, 2018, pp. 161–5.
186 R.P.D. Atkinson, D.W. Macdonald & R. Kamizola (2002) Dietary opportunism in side-striped jackals Canis adustus Sundevall, *Journal of Zoology*, 257, 129–39, p. 129.
187 Ibid.
188 Macdonald, Loveridge & Atkinson, 2004, p. 255.
189 Butler, du Toit & Bingham, 2004, p. 370.
190 Ginsberg & Macdonald, 1990, p. 12.
191 Ray, Hunter & Zigouris, 2005, pp. 100–02.
192 Castelló, 2018, pp. 142–3.
193 G. Giannatos (2004) *Conservation Action Plan for the Golden Jackal Canis aureus L. in Greece*. Athens: WWF Greece, p. 4.
194 Ginsburg & Macdonald, 1990, p. 12.
195 Castelló, 2018, pp. 132–43.
196 Ginsburg & Macdonald, 1990, p. 13.
197 N. Huisman (2017) First golden jackals arrived in France, *European Wilderness Society*, https://wilderness-society.org/first-golden-jackals-arrived-france/ accessed 26 January 2022.
198 C-R. Papp, O. C-tin Banea & A. Iuliana Szekely-Sitea (2013) Applied ecology and management aspects related to the golden jackal specific ecological system in Romania, Acta Musei Maramorosiensis IX, Sighetu Marmației, Ianuarie 2014 (Editia 2013), https://www.researchgate.net/publication/262011981 accessed 18 December 2020, no page numbers.
199 Ibid.
200 Ray, Hunter & Zigouris, 2005, pp. 103–4.
201 Z. Alif (2019) The golden jackal expands to Finland, *European Wilderness Society*, https://wilderness-society.org/the-golden-jackal-expands-to-finland/#:~:text=Recently%2C%20the%20first%20golden%20jackal,well%20as%20scavenge%20on%20carcasses accessed 26 January 2022.

202 C. Mircea Gherman & A. Daniel Mihalca (2017) A synoptic overview of golden jackal parasites reveals high diversity of species, *Parasites and Vectors*, 10, 419, 1–40, p. 1.

203 B. Yumnam et al. (2015) Phylogeography of the Golden Jackal (Canis aureus) in India, *PloS One*, 10, 9, 1–18 e0138497. doi: 10.1371/journal. pone.0138497 accessed 15 June 2021, p. 1.

204 Ibid., p. 2.

205 P.N.A. Lange, G. Lelieveld & H. de Knegt (2021) Diet composition of the golden jackal Canis aureus in south-east Europe – a review, *Mammal Review*, 51, 207–13, p. 208.

206 Papp, Banea & Szekely-Sitea, 2013, no page numbers. See, also, David W. Macdonald (1979) The flexible social system of the golden jackal, *Canis aureus, Behavioural Ecology and Sociobiology*, 5, 17–38.

207 Ginsburg & Macdonald, 1990, p. 13.

208 Nattrass, 2017, p. 13.

209 D.W. Macdonald & C. Sillero-Zubiri (2004) Wild canids – an introduction and dramatis personae, in D.W. Macdonald and C. Sillero-Zubiri (eds) *The Biology and Conservation of Wild Canids*, Oxford: Oxford University Press, 3–38, p. 5.

210 Lange, Lelieveld & de Knegt, p. 208.

211 Giannatos, 2004, p. 14.

212 Ibid.

213 Macdonald, 1979, p. 17.

214 Ibid., pp. 18–9.

215 Ibid., pp. 28–9.

216 Ibid., pp. 29–31.

217 Ibid., p. 36.

218 Ginsburg & Macdonald, 1990, pp. 12–3.

219 Papp, Banea & Szekely-Sitea, 2013, no page numbers.

220 Castelló, 2018, pp. 132–43.

221 C.M. Gherman & A.D. Mihalca (2017) A synoptic overview of golden jackal parasites reveals high diversity of species, *Parasites and Vectors*, 10, 419, 1–40, p. 1.

222 H. Tsunoda & M.U. Saito (2020) Variations in the trophic niches of the golden jackal Canis aureus across the Eurasian continent associated with biogeographic and anthropogenic factors, *Journal of Vertebrate Biology*, 69, 4, no page numbers.

223 Ibid.

224 Ibid.

225 Ibid.

226 Giannatos, 2004, p. 19.

227 Lange, Lelieveld & de Knegt, 2021, pp. 207–8.

228 Ibid., p. 211.

229 G.B. Schaller (1967) *The Deer and the Tiger: A Study of Wildlife in India*, Chicago, IL: University of Chicago Press, pp. 313–4.

230 F.J. Hill (1893) The jackal or lion provider, *Bombay Journal of the Natural History Society*, 8, pp. 306–7.

231 H. Shakespear (1860) *The Wild Sports of India*, Boston, MA: Ticknor and Fields, Kindle edition 2017, loc 806 and 878.

232 T. Kominos (2018) Preliminary findings on the impact of dogs on wild canid occupancy in western Greece, in Giannatos et al. (eds) 135–6, p. 135.

233 A-T. Bashta & L. Potish (2018) Current state and further expansion of the jackal in the Ukrainian Carpathian area, in Giannatos et al., p. 107.

234 A. Mohammadi (2017) Interspecific killing between wolves and golden jackals in Iran, *European Journal of Wildlife Research*, 63, 61, 1–5, p. 1.

235 Spassov and I. Acosta-Pankov (2019) Dispersal history of the golden jackal (*Canis aureus moreoticus* Geoffroy, 1835) in Europe and possible causes of its recent population explosion. *Biodiversity Data Journal*, 7, 1–22, p. 10.

236 Lange, Lelieveld & de Knegt, 2021, p. 211.

237 M. Shakarashvili et al. (2020) Population genetic structure and dispersal patterns of grey wolfs (Canis lupus) and golden jackals (Canis aureus) in Georgia, the Caucasus, *Journal of Zoology*, 312, 227–38, p. 227.

238 C. Home, Y.V. Bhatnagar & A.T. Vanak (2018) Canine Conundrum: domestic dogs as an invasive species and their impacts on wildlife in India, *Animal Conservation*, 275–82, p. 275.

239 Ibid., p. 276.

240 Giannatos, 2014, p. 24.

241 Giannatos, 2004, p. 13.

242 J. Knight (2006) *Waiting for Wolves in Japan. An Anthropological Study of People-Wildlife Relations*, Honolulu: University of Hawaii Press, p. 1.

243 K. Diós, V. Ulicsni & Z. Molnár (2018) The background of human-wildlife conflicts in connection with the golden jackal, in Giannatos et al., p. 98.

244 Ibid.

245 Ray, Hunter & Zigouris, 2005, pp. 103–4.

246 E.G. Raichev et al. (2013) The reliance of the golden jackal (Canis aureus) on anthropogenic foods in winter in central Bulgaria, *Mammal Study*, 38, 19–27, p. 19.

247 Ibid., p. 26.

248 Ibid.

249 Castelló, 2018, pp. 126–31. See, also, L. Hunter (2011) *Carnivores of the World*, Princeton, NJ: Princeton University Press, p. 104.

250 Castelló, 2018, p. 130.

251 Ibid., pp. 126–31.

252 Hugo and Jane van Lawick-Goodall (1970) *Innocent Killers*, London: Collins, pp. 105–48.

253 L. Campbell (2017) Wolves of the Atlas. Studying Africa's hidden wolf in the Moroccan Atlas Mountains *Wolf Print*, Autumn/Winter 2017, 14–17, p. 14.

254 W.V. Ferguson (1981) The systematic position of Canis aureus lupaster (*Carnivora: Canidae*) and the occurrence of Canis lupus in North Africa, Egypt and Sinai, *Mammalia*, 45(4), 459–65, p. 465.

255 Herodotus, trans. R. Waterfield (1988) *The Histories*, Oxford: Oxford World's Classics, Book Two, 67, p. 122. See, also, F.A. Machado and P. Teta (2020) Morphometric analysis of skull shape reveals unprecedented diversity of African Canidae, *Journal of Mammalogy*, 101, 2, 349–60, p. 349.

256 Cited by Campbell, 2017, p. 14. See also, K-P. Koepfli (2015) Genome-wide evidence reveals that African and Eurasian golden jackals are distinct species, *Current Biology*, 25, 2158–65.

257 E.K. Rueness et al. (2011) The cryptic African wolf: Canis aureus lupaster is not a golden jackal and is not endemic to Egypt, *PLoS One*, 6, 1, 1–5, p. 1.

258 Campbell, 2017, p. 15.

259 S. Viranta et al. (2017) Rediscovering a forgotten canid species Suvi Viranta1, *BMC Zoology*, 2–6, 1–9, p. 1–2. p. 2.

260 Rueness et al., 2011, p. 3.

261 Ibid., p. 4.

262 Personal correspondence with the author, 21 October 2020.

263 C.M. Stimpson et al. (2016) Middle Pleistocene vertebrate fossils from the Nefud Desert, Saudi Arabia: Implications for biogeography and palaeoecology, *Quaternary Science Review*, 143, 13–36, p. 13, pp. 23–4 and 32.

264 Viranta et al., 2017, p. 2.

265 Ibid., p. 5.

266 F.A. Machado and P. Teta (2020) Morphometric analysis of skull shape reveals unprecedented diversity of African Canidae, *Journal of Mammalogy*, 101, 2, 349–60, p. 350.

267 Macdonald & Sillero-Zubiri, 2004, p. 14.

268 Ginsburg & Macdonald, 1990, p. 13.
269 Castelló, 2018, pp. 126–31.
270 Moehlman, 1989, p. 152.
271 P.D. Moehlman, S. Temu & H. Hofer (2018) Cooperative breeding and feeding ecology of the African golden jackal, in G. Giannatos et al. (eds) *Proceedings of the 2nd International Jackal Symposium, Marathon Bay, Attiki Greece 2018.* Hellenic Zoological Archives. 9, November 2018, pp. 12–13.
272 A. Eddine et al. (2017) Diet composition of a newly recognized canid species, the African Golden Wolf (Canis anthus), in Northern Algeria Authors, *Annales Zoologici Fennici,* 54, 5–6, 347–56, p. 348.
273 Lamprecht, 1978, p. 214.
274 Ibid., p. 221.
275 Temu et al., 2016, p. 2.
276 Lamprecht, 1978, p. 218.
277 Ibid.
278 T.M. Gutema et al. (2018) Competition between sympatric wolf taxa: an example involving African and Ethiopian wolves. *Royal Society Open Science,* 5, 1–9, p. 2.
279 Ibid., p. 7.

2

ORIGINS AND EVOLUTION OF JACKALS AND GOLDEN WOLVES

My narrative of the evolution of jackals and golden wolves starts with early mammalian prototype carnivores, which (alongside surviving reptilian carnivores like crocodiles, alligators, monitor lizards, etc.) took their place in the food chain occupied by the dinosaurian carnivores, which became extinct c65mya. Many of the earliest mammalian carnivores are believed to have been primarily scavengers, though some may have hunted. Rat- or mouse-sized mammals, similar to modern small insectivorous mammals, lived alongside the dinosaurs from about 248mya to 65mya.[1] A squirrel-like carnivore, *Cimolestes*, appeared around 75–65mya, the time of the dinosaur extinction, and is the ancestor from which contemporary species labelled carnivores are descended.[2] As Macdonald explained,[3] the order Carnivora evolved over millions of years from *Cimolestes* from small animals preying on insects and small vertebrates into diverse species varying in size, behaviour, habitat, and diet. The order includes the Mustelidae or Mustelids, the family to which honey badgers belong, and the Canidae, which includes jackals and wolves.

Cimolestes split into two branches, the ancestors of modern carnivores and another group called the *creodonts*, which were much larger than *Cimolestes* and, between 55 and 35mya, were the dominant mammalian predators.[4] The ancestral carnivores proved more successful than the *creodonts*, which had included the strikingly named *Hyaenodon horribilis*, a wolf-sized meat-eater, and the *Megistotherium*, an 800kg hyena-like predator, probably "the largest mammalian land predator ever."[5] The evolving carnivores replaced the *creodonts* and were the top land predators in the northern hemisphere by 30–20mya. One reason for success was their adaptability, as they ate fruits and vegetables, as well as meat, and could survive declines in prey species in a way that the more specialised *creodonts* could not[6] – something that jackals and golden wolves developed to great effect, enabling them to vary diet according to availability.

DOI: 10.4324/9781003199793-3

The rise of the carnivores coincided with dramatic climate changes and the expansion of grassland savanna and open woodland areas, around 55–60mya. Grasses and grazing animals appear to have evolved in tandem.[7] By the Oligocene era, 33–23mya, the ancestors of most of the large mammals of modern African savannas, open woodlands, and denser forests had appeared. Their evolution was made possible by the volcanic and geological activity that reconnected Africa with Asia, formed the Rift Valley, and generated soils suitable for the growth of grasses and other vegetation that provided food for a diversity of insects, rodents and ungulates, and reptiles and birds which preyed on the insects and rodents. The insects, rodents, reptiles, birds, and ungulates were, in turn, preyed on by larger predators, including early canids, which evolved from the *Vulpavine* group of mammals, descended from the early carnivores. Increased availability of prey and of carcasses of animals preyed on by larger carnivores provided an abundant food source for the early canids.

The evidence suggests that canids were descended from early carnivores known as *miacids*, weasel-like, arboreal mammals larger than *Cimolestes*. The *miacids* evolved around 50–60mya in North America and dispersed to Europe and Asia. They hunted insects, small birds, birds' eggs, and small mammals.[8] The *miacid* descendants are varied – *Feloidea* (cat-like) groups of mammals, which include civets (*Viverridae*), cats (*Felidae*), hyenas (*Hyaenidae*), and mongooses (*Herpestidae*),[9] and the *Vulpavine* group, including the dogs (*Canidae*), bears (*Ursidae*), and weasels (*Mustelidae*).[10] The first canids developed from the *Miacidae* in North America more than 40mya.[11]

The *Canidae* evolved as terrestrial, highly mobile species with a varied diet, diverse foraging and hunting techniques, and a range of sizes.[12] The evolution of the first-known dog family species, in North America, occurred when their arboreal ancestors left the trees. For a large part of the history of the dog family, they were found in America before dispersing to Europe, Asia, and Africa around 7–8mya, when the land bridge between Alaska and north-east Eurasia formed.[13] They are the oldest of the currently living families within the order Carnivora.[14] The *Canidae* evolved three groups – the *Hesperocyoninae*, the *Borophaginae*, and the *Caninae*. The first was the base member of the entire dog family, but its species became extinct during or after 15mya.[15] The *Borophaginae* evolved numerous species, which were endemic to North America from the junction of the Oligocene and Miocene (around 33–32mya) through to the end of the Pliocene or very early Pleistocene (ending 2.6–2mya).[16] Some of the species developed strong bone-crushing teeth, and they may have paralleled the scavenging habits of early hyenas, though many were fox- or coyote-sized.[17] The surviving sub-family of dogs, the *Caninae*, may share a common ancestry with the *Borophaginae*, probably because of similarities in dentition.[18] The early dogs developed teeth that could crush as well as cut, and they had longer muzzles to accommodate their evolving dentition. Diet would have included small mammals, insects, eggs, and fruits.[19] As the *Caninae* evolved and larger species appeared, they began to develop strong jaw muscles and jaws large enough to accommodate 42 teeth. This enabled them to kill prey and tear apart meat. Many of the species were light with long legs and paws adapted for running, enabling them to catch fast prey by pursuit and to evade larger, potentially lethal carnivores.[20]

A diversity of species evolved, the earliest being the *Leptocyon*, a small, fox-like animal which appeared in the early Oligocene (34–32mya)[21] but became extinct about 12–8mya. In the late Miocene, *Caninae* species evolved from the *Leptocyon* base – the first being similar to present-day raccoons, mongooses, civets, and genets, rather than resembling modern canid species.[22] The ecological niche *Leptocyon* had filled was taken by the first foxes, the *Vulpini* which produced *Vulpes* and *Urocyon*, around 9–5mya (and much later, *Otocyon*, the bat-eared foxes of Africa – which emerged in the Pliocene).[23] The *Urocyon* produced the grey fox (*Urocyon cinereoargenteus*) of the southern USA and Central America.[24] What has been labelled a transitional taxon, *Eucyon*, evolved around 9–12mya. It was a jackal-sized canid different from the other canids in dentition and facial bone structure.[25] Around 7–5mya, *canids* spread from North America into Eurasia over the land bridge across the Bering Straits. This occurred just as the climate was cooling and in Eurasia and Africa woodlands were retreating and grasslands expanding. These conditions suited the evolving, longer-legged canids, and numerous species evolved from common ancestors to become wolves, jackals, wild dogs, and foxes.[26] During the Miocene, canids split into fox-like and wolf-like species. The wolf lineage evolved to include the black-backed jackal, side-striped jackal, the Eurasian golden jackal, and the African golden wolf. The divergence of the different species has meant that the still living *Canidae* are divided into five groupings – the wolf-like (which includes wolves, jackals, coyotes, dholes, and African wild dogs), South American foxes, the bush dog/maned wolf canids, the red fox group, and the raccoon dog, bat-eared fox, and grey fox branch.[27]

The arrival of canid species in Eurasia from the Americas after 7ma, and the evolution of large, fast species, such as wolves, capable of hunting large prey and surviving in forests, open woodlands, and grasslands, may have contributed to the demise of competing carnivores there, notably European hyenas.[28] The first canids in Eurasia, though, are likely to have been the jackal-sized *Eucyon*[29] – ten species have been identified with a wide distribution including China, Mongolia, Kazakhstan, France, Italy, Kenya, Ethiopia, Canada, Arizona, Nevada, Nebraska, New Mexico, Washington, Wyoming, and Texas.[30] *Eucyon* moved into Europe, East Asia, and, by the late Miocene, Africa, with *Eucyon*, *Eucyonulpes*, and early *Canis* species all playing their part in the dispersal across Europe, Asia, and Africa.[31] Werdelin et al. described *Eucyon* as a genus of jackal-sized canids descended from common ancestors of the jackals and golden wolves that predated the appearance of true *Canis* species in North America, Eurasia, and Africa; they can be placed on the stem lineage to *Canis* and *Lupulella* (the side-striped and black-backed jackals).[32] African finds of *Eucyon* include *Eucyon intrepidus* from the late Miocene in Kenya and *Eucyon wokari* from Aramis in Ethiopia.[33] Werdelin et al.'s paper identifies a new *Eucyon* species from Woranso-Mille, Afar region, Ethiopia, which they believe "may be the youngest *Eucyon* yet described from the

African continent."[34] They examined the bones found and concluded this *Eucyon* was of the size of a medium-sized black-backed jackal (*Lupulella mesomelas*) and labelled it *Eucyon kuta*.[35] They believe that

> [n]ot enough of this new species is preserved to indicate anything more than a possibly jackal-like ecology. However, it is clear that even though modern jackals are not yet present in the middle Pliocene fossil record, there is at least one genus that is a candidate for occupying a similar niche.[36]

At the end of the Miocene and beginning of the Pliocene, the first members of the *Canis* genus had appeared in North America around 5–6ma – again jackal-sized. They dispersed into Eurasia in the early Pliocene and colonised most of Asia, Europe, and Africa, evolving into a variety of species, including wolves and jackals. The Pliocene and early Pleistocene was a "critical period for the expansions of the *canids*." In this period, they came to inhabit Asia, Europe, and Africa, in addition to North America, between 5 and 7ma.[37] O'Regan et al. said that the appearance of *Canis* species in the Pliocene, often referred to as "the wolf event," involved at least four different taxa.[38] They recorded that earliest known *Canis*,

> even if not demonstrably a "wolf" … from Vialette in the Massif Central of France, dated to 3.14 Ma … [while] Members of the African hunting dog genus *Lycaon* may be the only instance of simultaneous dispersal into Africa and Europe from Asia. The earliest find of the genus is in China at 2.4 Ma, followed by 1.9 Ma in Europe and 1.8 Ma in north Africa.[39]

The evolution from this period of a diversity of dog species, mostly generalists feeding on a wide variety of foods including fruit and vegetable matter, enabled them to spread widely in a great diversity of habitats from Arctic tundra to thick forests and deserts. They hunted, foraged, and scavenged at carcasses of animals hunted by larger carnivores or which died of natural causes, much as jackals and golden wolves still do.[40] The *Canis* species in Europe, Asia, and then Africa massively expanded their ranges and diversified as new species in the late Pliocene and early Pleistocene, "resulting in multiple, closely-related species in Europe, Africa and Asia."[41] The species, closely linked genetically to the grey wolf, evolved into the Eurasian golden jackal and the African species of golden jackal now designated as the golden wolf.[42] There is evidence of some hybridisation between golden jackal and grey wolf in the Balkans and possibly in the Caucasus.[43] Dating canid dispersals into Africa and their diversification has been more difficult than with Eurasian and North American species, given the relative poverty of the fossil record – but discoveries of the remains of a small fox, *Vulpes riffautae* in north-western Chad, and of a *Eucyon* species – *intrepidus* – in Kenya's Baringo district, have suggested the presence of the former around 7mya and the latter from about 6.1mya.[44] Fossil records indicate the presence of canids in North, East, and Central Africa as early as late Miocene (11.63–5.333ma). Fossil evidence from East and southern Africa indicates that black-backed jackals, which may have evolved between 5 and 2.5ma, have been present in those regions for 2–3 million years.[45]

Yumnam et al. believe that the expansion of the range of the golden jackal in India can be put much later than the known presence of black-backed jackals in Africa, at around 37,000BP (in a possible range of 16–71,250BP). They estimated that the expansion of the golden jackal in south-east Europe, Israel, and elsewhere in Eurasia can be dated later at around 20,548BP (in a range of 2,649–23,973BP).[46] The timing of their population expansion was linked to interglacial cycles of the late Pleistocene, with the golden jackals, which appeared in a warmer period, shrinking back into limited ranges and surviving in areas less affected by cold periods, before expanding back from these refuges as the northern hemisphere ice sheets retreated.[47] Even before the end of the glacial period (aka ice age), 10–15,000ka, *Canis* species had gained access to the whole of the Americas, Eurasia, and Africa and were continuing to diversify.[48] A greater dispersal of canids occurred at the end of the ice age when the glacial barrier that had developed between arctic latitudes in the northern hemisphere and regions to the south started to melt, with many Eurasian species both of *canids*, notably the grey wolf (*Canis lupus*), and of prey species including elk, caribou, mountain sheep, goats, and bison moved south.[49]

As they and their prey dispersed and in some areas dense forest thinned out or, in Africa, became savanna, wolf-like canids developed social hunting – whether in pairs as black-backed jackals generally do or larger packs as with wolves (*Canis lupus*), African wild dogs (*Lycaon pictus*), and Asian dholes. These wolf-like species were from the lupine branch of canids which evolved into several lineages, with one giving rise to the black-backed and side-striped jackals and another to the lineage which produced the grey wolf, coyote, dhole, Ethiopian wolf, and the Eurasian golden jackal and the African golden wolf.[50] The jackals and the African golden wolf have great similarities in diet, habitat, and behaviour. The golden jackals/golden wolves are now believed to be sister taxa with the grey wolf, coyote, and Ethiopian wolf, while the black-backed jackal and side-striped jackal are related taxa but not in the same branch within the canids as the wolves. The black-backed jackal is the most adapted to arid regions, coastal deserts, savannas, and woodland savanna mosaics, as well as farmlands, while the side-striped is also found in thicker woodland and denser scrub; the African golden wolf in arid, semi-desert, and rocky regions; and the Eurasian jackals associated with wetlands, lake and river margins, scrublands, and outskirts of urban areas.[51] All the jackal species have adapted to human settlement and take advantage of the foraging opportunities it presents.

In Europe, *Canis arnensis* (Arno River dog) was endemic to Mediterranean Europe during the early Pleistocene around 2.6–0.75ma. Fossils suggest it was a small jackal-like dog. *Canis arnensis*'s anatomy and morphology indicate it was closely related to the extant Eurasian golden jackal (*Canis aureus*) rather than to now-extinct wolves such as the Etruscan wolf (*Canis etruscus*) of the time, which via a later but also extinct wolf, *Canis mosbachensis*, was the ancestor of the modern wolf, *Canis lupus*. The Arno River dog was probably the ancestor of modern jackals.[52] Sufficient remains have been found to suggest it was present in reasonable number in the Italian Apennines and parts of the Balkans around 1.7–1.0ma.[53] Another possible link in the ancestral chain of jackals and wolves is *Canis accitanus*, a small canid whose fossils have been found in the Guadix Basin, near Granada in Spain. Its skull has similarities with those of Eurasian golden jackals and black-backed jackals, which suggest that like jackals it fed on carrion and needed the dentition for stripping flesh from bones and grinding food.[54]

Looking at the evolution of Africa's jackals and golden wolves, the expansion of savanna and savanna-woodland in Africa created the conditions for the substantial growth in numbers and diversification of species of herbivores, rodents, and reptiles that would provide a prey base and scavenging opportunities for a range of carnivores, from small species like mongooses and civets via honey badgers to foxes, jackals, wolves, wild dogs, hyenas, and the large cats. Grasses and grazing animals appear to have evolved in tandem.[55] By 20–30ma, the ancestors of most of the large mammals of modern African savannas had appeared, helped by volcanic and geological activity that formed the Rift Valley and generated soils producing abundant grasses. The hoofed mammals found in Africa included pigs, giraffes, buffalo, wildebeest, hartebeest, reedbuck, gazelles, and smaller antelopes.[56] This enabled the evolution and dispersal across suitable habitats of carnivores, with the appearance over a period of tens of millions of years of the jackals, golden wolves, wild dogs, Ethiopian wolves (*Canis simensis*), hyenas (*Hyaenidae*), and cats (*Felidae*).[57] One of the ancestors of the African jackals appears to be a new medium-sized canid, *Eucyon khoikhoi*, whose remains have been found at the early Pliocene site of Langebaanweg 'E' Quarry (South Africa). Analysis of the remains "places *E. khoikhoi* as the most basal taxon of an African clade composed of number of now extinct *canids*" and of early examples of the side-striped jackal.[58] This *Eucyon* species is believed to be related to the East African fossils of *Eucyon intrepidus*, "whose dentition suggests a closer relation with the Lupulella group," which produced the black-backed and side-striped jackals.[59] Fossil evidence shows that as the African canids diversified and dispersed, black-backed jackals evolved and were present in South Africa (with finds at Sterkfontein – about 2.7mya; Swartkrans – 1.8–1mya; and Kromdraai – about 2ma) and East Africa (Olduvai I and II, Ngurusi, and East Turkana) in the early Pleistocene (2.6ma–781ka); side-striped jackals in Zambia and East Africa in the late Pleistocene (126–11.7ka), while fossils described as Eurasian jackals, but probably African golden wolves, have been found in North Africa and dated as late Pleistocene.[60] Kingdon wrote that black-backed jackal fossils show its presence "throughout the Pleistocene" and that "it would appear to be an exceptionally stable and ancient form."[61]

Other North African fossil discoveries suggest that an extinct ancestor of the black-backed and side-striped jackals *Lupulella* lineage was present there around

1.8mya–774ka. The finds were labelled *Lupulella mohibi*, part of a group of species classified as primitive *Lupulella* jackals.[62] The species has not been found anywhere other than north-western Africa, despite this is more closely linked to the current *Lupulella* species than to the modern North African golden wolf. Bones of a related extinct species – *Lupulella paralius* – have been found near Casablanca in Morocco and are similarly most closely related to black-backed and side-striped jackals.[63] Gaubert et al. wrote that the canids of North Africa, previously categorised as golden jackals, were, as Aristotle and other observers up to the 19th century have called them, wolves and that *Canis lupaster lupaster*, the North African golden wolf as we now know it, has been present in Africa since the middle to late Pleistocene (between 781,000 and 11,700 years ago).[64]

Jackals, golden wolves, and humans – evolution and the roots of conflict

The evolution of *Homo sapiens* from hominid and hominin predecessors, to a great extent, paralleled carnivore evolution in Africa. Hominid ecology from the Pliocene to the Holocene was affected by changes in climate, vegetation, and geology. In the early period when forest was the dominant vegetation across Africa, hominid distribution and survival relied heavily on "potable water, and animal-based food resources and plant-based food resources, with the plant diet providing more food than animal diet."[65] The early hominids were chimp-sized and needed safe sleeping sites, confining them to forests. They were preyed on by dirk- and scimitar-toothed cats and probably large ancestors of the spotted, striped, and brown hyenas. Jackals would have been no threat to hominids and, later, hominins, though it is very possible that early, forest-dwelling jackals (probably most close in diet and habitat to side-striped jackals) may have competed with them in scavenging carcasses of ungulates and other prey killed by larger carnivores.

Geological and climate changes, which produced drier upland areas and the spread of open woodland and grassland, encouraged the evolution of hominids adapted to such terrain in East and southern Africa. These habitat changes affected the evolution of jackals, with ancestors of the modern black-backed jackals thriving in savanna and open woodland areas, in a similar way to the evolving hominids. The fossil evidence leads archaeologists and anthropologists to believe that in East Africa, the hominid line derived from arboreal ancestors who inhabited forests that shrank during climate changes. These key changes in the environment occurred first during the Miocene (15.5–12.5ma) and then the Pliocene-Pleistocene period (3.0–2.5mya). Volcanic activity enriched soils, providing more nutritious grasses over wide areas. The improved availability and quality of plant food supported a wide range of herbivores, an abundant source of food for hunters of small prey and scavengers of larger carcasses, like jackals and golden wolves.

The two periods of cooling and drying were followed by what Owen-Smith calls "turnover pulses" in large mammalian herbivores taking place between 10 and 5mya and around 2.5ma.[66] The first saw the development of the modern bovid genera and the first *australopithecines*, which evolved from the hominids. The presence of these hominins, who evolved from the older more ape-like hominids, in savanna and open woodland at the time of the second turnover pulse encouraged changes in their meat-eating and meat acquisition, just as canids were adapting to the new habitat. By then, archaeological evidence indicates meat-eating by hominins including

> unmistakable evidence for at least a partial focus on tool-assisted consumption of medium to larger-sized mammals at 2.5–2.6 Ma (millions of years ago) ... This fundamental shift does not simply represent a change in diet, but also a change in ... habitat preferences, activity patterns, population size and structure, social behavior, predator avoidance, technology, and cognitive capabilities.[67]

It may also have "forced increased and novel interactions between hominins and *carnivores*, including competition for these carcasses ... and enhanced *predation* risk from sympatric carnivores."[68] The second turnover involved the thinning out of bovid species, with those unable to adapt becoming extinct, and the divergence from other hominids of "robust australopithecines" adaptable to the open environments and the evolution of the more advanced hominins.[69] The growth in open habitat prompted changes in social organisation through the development of social/pack hunting by hominins and other carnivores, though little is known about the effects on jackals or wolves, though it seems unlikely that their foraging and hunting behaviour was greatly different from that of modern jackals, beyond the absence of any killing of domestic stock or scavenging from human waste middens.

The opening up of savanna areas and decline in dense woodlands affected the diversity of the fauna of eastern and southern Africa, with herbivore and carnivore megafauna disappearing, but a diversity of species close to what exists now surviving. Open woodland, low tree and shrub bush, and grassland came to cover about 40% of Africa's land surface. This supported a diversity of herbivores and is an explanatory factor for why Africa avoided the more substantial extinctions, which affected North America and Eurasia. Owen-Smith estimates that during the megafaunal extinctions at the end of the Pleistocene between 15,000 and 10,000 years ago, the Americas lost 75% of genera, Europe and Australia 45%, and Africa only 13%.[70] The carnivore extinctions in the Miocene led to a turnover of carnivore taxa and the development of wolf-like and fox-like species, as well as large cats and hyenas, to replace carnivore taxa that became extinct. As Purvis et al. noted, the evolution of new canid species saw developments in dentition (such as carnassial teeth more efficient at removing and slicing up flesh), improved coursing abilities, and hunting tactics.[71]

During the latter part of this period of carnivore turnover, it is likely that after 2.6mya hominins entered into the *carnivore* guild, through their increased meat consumption. While they may have competed directly with large carnivores such as lions and hyenas both as hunters and scavengers, it is unlikely that there was a serious competitive element in the interactions between hominins and early species of jackals/wolves, other than hominins being capable of stripping carcasses and transporting large sections from the kill site, so leaving less for the jackals to scavenge. Hominins would have been very capable of driving jackals away, even without weapons, and of killing jackals, possibly for food but later in hominin development for their skins – though there is little fossil or other archaeological evidence to support my contention until we get to the last 2 million years when it is believed hominins may have started making clothes from animal skins.[72]

The black-backed and side-striped jackals and golden wolves or their immediate ancestors in East Africa would have been present at the time of the evolution of ape-like hominids into the australopithecines, notably *Australopithecus afarensis* (appearing 3.7–3.5ma) and later *Homo habilis* (2.4–5ma) and *Homo erectus* (2ma). Direct competition between these hominins and jackals would have been minimal, with the hominins more than capable of taking carcasses from which the jackals were scavenging and jackals posing no threat to the lives of hominins. It is very possible, though, that jackals would have been attracted to places where hominin hunter/scavengers transported disarticulated sections of carcasses to be consumed by a wider group than just the hunters. Based on studies from Olduvai in Tanzania (2–0.5ma) and Koobi Fora in Kenya (1.9–1.3ma), Bunn suggests that scavenging carnivores, which would have included jackals, would have been attracted to areas where hominin transport of parts of carcasses would have created bone assemblages and waste sites.[73] This is supported by Shipman's studies at Olduvai which revealed that bones found at hominin sites had been chewed by animals ranging in size from jackals to hyenas, and which also suggest that hominins at this time (1–2ma) were moving to a recognisable hunter-gatherer mode of subsistence.[74] The diversity of African grazers and browsers which survived the period of extinctions enabled large predators to flourish, helped hominids and then hominins to multiply and evolve, and enabled canid species to thrive – notably jackals, golden wolves, and wild dogs. In Eurasia, wolves, dholes, and jackals were also able to take advantage of changing habitat, unlike some of the specialised mega-carnivores, which disappeared, such as the dirk- and scimitar-toothed cats, which disappeared from Africa and Eurasia between 1.5ma and 900ka.[75]

Around 2.6mya hominins had begun to fashion tools from stone rather than just using unworked stones and branches for hunting or driving predators and scavengers from kills.[76] They were named *Homo habilis* by Louis Leakey and identified as part of a stone-tool making culture from 2.5 to 1.5ma, following tool and other fossil discoveries at Olduvai Gorge, Tanzania. The making of tools and weapons made the hunting of large herbivores and competitive scavenging possible with more chance of success and less risk of death or injury.[77] Jackals and golden wolves would not have been serious competitors like the larger carnivores and may have opportunistically benefitted by feeding from the remains of hominin kills that weren't carried by to home sites by them and, very likely, by scavenging remains from around their settlements. Around 1.8ma, the species *Homo ergaster* evolved, with the ability to make and use more sophisticated tools, and Iliffe says *Homo ergaster* evolved into the more modern human species, evidence for this coming from Ethiopia's Awash Valley dated at 160,000 years ago.[78] Recent finds from the Jebel Irhoud in Morocco suggest that the evolution of *Homo sapiens* from *Homo heidelbergensis* or *rhodesiensis* occurred around 315,000 years ago (fitting in with the pattern derived from fossil evidence in East Africa of *Homo sapiens* origins between 400,000 and 200,000 years ago).[79] DNA-based research on the skeletal remains of a boy believed to have died 2,000 years ago in South Africa has helped researchers to "recalculate the time at which humans like us – *H. sapiens* – split or branched from pre-modern human groups to between 350 000 and 260 000 years ago."[80] The latest research in the Omo Kibish region of Ethiopia, by Vidal et al., also suggests that *Homo sapiens* evolved around or after 350,000 years.[81] "Omo I is the oldest *Homo sapiens* with unequivocal modern human traits," according to Vidal's co-researcher/author Oppenheimer.[82] The discoveries at Jebel Irhoud in Morocco indicate that golden wolves of northern Africa would have coexisted with the evolving *Homo* species at the time, and the finds of hominin and jackal remains in east and southern Africa in the regions central to human evolution confirm coexistence of early humans with golden wolves and African jackals. At this stage of human development, people remained hunter-gatherers with all food coming from wild sources.[83] As the hunting and defensive capabilities of *Homo* species advanced and the population expanded, they dispersed, spreading to Eurasia and southern Asia.

Jackals would have lived alongside the expanding and dispersing hominin communities but would have been peripheral as either competitors or potential prey. Golden wolves and jackals would not have been hunted for food, though they may have been killed for their skins to use a bedding or simple capes. The later chapters will detail the use of jackals to provide skins/furs for clothing, decoration, or bedding, but there is evidence from this early period of coexistence of humans and jackals and golden wolves and human utilisation of them and other canids. Excavations near Palmyra in Syria found bone assemblages in caves believed to have been created by early humans, with finds dated between 300,000 and 50,000 years BP. The bones included those of early *Homo sapiens* as well as jackals, wolves, foxes, and porcupines. It is not clear whether the canid bones were from animals killed by humans or whether at different times the caves were used by humans and canid species at shelters.[84]

Blumenschine in his study on hominid/hominin scavenging and competition for carcasses with other predators does not include jackals, suggesting they were very peripheral competitors for scavenging or hunting humans. He says calculations based on observation of scavenging from carcasses in the Seronera region of Tanzania's Serengeti National Park indicated that jackals took only a small amount from carcasses and made marginally "immeasurable contributions to carcass consumption."[85]

Jackals survive the mass extinctions

When the mass extinctions of megafauna and many carnivore species extinctions occurred in the late Pleistocene in the Americas, Eurasia, and, to a much lesser extent, Africa, the jackals and golden wolves seem to have been unaffected. The causes of the mass extinctions are hotly debated with potential extinction hypotheses including climate change, human overkill, or a combination of human/climatic effects during interglacial warming there. As De Santis et al. contend, "While large carnivores like *S. fatalis* [Smilodon] and *P. atrox* are unlikely to have been directly hunted to extinction by humans, they were likely vulnerable given competition with humans for prey species."[86] There is no definitive version of the extent to which megafaunal extinctions can be laid at the door of humans. The strength of causal factors varied across regions, but human activities, notably hunting, were part of the mix.

Africa retained more species than other continents during this time and supported a range of prey and carnivore species, including the two jackal species and the golden wolf. The major difference between Africa and the rest of the world was the gradual evolution of efficient hunting by hominins in Africa living alongside their prey and competing members of the carnivore guild (in which one should include jackals even though their size and prey do not put themselves in a directly comparable carnivore profile as lions, leopards, etc.), which would have allowed prey to evolve anti-predator behaviour as human hunting abilities improved. Extinctions were slower in Eurasia than in North and South America, possibly because again, the small human population enabled time for evolution of avoidance or defence strategies by prey. When efficient hunters and their dogs arrived in Australia and the Americas, they increased and dispersed and honed and developed hunting skills of which prey species had no experience and no defensive strategies. Coinciding with climate and vegetational change, human hunting rendered prey species particularly vulnerable – far more so than in Africa and Eurasia.

It is worth noting in relation to extinctions that affected all the carnivores, that the canids, with their adaptability in terms of diet and ability to survive in a diversity of habitats, were able to survive with a large range of different species taking full advantage both of the disappearance of many larger carnivores and their own ecological plasticity, something which the jackal species continue to demonstrate well in their adaptability to the massively increased human presence and the anthropogenic effects this has brought about.[87] This is not to say that jackal species did not disappear. Fossil evidence shows that predecessors of the black-backed jackal were found in North, East, and southern Africa. In South Africa, excavations have produced evidence of two extinct jackals called *Canis terblanchei* and *Canis brevirostris*, the latter often referred to as the short-faced jackal.[88] Van Valkenburgh recorded the presence of both *Canis terblanchei* and *Canis brevirostris* in the Serengeti region from before 2.35–2ma but disappearing before 1.5mya, while black-backed jackals and side-striped jackals appeared there before much earlier (see above) and East African golden wolves just after 1ma.[89]

Jackals, golden wolves, and early humans in Africa

Hominins in southern Africa developed toolmaking in the 1.7–1.1ma period and in North Africa around 1.5–1.1mya. Scavenging and/or hunting yielded not only meat but also bone, horn, and antler that could be used for weapon-making. Around 1.0mya, human is believed to have started utilising fire for warmth, protection from predators at night, cooking, and the production of fire-hardened wooden weapons for defence and hunting. Herbivore remains in fossil finds relating to hominin hunting and food consumption from after 1.8mya rose from about 15–25% of assemblages to 45%.[90, 91] At some stage, snares, pitfalls, and game pits would have been added to the array of hunting technologies, as shown in bone assemblages and artefacts from early San settlements in southern Africa.[92]

The world's oldest known spears, some 400,000 years old and from Germany, enabled early *Homo* species to hunt large animals and kill predators. Hafted weapons with sharp points appear in the African archaeological record 100,000–200,000 years ago.[93] There is evidence of "an inexorable progression" in the relationship of humans to other carnivores: from being prey to being scavengers (passive or competitive) and competitive hunters across this period.[94] Humans, as Hart and Sussman neatly summarise it, could scavenge without being primarily scavengers and hunt without being fully hunters – the flexibility of diet and methods of obtaining protein were huge advantages,[95] as over time they have proved to be for the equally adaptable jackals and golden wolves, up until real conflict with humans developed as the latter domesticated, bred, and consumed sheep, goats, cattle, and poultry, when serious conflict developed, with jackals adapting to scavenge livestock carcasses and kill young livestock, sheep, goats, and poultry. Humans quickly declared war on them and other carnivores.

Discoveries of human and animal remains at many archaeological sites in East, and particularly, southern Africa show that jackals lived in close proximity to early humans – jackals' bones being found in bone assemblages at sites of human settlement. Fossils uncovered at Cooper's site, Kromdraai, Sterkfontein, and Swartkrans in Gauteng, South Africa, indicate the presence not only of black-backed jackals but also of the two extinct jackals[96] – *Canis terblanchei* at Cooper's and *Canis brevirostris* at Sterkfontein. Brain said that the extant South African black-backed jackal was common to the sites at Sterkfontein 4, Swartkrans, and Kromdraai A and B.[97] At Sterkfontein 4, jackal remains were found in the same excavation site as the hominin, *Australopithecus africanus*. They have been dated to 2.6–2.0ma. A black-backed jackal mandible was found at Minnaar's Cave alongside the skull identified as *Australopithecus transvaalensis*, which is a label that has been used for *Australopithecus africanus*.[98]

According to Ewer, jackals were the most abundant canids among fossil assemblages in Gauteng, which fits with the ubiquitous distribution in South Africa prior to European extermination campaigns from the 18th century onwards. Specimens of extinct jackals found there were compared with skulls and skeletons of black-backed jackals and Sundevall side-striped jackals (*Lupulella adustus adustus*)

– they were found to overlap in many characteristics but had differences that made it clear they were earlier species.[99] *Canis terblanchei* was closer to the Sundevall side-striped jackal than to the black-backed, notably in dentition and size.[100] Ewer believed that earlier studies were correct in their labelling of *Canis terblanchei* as a species distinct from any living jackal.[101] She also refers to the finding of remains of a third extinct jackal, labelled *Canis antiquus*, in excavations at Sterkfontein, Swartkrans, and Kromdraai, dated at 2.5–2ma and very similar to the modern black-backed jackal in size but with some physiological differences, though not enough, Ewer says, to definitively call it a separate species, and it could be a now–extinct sub-species of black-backed jackal that was labelled *Canis mesomelas pappos*.[102]

Some physical anthropologists attribute some of the earliest discovered fossils to modern *Homo sapiens* who came from the Klasies River mouth, a complex located on the southern coast of the Cape Province, roughly 130km west of Port Elizabeth, and Border Cave on the Natal-Swaziland border. Phillipson has dated some of the evidence of human habitation at the Border Cave at more than 200,000 years, while Klasies River finds suggest the earliest occupation there was over 120,000 years ago. These were groups that varied their means of subsistence, foraging/hunting methods, and tool use according to the different environments and resources available across the region – with geology, vegetation, climate, and species availability affecting diet, group organisation, etc. The evidence of tools and other findings from excavations, from around 80,000 years leading up to 1,000 years ago, "both terrestrial and marine creatures were exploited for food … large land animals such as eland and buffalo dominate the associated faunal remains in the sites referred to above."[103] The ability to kill or scavenge from the carcasses of very large herbivores shows a developing scavenging and hunting capability. The consumption of large animals would have resulted in the production of remains around human habitations that would undoubtedly have attracted scavengers like jackals and hyenas, both of which have demonstrated the ability to adapt to human presence.

Sterkfontein area 5 and the nearby Lincoln's Cave excavations have a range of human and mammal finds, most of which are dated at 252–115ka, but the presence of remains of *Homo ergaster*, who lived around 1.9–1.5ma, suggests the finds could be considerably older or covering a wide timescale; the "occurrence of typical Middle Stone Age [280ka–50–25ka] tool forms with typical Early Acheulean [the period starting around 1.7ma] artifacts strongly suggests mixing of older and younger material in the Lincoln Cave."[104] The remains also include at least 11 *Canis mesomelas* (now renamed *Lupulella mesomelas*), the southern black-backed jackal. They were the most numerous carnivore remains at this site.[105] The bones of small canids and *viverrids* are far more abundant in later deposits, which Reynolds et al. suggest that canids like black-backed jackals may have varied their use of areas of Sterkfontein through time, most likely for denning purposes.[106] Excavations at Plover's Lake, near Sterkfontein, Swartkrans, and Kromdraai dating between 62.9 and 88.7ka, revealed large numbers of black-backed jackal remains in flowstone deposits in caves, alongside human, hyena (brown and spotted), and a mass of ungulate and, in particular, hyrax bones.[107] Hyrax would have been a likely prey for jackals and other mammalian carnivores and raptors.

At Pinnacle Point in South Africa's Western Cape, on the coast east near Mosselbaai, 190 black-backed jackal bones from at least nine individual remains were found during excavations of fossil assemblages in the Cape fynbos region (a belt of natural shrubland or heathland vegetation inland from the southern Cape coast), dating between 195 and 130ka.[108] Excavations on the southern bank of the Verlorevlei River, on the Western Cape coast north of St Helen's Bay, found a limited number of black-backed jackal remains in bone assemblages which could have been from the Middle Stone Age in Africa (250–50/25ka) or the Later Stone Age (starting around 50–25ka). They were found along with reed and wooden artefacts and faunal remains, including numerous bones of domestic sheep, though the latter will be from much more recent times.[109] Black-backed jackal remains, and those of honey badgers, dating back to the Middle Stone Age in Africa, were also found at the Klasies River mouth site.[110] At the Swartklip archaeological site, west of Mosselbaai in the Western Cape, the remains of large black-backed jackals, bigger than modern ones, suggested to Cruz-Uribe that they dated from a period when the climate was far cooler prompting an increase in body size to cope with a colder climate – either during an interglacial period when temperatures had not risen again or to a glacial period of perhaps 115,000–11,700 years ago.[111] At all the sites mentioned, black-backed jackals, to judge by the numerous remains found, were abundant, and there is evidence that juvenile jackals were killed and eaten by brown hyenas, which were believed to be responsible for many of the bone assemblages.[112]

In southern Africa, the climate and vegetation fluctuated, under the influence of the glacial periods in the northern hemisphere. During the last glacial maximum (LGM) in the north, for example, cooler conditions prevailed in Africa, as far south as the Cape. The cooler conditions during glacial periods were also drier, and Owen-Smith pointed out that during the LGM about 20,000 years ago, conditions became more arid "over most of southern Africa."[113] Human communities expanded and continued to live in semi-nomadic existences across most of Africa at this time, with permanent settlements developing later. It was only towards the end of the last glacial period (lasting from about 115,000 to 11,700 years ago) that human settlement became more permanent as populations expanded. Evidence obtained from the Nile Valley suggests human expansion 20,000–11,000 years ago, with communities there and perhaps also in North Africa engaging in intensive exploitation of selected plants and animals. More sophisticated tools were being developed, such as the grindstones found at Wadi Kubbanya near Aswan in Egypt,[114] which suggested larger-scale harvesting of edible seeds and tubers and also a more settled lifestyle centred on food resources.

In much of eastern and southern Africa, the climate across this period, and the spells of aridity, reinforced the spread of savanna and open woodland habitats that had encouraged the large numbers and diversity of herbivores, which, in turn, supported evolution of carnivore species and hominins into the species we recognise today.[115] The expansion of grassland and open woodland areas was to provide the habitat most suitable for pastoralism, when it spread from the Nile Valley and north-eastern Africa into eastern and then central and southern

Africa, around 5,000 years ago, and enabled the Bantu migration from west and west-central Africa about 3,000 years ago.[116] Jackals, especially black-backed, thrived in the savanna and open woodland habitats, with their wide range of small herbivore, rodent, avian, and insect prey and access to large carcasses from kills by other carnivores. In areas that retained slightly denser, deciduous wood-lands (like miombo) in parts of southern and eastern Africa and the Horn of Africa, side-striped jackals were better adapted and black-backed rare or absent.

There has been little focused study on jackals in this period, and works on faunal archaeology mention them generally in passing, but extensive research into the behaviour of jackals and golden wolves around human settlements and in long-cultivated areas suggests that the plasticity of diet and behaviour would have meant that they adapted well to growing human populations and the edible waste they produced and, when permanent settlements began to develop, would have adapted well to both the opportunities and the dangers of sedentary hu-man lifestyles and both pastoralism and crop farming. This would not have been without conflict, but, as later chapters will show, jackals have been remarkably resilient in surviving the most concerted persecution by humans.

In southern Africa, human expansion first involved the San and Khoikhoi hunter-gatherers, though the latter would develop into pastoralists who were in-volved in livestock husbandry, developed through contact with Bantu migrants, with hunting. Thousands of generations of Stone Age hunter-gatherers populated the South African landscape. The San are the best model we have for the hunter-gatherer lifestyle that saw so many generations through the Stone Age, and it is tempting to say that the history of the Later Stone Age is the history of the San. This can only be done at a very broad level of generalisation, but evidence does point to a "San" history.[117] Human remains from the last 10,000 years excavated by archaeologists are broadly similar to those of the 19th- and 20th-century San people. The weapons, digging tools, and other implements of the more modern San people are similar to artefacts found and dated back to Later Stone Age hunter-gatherers.[118] Following their contacts with Bantu migrants, who farmed and kept livestock, who were moving south and interacting with them, the Khoik-hoi developed pastoralism, and it was Khoikhoi herders who brought sheep and cattle into South Africa in the past 2,000 years. The Khoikhoi had adopted the keeping of dogs for hunting and for guarding livestock. These came into contact with black-backed jackals and other jackal species as hunter-gatherers dispersed through east and in southern Africa.[119] The western and south-western regions of South Africa were peopled over millennia by San and Khoikhoi, with the lat-ter gradually introducing cattle, sheep, and dogs, which they had acquired prior to their movement into the region, while the eastern part of the region was later settled by Bantu migrants, bringing both livestock and crop cultivation.[120]

Eurasia, West Asia, and South Asia

There are huge gaps in the archaeological, oral, and written accounts and a need to try to interpret what data is available in the light of what we currently know about the sub-species of Eurasian jackals and their coexistence with humans. Archaeological evidence from North Africa, the Levant, and the Arabian peninsula have given some clues, but it is a matter of piecing together scattered bits of information and trying to form a reasonable narrative. India and neighbouring regions in South and Southeast Asia present particular problems, with few written or other records to shed light on the prehistoric period.[121] The result is an attempt to reconstruct the past by piecing together a historical account from "random inscriptions, titbits of oral tradition, literary compositions and religious texts" to add to what can be ascertained from archaeological research.[122]

In South Asia it is known, though, that the earliest humans, *Homo habilis* or *Homo erectus*, appeared in the Salt Range (Pakistan) and the Siwaliks (India) about 2mya.[123] Petraglia recorded in the Riwat district of the Soan Valley in Pakistan that flaked stone tools excavated there represent "one of the best potential cases for a Late Pliocene presence of hominins in the subcontinent. The Riwat locality contains a small number of flaked pieces that are from a boulder conglomerate context dating to ca. 1.9 Ma [million years ago] or more."[124] He believes this region of Pakistan and the Siwaliks of the Nepal/India border and Narmada Valley in India are the most likely sites of early colonisation by early humans of the Indian subcontinent.[125] The colonisation of the region may have been sporadic and, in many areas, non-permanent, to begin with, and more concerted settlement on a permanent basis may have occurred later with more advanced *Homo* species. Petraglia concluded,

> [E]vidence indicates a dispersal into the subcontinent after 700 Ka [thousand years ago] … Such chronological and technological evidence would coincide with the presence of *Homo heidelbergensis* in India. However, given that the Narmada specimen shares many attributes with *Homo erectus*, the presence of this species, or more than one archaic hominin population in South Asia cannot be discounted.[126]

Much later, after the initial movement into Pakistan and north-western India, at around 75ka, *Homo sapiens* moved into western India and spread across the whole subcontinent over the next 40,000–50,000 years. Their survival and expansion, and that of the indigenous fauna, were affected by waves of climatic change during the northern hemisphere glacial and interglacial periods, which, in turn, affected vegetation and sources of plant-based foods.[127]

In the last 50,000 years, as already noted above, a substantial proportion of global megafauna became extinct, varying in the extent of species loss from continent to continent. But tropical Asia retained significant populations of megafaunal species. As Jukar et al. concluded,

> [F]aunal dynamics are largely understudied in South Asia … The lack of a taxonomically updated dataset of late Quaternary terrestrial vertebrates in the Indian Subcontinent and associated geochronological information have prevented a comprehensive study of the pattern and magnitude of extinction, and as a consequence, the ability to evaluate the potential impact of hominins and climate change on the fauna.[128]

This includes data on the survival and widespread distribution of jackal sub-species in South Asia and Indochina. Despite the dispersal of early *Homo* species and then *Homo sapiens* from Africa, there was a comparatively low loss of predators in the Indian subcontinent and adjacent regions of Southeast Asia. This may be related mainly to the driver of loss there being climate change, with relatively little evidence of human influences when modern humans first peopled the region and climatic changes the most likely cause of the extinctions that occurred.[129] Early humans and the established fauna coexisted without huge loss due to human hunting or land use, at first. Cohabitation and coevolution between humans and predators may explain the survival of tigers, lions, leopards, bears, and wolves in India and other parts of Asia, when they disappeared in parts of Europe and much of West Asia. And, of course, in most areas it inhabited before, during and after the megafaunal extinctions, the jackal has shown great adaptability, even after humans started keeping livestock and persecuted carnivores as stock killers. It survived in south-eastern Europe and much of West Asia, when larger carnivores disappeared, often at the hand of man. In the late Pleistocene, some habitat and wildlife changes occurred across northern and western Asia – from the Thar Desert of India, through Baluchistan in Pakistan, into Iran, up into Central Asia and Mongolia, and west towards the Caspian Sea – as a result of post-glacial drying and cooling, leading to extensive desertification across large areas of this region.[130] Jackals are capable of surviving in arid regions, but the spread of large areas of desert, often with few freshwater sources and saline-dominated depressions and pans, may have had some effect on golden jackal ranges and numbers as the deserts spread and hunting/foraging options declined.

In scrub and arid regions of West Asia and the Middle East, during wetter and cooler periods of the middle and late Pleistocene, fauna survived and dispersed between Africa, West Asia, and South Asia, with the Levant and parts of Arabia as a corridor for movement for wildlife and for hominins.[131] They moved through the region into the rest of Asia but also down into the Arabian peninsula. The wider West Asia region was populated by gazelles, antelopes, deer, camels, equines, pig species, wild goats, and sheep and wild *species* of cattle. Canids were present with the genera *Canis, Lycaon*, and *Otocyon* all represented.[132] The golden jackal was clearly present in Eurasia at the time, and remains have been found at Gesher Benot Ya'aqov in Israel and Dursunlu in Turkey, while in Arabia, numerous canid cranial and postcranial remains were uncovered from Ti's al Ghadah, the majority of which can be assigned to a medium-sized *Canis* species not yet completely identified. There are indications that it could be from a golden jackal or possibly an ancestral golden wolf, which researchers have referred to as *Canis anthus*.[133] Clear evidence from the Levant and areas to the north show the presence of the golden jackal.[134]

Fossil remains from the Balkans, through Turkey, the Levant, Palestine, and east to India, show the presence across the region of golden jackals during the late Pleistocene from 126,000 to 11,700 years ago. Kurtén reported that *Canis lupaster* remains, larger than *Canis aureus*, golden jackals, were only found in North

Africa, but at the time, he wrote, the North African species were still labelled as jackals, though he often calls them wolf-jackals, and he says there is evidence of them having been in the Levant and Palestine in what he refers to as "ancient times" but not the present. He believes the few found there were at the extreme northern edge of their range at the time.[135] He further suggests that they disappeared from the region and were confined to Africa because of climatic changes during the late glacial periods and competition from the Syrian or Persian/Indian wolf, while the golden jackal survived and most of its fossils found are from the post-glacial period.[136]

Fossils of jackals found in parts of Eurasia give indications of the dispersal of jackal species in periods of glaciation and particularly interglacial periods. Those of the golden jackal, *Canis aureus*, have been found in southern Russia, the Crimea, and the Caucasus, dating to the late Pleistocene (126–11.7ka) and early Holocene (from 11.7ka).[137] The lowlands at the base of the Caucasus may have been a dispersal route for many species, including canids and hominins between 300,000 years BP up to the start of the Holocene. Golden jackal remains have been found in the Azokh Caves in the Caucasus, which have fossil remains relating to hominins and fauna, which date to 10,000–18,000 years BP.[138] The caves also contained bones of a variety of prey animals from deer, through antelope, gazelle, to wild sheep and goats. Other carnivore remains found there are of wolf, badger, and hyena.[139]

In the European part of their range, golden jackals were resident in south-central and south-east Europe, the Balkans, Bulgaria, Romania, and possibly Hungary, and into parts of what are now of southern Russia, into Central Asia, Turkey, and the regions noted above in West Asia, into South Asia (including Sri Lanka), and Indochina.[140] They were able to thrive in a wide variety of habitats including grasslands, scrubs, forests, and mangroves.[141] The climatic shift to warmer and wetter climatic conditions (estimated over one or two centuries around 11,500 years ago) and the migration and settlement by relatively small, low-density populations of hunter-gatherers are likely to have had little effect of adaptable jackals, though surprisingly, they did lead to the disappearance of the spotted hyena from Asia and the shrinking of the range of the striped hyena.[142] Larger carnivores, notably cave hyenas and lions, disappeared from most of Europe. In terms of human presence and dispersal in Europe, *Homo antecessor* (1.2ma–500,000 years ago) reached Europe around 780,000 years ago.[143] *Homo heidelbergensis* and *Homo neanderthalensis* emerged in Eurasia between 350,000 and 600,000 years ago in the middle Pleistocene. Modern humans (*Homo sapiens*) arrived in Europe during the late Pleistocene, between 45,000 and 40,000 years ago.[144]

Surprisingly, "the understanding of historic development of jackal populations in Europe is lacking." One of the hypotheses suggested that the European population goes back to the introduction of jackals from northern Africa in the 15th century. This was later rejected on the basis of morphology, but the origin of most of the European population remains unknown.[145] It is known that golden jackals were present along the eastern Mediterranean and Adriatic coasts of Croatia and Greece 7,000–6,500 years BP, but they remained absent or unrecorded in much of Europe until the 19th century and then expanded in the 20th century.[146]

Climate change and consequent changes in vegetation reduced the more open habitats favoured by many carnivores and their prey in Eurasia at the end of the last glacial period. The decline in large prey hunted by other carnivores (such as cave lions and hyenas, both of which became extinct) would have resulted in a reduction in carcasses from which jackals would have scavenged and could have resulted in a diminution of the range and also in population size, though the dietary adaptability of jackals and their ability to exist in a variety of habitats in temperate and tropical regions may explain why they survived in some southern areas, while Europe's large carnivores like lions and hyenas, except wolves, lynx, and bears, disappeared.

Notes

1 X. Wang & R.H. Tedford (2008) *Dogs Their Fossil Relatives and Evolutionary History*, New York: Columbia University Press, p. 8.
2 D. Macdonald (1992) *The Velvet Claw. A Natural History of the Carnivores*, London: BBC, pp. 22–3.
3 Ibid.
4 Macdonald, 1992, p. 23.
5 Ibid., p. 23.
6 Ibid., p. 26.
7 R.S. Reid (2012) *Savannas of Our Birth. People, Wildlife and Change in East Africa*, Berkeley: University of California Press, 2012, p. 83.
8 B. Strauu (2019) 40 Million Years of Dog Evolution, https://www.thoughtco.com/prehistoric-dogs-1093301 accessed 13 January 2022.
9 Evolution of ancestral carnivores leading to the modern wild cat family, http://www.catsurvivaltrust.org/evolution.aspx accessed 28 September 2020.
10 Macdonald, 1992, p. 34; and G.D. Welsey-Hunt & J.J. Flynn (2005) Phylogeny of the Carnivora: basal relationships among the Carnivoramorphans, and assessment of the position of 'Miacoidea' relative to Carnivora. *Journal of Systematic Palaeontology*, 3, 1, 1–28, pp. 9–12.
11 X. Wang et al. (2004a) Ancestry. Evolutionary history, molecular systematics and evolutionary ecology of canidae, in D.W. Macdonald & C. Sillero-Zubiri (2004) *The Biology and Conservation of Wild Canids*, Oxford: Oxford University Press, 38–54, p. 39.
12 D. Macdonald (1992) *The Velvet Claw. A Natural History of the Carnivores*, London: BBC, p. 78.
13 X. Wang & R.H. Tedford (2008) *Dogs Their Fossil Relatives and Evolutionary History*, New York: Columbia University Press, p. 8.
14 X. Wang et al., 2004a, p. 41.
15 Ibid., p. 23.
16 Ibid., p. 33.
17 X. Wang et al. (2004b) Phylogeny, classification and evolutionary ecology of the Canidae, in C. Sillero-Zubiri, M. Hoffmann & D.W. Macdonald (eds) *Canids: Foxes, Wolves, Jackals and Dogs*, Gland, Switzerland: IUCN, 8–20, p. 8.
18 Ibid.
19 Macdonald, 1992, p. 79.

20 L.J. Rogers & G. Kaplan (2003) *Spirit of the Wild Dog*, Crows Nest: Allen and Unwin, pp. 2–3.
21 Xiaoming & Tedford, 2008, p. 49.
22 Rogers & Kaplan, 2003, p. 7.
23 Xiaoming et al., 2004b, p. 8.
24 J.E. Castelló (2018) *Canids of the World*, Princeton, NJ: Princeton University Press/ Princeton Field Guides, p. 274.
25 Xiaoming et al., 2004a, p. 45.
26 Rogers & Kaplan, 2003, p. 8.
27 Xiaoming et al., 2004a, p. 49.
28 K. Somerville (2021) *Humans and Hyenas. Monster or Misunderstood*, Abingdon, Oxon: Routledge/Earthscan, p. 41.
29 Ibid., p. 20.
30 Eucyon, http://www.prehistoric-wildlife.com/species/e/eucyon.html accessed 29 September 2020.
31 L. Rook et al. (2017) The Kvabebi Canidae record revisited (late Pliocene, Sighnaghi, eastern Georgia), *Journal of Paleontology*, 91, 6, 1258–71, p. 1268.
32 L. Werdelin, M.E. Lewis & Y.H. Selassie (2015) A critical review of African species of Eucyon *Mammalaia; Carnivora;* Canidae with a new species from the Pliocene of the Woranso-Mille area, Afar region, Ethiopia, *Papers in Paleontology*, 1, 1, 33–40, p. 33.
33 Ibid.
34 Ibid.
35 Ibid., pp. 38–9.
36 Ibid., p. 39.
37 Xiaoming & Tedford, 2008, p. 133.
38 H.J. O'Regan, et al. (2009) Hominins without fellow travellers? First appearances and inferred dispersals of..., *Quaternary Science Reviews*, doi:10.1016/j.quascirev.2009.11.028, accessed 27 December 2021, 1–10, p. 7.
39 Ibid.
40 Macdonald, 1992, p. 94.
41 Xiaoming et al., 2004, p. 11; and B. Kurten (1984) Geographic differentiation in the Rancholabrean dire wolf (*Canis di5rus Leidy*) in North America, In Genoways, H.H. and Dawson, Mr. R. (ed) *Contributions in Quaternary Vertebrate Paleontology: A Volume in Memorial to John E. Guilday*, Special Publication N. 8, Pittsburgh: Carnegie Museum of Natural History, pp. 218–27.
42 B. Yumnam et al. (2015) Phylogeography of the Golden Jackal (Canis aureus) in India, *PloS One*, 10, 9, 1–18 e0138497. doi:10.1371/journal. pone.0138497, accessed 15 June 2021, p. 2. See also K.P. Koepfli et al. (2015) Genome-wide evidence reveals that African and Eurasian golden jackals are distinct species, *Current Biology*. http:// dx.doi.org/10.1016/j.cub.2015.06.060, accessed 15 June 2021.
43 Yumnam et al., 2015, p. 2.
44 Xiaoming & Tedford, 2008, pp. 133 and 144–5.
45 N. Nattrass et al. (2017) Understanding the black-backed jackal, *CSSR Working Paper*, No. 399, Institute for Communities and Wildlife in Africa, http://cssr.uct.ac.za/ pub/wp/399, accessed 4 June 2021, p. 1.
46 Yumnam et al., 2015, p. 14.
47 Ibid.
48 Xiaoming & Tedford, 2008, p. 60.
49 Ibid., p. 61.
50 Macdonald, 1992, pp. 99–101.
51 Castelló, 2018, pp. 126–43, 152–65.
52 See, Jean-Philip & M. Boudade-Maligne (2011) Quaternary small to large canids in Europe: Taxonomic status and biochronological contribution, *Quaternary International*, 243, 1, pp. 171–82.

53 J. Rodríguez & A. Mateos (2018) Carrying capacity, carnivoran richness and hominin survival in Europe, *Journal of Human Evolution*, 118, 72–88, p. 81.

54 G. Garrido & A. Arribas (2008) *Canis accitanusnov. sp.*, a new small dog (Canidae, Carnivora, Mammalia) from the Fonelas P-1 Plio-Pleistocene site (Guadix basin, Granada, Spain), *Geobios*, 41, 6, 751–61, p. 758.

55 R.S. Reid (2012) *Savannas of Our Birth. People, Wildlife and Change in East Africa*, Berkeley: University of California Press, 2012, p. 83.

56 Ibid., p. 83.

57 R.C. Bigalke (1978) Present-day mammals of Africa, in V.J. Maglio & H.B.S. Cooke (eds) *Evolution of African Mammals*, Cambridge, MA: Harvard University Press, 1–16, p. 7.

58 A. Valenciano, J. Morales & R. Govender (2021) *Eucyon khoikhoi* sp. nov. (Carnivora: Canidae) from Langebaanweg 'E' Quarry (early Pliocene, South Africa): the most complete African canini from the Mio-Pliocene, *Zoological Journal of the Linnean Society*, XX, 1–29, p. 1.

59 Ibid.

60 R.J.G. Savage (1978) Carnivora, in Maglio and Cooke, 249–67, p. 254.

61 J. Kingdon (1979) *East African Mammals an Atlas of Evolution in Africa, V IIIA*. London: Academic Press, p. 31.

62 D. Geraads (2016) Pleistocene Carnivora (Mammalia) from Tighennif (Ternifine), Algeria, *Geobios*, 49, 445–58, pp. 455–6.

63 D. Geraads (2011) A revision of the fossil Canidae (Mammalia) of north-western Africa, *Paleontology*, 54, 2, 429–46, p. 432.

64 P. Gaubert et al. (2012) Reviving the African Wolf Canis lupus lupaster in North and West Africa: A Mitochondrial Lineage Ranging More than 6,000 km Wide. *PLoS One*, 7, 8, p. 6.

65 E.M. O'Brien & C.R. Peters (1999) Landforms, climate, ecographic mosaics, and the potential for hominid diversity in Pliocene Africa, in Timothy Bromage and Friedemann Schrenk (eds), *African Biogeography, Climate Change, and Human Evolution*, New York: Oxford University Press, 115–37, pp. 134–5.

66 N. Owen-Smith (1999) Ecological links between African Savanna environments, climate change, and early hominid evolution, in Bromage and Schrenk, 138–49, p. 138.

67 B. Pobiner (2015) New actualistic data on the ecology and energetics of hominin scavenging opportunities, *Journal of Human Evolution*, 80, 1–16.

68 Ibid.

69 Owen-Smith, 1999, p. 138.

70 Owen-Smith, 1999, p. 148.

71 A. Purvis, G.M. Mace & J.L. Gittelman (2001) Past and future carnivore extinctions: a phylogenetic perspective, in J.L. Gittelman et al. (eds) *Carnivore Conservation*, Cambridge: Cambridge University Press, 11–34, p. 14.

72 A. Roberts (2018) *Evolution. The Human Story*, London: Dorling Kindersley, p. 180.

73 H.T. Bunn (1983) Evidence on the diet and subsistence patterns of Plio-Pleistocene hominids at Koobi For a, Kenya, and Olduvai Gorge, Tanzania, in Juliet Clutton-Brock and Caroline Grigson (eds) *Animals and Archaeology. 1. Hunters and Their Prey*, Oxford: BAR International Series 163, 21–30, p. 28.

74 P. Shipman (1983) Early hominid lifestyle: Hunting and gathering or foraging and scavenging? in Clutton-Brock and Grigson (eds), 31–49, p. 32.

75 K. Somerville (2020) *Humans and Lions Conflict Conservation and Coexistence*, London: Routledge/Earthscan, pp. 8–10.

76 C. Ehret (2016) *The Civilizations of Africa. A History to 1800*, Charlottesville: University of Virginia Press, pp. 16–7.

77 L.S.B. Leakey (1961) *The Progress and Evolution of Man in Africa*, Oxford: Oxford University Press, 9 and 40.

78 J. Iliffe (2007) *Africans. The History of a Continent*, Cambridge: Cambridge University Press, 2nd edition, pp. 7–8.

79 J-J. Hublin et al. (2017) New fossils from Jebel Irhoud, Morocco and the pan-African origin of Homo sapiens, *Nature* 546, 289–92, 08 June 2017, p. 289.
80 M. Lombard (2017) Ancient DNA increases the genetic time of modern humans, *The Conversation*, 13 October 2017, https://theconversation.com/ancient-dna-increases-the-genetic-time-depth-of-modern-humans-84716?utm_medium=email&utm_campaign=Latest%20from%20The%20Conversation%20for%20October%2012%202017%20-%2085447073&utm_content=Latest%20from%20The%20Conversation%20for%20October%2012%202017%20-%2085447073+CID_0149bdeb7533904b84a09611a069281e&utm_source=campaign_monitor_africa&utm_term=Ancient%20DNA%20increases%20the%20genetic%20time%20depth%20of%20modern%20humans accessed 13 November 2017.
81 Vidal et al., 2022, no page numbers.
82 W. Dunham (2022) Older date for Ethiopian fossils sheds light on rise of Homo sapiens, *Reuters*, 13 January 2022, Older date for Ethiopian fossils sheds light on rise of Homo sapiens | Reuters, accessed 13 January 2022.
83 Phillipson, 2005, p. 92.
84 S. Payne (1983) Bones from cave sites: Who ate what? Problems and a case study, in Clutton-Brock and Grigson, 149–62, p. 149.
85 R.J. Blumenschine (1986) *Early Hominid Scavenging Opportunities. Implications of Carcass Availability in the Serengeti and Ngorongoro Ecosystems*, Oxford: BAR International Series 283, p. 30.
86 L.R.G. DeSantis et al. (2017) Implications of diet for the extinction of saber-toothed cats and American Lions, *PloS ONE*, December 2012, 7, (12), p. 1.
87 N. Nattrass, M. Drouilly & M. Justin O'Riain (2020) Learning from science and history about black-backed jackals *Canis mesomelas* and their conflict with sheep farmers in South Africa, *Mammal Review*, 50, 101–11, p. 104.
88 C.K. Brain (1981) *The Hunters or the Hunted? An Introduction to African Cave Taphonomy*, Chicago, IL: University of Chicago Press, p. 164.
89 B. Van Valkenburgh (1988) Trophic diversity in past and present guilds of large predatory mammals, *Paleobiology*, 14(2), 155–73, p. 166.
90 L. Barham & P. Mitchell (2008) *The First Africans. African Archaeology from the Earliest Toolmakers to Most Recent Foragers*, Cambridge: Cambridge University Press, Kindle edition, p. 95.
91 Clark, 1976, pp. 24, 146.
92 G. Mokhtar (ed) (1990) *General History of Africa. II Ancient Civilizations of Africa*, London: James Currey/UNESCO, 1990, p. 353.
93 Reid, 2012, p. 93.
94 D. Hart & R.W. Sussman, *Man the Hunted. Primates, Predators, and Human Evolution*, Philadelphia, PA: Westview Press, 2009, p. xi and p. 5.
95 Ibid., p. 15.
96 Brain, 1981, p. 164.
97 Ibid.
98 Ibid.
99 R.F. Ewer (1956) The fossil carnivores of the Transvaal caves: Canidae, *Journal of Zoology*, January 1956, 97–119, p. 97.
100 Ibid., pp. 99–101.
101 Ibid., p. 104.
102 Ibid., p. 113.
103 Ibid., p. 99.
104 S.C. Reynolds, R.J. Clarke & K.A. Kuman (2007) The view from the Lincoln Cave: mid- to late Pleistocene fossil deposits from Sterkfontein hominid site, South Africa, *Journal of Human Evolution*, 53, 260–71, p. 269.
105 Ibid., p. 263.
106 Ibid., p. 270.

107 D.J. de Ruiter (2008) Faunal assemblage composition and paleoenvironment of Plovers Lake, a Middle Stone Age locality in Gauteng Province, South Africa, *Journal of Human Evolution*, 55, 1102–17, p. 1106.

108 A.L. Rector & K.E. Reed (2010) Middle and late Pleistocene faunas of Pinnacle Point and their paleoecological implications, *Journal of Human Evolution*, 59, 340–57, pp. 340 and 344.

109 T.E. Steele & R.G. Klein (2013) The Middle and Later Stone Age faunal remains from Diepkloof Rock Shelter, Western Cape, South Africa, *Journal of Archaeological Science*, 40, 3453–62, p. 3453.

110 R.G. Klein (1976) The mammalian fauna of the Klasies River Mouth Sites, Southern Cape Province, South Africa, *South African Archaeological Bulletin*, 31, 123/124, 75–98, p. 77.

111 K. Cruz-Uribe (1988) The use and meaning of species diversity and richness in archaeological faunas, *Journal of Archaeological Science*, 15, 179–96, pp. 185–6.

112 R.G. Klein & K. Cruz-Uribe (1984) *The Analysis of Animal Bones from Archaeological Sites*, Chicago, IL: University of Chicago Press, pp. 82–5.

113 N. Owen-Smith (2021) *Only in Africa. The Ecology of Human Evolution*, Cambridge: Cambridge University Press, p. 24.

114 Ibid., p. 147.

115 Ibid., p. 141.

116 Ibid., 2021, p. 318.

117 South African History Online (no date) The San, https://www.sahistory.org.za/article/san accessed 24 August 2021.

118 Ibid.

119 K. Ann Horsburgh (2008) Wild or domesticated? An ancient DNA approach to canid species identification in South Africa's Western Cape Province, *Journal of Archaeological Science*, 35, 1474–80, p. 1474.

120 Ibid.

121 E.C. Majumder (1950) The vedic age, in R.C. Majumder (ed) *The History and Culture of the Indian People, Vol 1*, Bombay: Bharatiya Vidya Bhavan, p. 47.

122 J. Keay (2010) *India a History from the Earliest Civilisations to the Boom of the Twenty-First Century*, London: Harper Collins, Updated edition, pp. xvii–xviii.

123 I. Habib (2001) *People's History of India: Prehistory*, New Delhi: Tulika Books, p. 7.

124 M.D. Petraglia (2010) The early paleolithic of the Indian subcontinent: Hominin colonization, dispersals and occupation history, in J.G. Fleagle et al. (eds) *Out of Africa I: The First Hominin Colonization of Eurasia, Vertebrate Paleobiology and Paleoanthropology*, Heidelberg: Springer Dordrecht, 165–79, p. 168.

125 Ibid.

126 Ibid., p. 175.

127 Habib, 2001, pp. 14–5.

128 A.M. Jukar et al. (2020) Late quaternary extinctions in the Indian Subcontinent, *Palaeogeography, Palaeoclimatology, Palaeoecology*, https://doi.org/10.1016/j.palaeo.2020.110137 accessed 14 December 2020, 1–11, p. 1.

129 Ibid., pp. 1–2.

130 D.N. Wadia (1960) *The Post-glacial Desertification of Central Asia and the Evolution of the Arid Zone of Asia*, New Delhi: National Institute of Sciences of India, p. 1.

131 M. Stewart et al. (2017) Middle and Late Pleistocene mammal fossils of Arabia and surrounding regions: Implications for biogeography and hominin dispersals, *Quaternary International*, https://doi.org/10.1016/j.quaint.2017.11.052 accessed 19 January 2022, 1–18, p. 2.

132 Ibid., p. 8.

133 Ibid.

134 Ibid., p. 10.

135 B. Kurtén (1965) *The Carnivora of the Palestine Caves*, Helsinki: Acta Zoologica Fennica, 107, 1–74, pp. 41–2.

136 Ibid., p. 60.
137 N.K. Vereshchagin & G.F. Baryshnikov (1984) Quaternary mammalian extinctions in Northern Eurasia, in P.S. Martin & R.G. Klein (eds), *Quaternary Extinctions. A Prehistoric Revolution* I Tucson, Arizona: University of Arizona Press, pp. 483–516, p. 487.
138 Y. Fernandez-Jalvo et al. (eds) (2016) *Azokh Cave and the Transcaucasian Corridor*, Stuttgart: Springer, p. 1.
139 Ibid., p. 105.
140 Castelló, 2018, pp. 132–43.
141 Castelló, 2018, pp. 138–43.
142 Somerville, 2021, p. 37.
143 Roberts, 2018, p. 130.
144 Ibid., p. 189.
145 R. Rutkowski (2015) A European concern? Genetic structure and expansion of golden jackals (Canis aureus) in Europe and the Caucasus. *PLoS ONE*, 10, 11, doi:10.1371/journal.pone.0141236, accessed 8 December 2020, 1–22, p. 3.
146 Ibid.

3

FROM THE END OF THE PLEISTOCENE TO THE START OF THE COMMON ERA (CE)

In the last 10,000 years of the Pleistocene, from around 20,000BCE, what Reid called the "soft boundary" between humans, wildlife, and the environment, with people living in small communities gathering wild plant foods and hunting/scavenging for meat, began evolving. Over the next 5,000–10,000 years, the "soft boundary" gradually developed into a "hard boundary" through the domestication of animals and spread of the cultivation of plants for food, the latter developing from the regular harvesting of wild cereals, vegetables, tubers, and fruits.[1]

Initially, the soft boundary consisted of human defence from other predators like hyenas and lions, people hunting wild prey; scavenging animal carcasses in competition with other carnivores like jackals, hyenas, and lions; foraging for wild cereals, fruits, tubers, and honey; and gathering wood for fires. The wildlife side of the boundary involved predation by carnivores, competition within the carnivore guild for prey and scavenging opportunities, browsing of plants by ungulates, and foraging for fruits. It would have included competition with omnivores/carnivores like honey badgers and jackals which also foraged for bee nests, wild plants, seeds, fruit, nuts, and water-bearing vegetables like melons in arid areas. There was no territorial boundary between human and wildlife worlds, and they shared the same habitat and resources.

Using Reid's model, this then evolved into a mixed boundary with people beginning to cultivate on a small scale and keep livestock, while still hunting, foraging, gathering wood and water, and transporting it to settlements, instead of drinking water where it was found in streams, rivers, or lakes. At some stage in this process, they started establishing hives to encourage bees and then harvest honey – creating an early point of competition and possibly conflict with badgers (see later chapters for greater detail of this conflict). Then came the hard border, with clearing of woodland or bush, fencing of cultivated land, attempts

DOI: 10.4324/9781003199793-4

to exclude wildlife from crop fields or enclosures for livestock, protection of resources, and the beginnings of irrigation. As human society and forms of subsistence developed, so the boundary between farmed/settled areas and the habitat of wildlife became harder and more exclusionary, with killing of ungulates and other mammalian, avian, reptile, or insect herbivores, and killing of or driving out of predators that threatened livestock.[2] As the narrative will highlight, jackals proved adept at both avoidance of contact and conflict which, combined with the ability to break through human–created boundaries and the plasticity of diet, enabled them to survive through their ability to both hunt small prey, eat fruit and other vegetable matter, scavenge carcasses, and benefit from the detritus of human settlements. They also proved resilient in terms of breeding and dispersal in the face of persecution.

Jackals adapted to human domestication of wild sheep, goats, cattle, and poultry. This will have presented both opportunities and dangers. Across the ranges of the golden jackal, golden wolf, and the sub-Saharan jackal species, expansion of human settlements and pastoral and other agricultural activities created new food sources for carnivores already used to scavenging from the leftovers at hunter-gatherer camps. The dietary and foraging adaptability of the jackals meant that the hard borders of human activity were still porous; more so for them than for larger, dangerous predators like lions, tigers, leopards, and hyenas. This created conflict and persecution. As Carruthers and Nattrass identified,

> It is a truism that livestock-keepers from time immemorial have felt the need to protect their flocks and herds from predators to which all vulnerable animals are prey. In Africa, large, or apex, predatory carnivores abounded in bygone eras and over wide areas. Therefore, from the dawn of pastoralism on the continent it has been necessary to provide protection from wild predators for domestic livestock.[3]

The growth of antagonism towards predators, including jackals which were to become seen as a major threat to goats, sheep, young cattle, and poultry, was found across the regions of the world into which humans had dispersed and kept livestock and in which jackals were found.

The beginnings of cultivation and pastoralism by humans

Hunter-gatherers around the Tigris and Euphrates Rivers, in modern-day Iraq, and then the eastern Mediterranean littoral, are believed to have been the first to have created more permanent settlements as they harvested abundant wild cereals, nuts, pulses, and fruits, which they supplemented with hunting, fishing, and foraging for foods like honey.[4] The evolution from harvesting wild plant foods into early forms of cereal and pulse cultivation probably took place independently in regions with suitable climates and soils. There is evidence that this form of food production had begun in the Middle East and West Asia by

9,600BCE, while sheep and goat domestication occurred between 9,000 and 5,500BCE in West Asia. The domesticated sheep and goats found in Africa are all likely to have descended West Asian breeds, as there is no evidence of Barbary or other wild sheep or goats in Africa having given rise to domesticated stock.[5] Wild ibex and mouflon were probably ancestors of domesticated goats and sheep.[6] After 7,000BCE, communities growing cereals and herding livestock moved into northern Mesopotamia.[7] These practices became established across Mesopotamia and in Persia, with people growing primitive varieties of wheat and barley developed from wild varieties.[8] The new farming communities based on cultivation and livestock expanded and dispersed into wider regions of the eastern Mediterranean and later into north-east Africa via the Nile Delta, with pastoralism particularly important in many areas.[9] These regions had widely distributed populations of jackals and golden wolves, for whom scavenging from human settlements and livestock carcasses would have supplemented their natural diet of wild ungulate remains, smaller mammals, birds, reptiles and insects they hunted, and the seeds and fruits they foraged. This brought them into conflict with people, who would often have presumed that jackals consuming a dead lamb, sheep, kid, goat, or calf had killed it in the first place. Aynard in his survey of animals in ancient Mesopotamia listed a diversity of predators, including golden jackals, lions, leopards, cheetah, wolves, and foxes.[10]

The new production methods spread into southern Europe from Anatolia around 6,000–700BCE, where cultivation of plants foods was supplemented by domestication of animals, including wild boar and wild cattle. Arable farming and pastoralism spread eastwards, or in some areas developed independently, in Asia, with evidence of both cultivation and pastoralism in around the Caspian Sea by 6,000BCE.[11] In the Indian sub-continent, some regions were suitable for the development of agriculture-based societies, while others, like the dense humid forests of the north-east, the Eastern and Western Ghats of central-southern India, and other mountainous or thickly forested regions, retained hunter-gatherer and basic shifting agricultural systems.[12] Shifting then settled agriculture developed between 10,000 and 4,000 years ago, according to Gadgil and Guha, with settlement and population expansion taking off c4,000BP.[13] Farmers and pastoralists spread across India over several thousand years as livestock husbandry, based particularly on zebu cattle, and the cultivation of rice, gram, beans, etc. enabled populations to grow.[14] The clearing of land and need to protect livestock will have excluded wildlife from fields used for cultivation and created greater conflict with predators. But well into the Common Era, wildlife remained diverse, widely distributed, and with sufficient wild prey or their carcasses providing food for tigers, leopards, dholes, jackals, and hyenas.

Jackals in India, West Asia, south-east Europe, and Africa adapted to life around human settlements, feeding opportunistically on waste produced by humans and on the carcasses of livestock that died from disease or other natural causes. The development of human food production – from cultivation of formerly wild grains and domestication of wild goats, sheep, and cattle to increasingly advanced land clearance, crop production, pest control, and predator removal – had a huge effect on the environment, on wildlife, the ungulates on which carnivores depended either for prey or carcasses from which to scavenge, and so on, the predators themselves. But across their ranges, the various jackal species were to prove particularly adept at slipping through the increasingly hard boundaries, with their fences, cleared land, and growing human settlements, to continue to thrive by taking advantage of new foraging and hunting opportunities to add to existing ones. The increasing sophistication of hunting and anti-predator weapons (from rocks, stone axes, and spearheads to metal weapons and projectile weapons) all affected the ranges, distribution within the ranges, and the survivability of jackal populations, though not to the extent they did for larger predators, such as lions or tigers, which were a threat to human life as well as livestock.[15]

Human development in Africa and consequences for carnivores

North Africa and Egypt

Evidence of the activities of hunter-gatherer communities in the Nile Valley, as far south as Wadi Kubbaniya, dates back to 18,000BP, about 12,000 years before domestication of animals started in Egypt.[16] The evidence from Wadi Kubbaniya, and elsewhere in the Aswan region, suggests communities engaged in intensive exploitation of selected plants and animals. The grindstones found at Wadi Kubbanya indicated use of efficient tools for obtaining food from plants,[17] with the large-scale harvesting of edible seeds and tubers leading to a settled lifestyle. This started a move to a mixed boundary and then a progressively harder boundary between human and natural worlds. Around 18,000–9,000 years ago, archaeological evidence suggests people began intensive exploitation of tubers and fish all along the Nile Valley of Egypt.[18] By 15,000BCE, communities along the northern Nile were harvesting wild cereals.[19] In the 15,000–11,000BCE period, evidence from sites excavated in the Nubian Nile Valley, in what is now southern Egypt and northern Sudan, indicated hunting of wild cattle and other large ungulates, fishing, and extensive exploitation of wild plant foods.[20] At hunter-gatherer sites in the region, fossil finds include canid bones which are believed to result from people killing canid species, most likely for their pelts and perhaps claws and teeth.[21] Golden wolves and foxes were the canid species most likely to be found in the area. The bones found in these excavations included hartebeest and gazelle remains. These would have been hunted by people and by the resident populations of lions, cheetahs, leopards, wild dogs, and hyenas. This would have produced carcasses for golden wolves to scavenge, including from hunter-gatherer camps. To the west of the Nile, in north-west Sudan and Chad, people pursued livelihoods based on foraging, hunting, and fishing.[22] They benefitted from changes in the climate around 12,000 years ago, which led to higher rainfall, raised temperatures, higher river levels, and the creation of lakes, the latter two resulting partly from the meltwater from melting glaciers in the Ethiopian highlands. Iliffe wrote that in the period of 12,000–7,500 years ago, many communities moved away from river valleys to higher areas of grassland and open woodland.[23] Much of what is now the Sahara Desert received sufficient rainfall to provide vegetation for ungulates and permanent water, until a long arid period turned it to an inhospitable desert.[24]

The growing human communities of north-east Africa hunted using more sophisticated spears, bows, and arrows, as weapon-making and toolmaking capabilities improved. This enabled population expansion in the Nile Valley (as it had in Mesopotamia and West Asia) and adjacent highland regions.[25] In the Aswan region of southern Egypt, evidence of substantial exploitation of the tubers of nut grass and the use of millstones to grind them down suggests improving technology.[26] Artefacts found in the Nubian Nile region show local specialisation in tool development and the production of stone weapons that enabled the hunting of wild cattle and other large ungulates.[27] The butchering of carcasses where the ungulates were killed or back at settlements, when portions of carcasses were transported back, would have created growing scavenging opportunities for golden wolves. In highland areas adjacent to the Nile Valley, more permanent settlements had developed around 9,000–8,500BP. Stone shelters discovered by archaeologists and "intensive exploitation" of wild Barbary sheep by human hunters took place, according to the fossil records from the Acacus highlands in south-western Libya.[28] African golden wolves may also have hunted the lambs of Barbary sheep and scavenged from the carcasses of sheep killed by people or other large predators, such as lions and leopards.

The increased rainfall in North Africa, the Nile Valley, and into the Horn and East Africa assisted not only the expansion of human populations, the move to permanent settlements, but also the flourishing of a diversity of ungulates, small mammals, birds and reptiles, and the large, medium, and small carnivores that preyed on them. But in the areas suitable for cultivation and pastoralism, human activity slowly but surely began to change the landscape and to develop Reid's hard boundary, as the evolving farming communities sought to protect crops and livestock. As Eddine et al. pointed out,

> In North Africa, the diffusion of earlier Neolithic technology arrived approximately 9000–7000 years BP, during the Holocene Climatic Optimum, when a marked climatic shift changed arid desert conditions into savannah-like environments, fostering the establishment of human settlements and the regional development of pastoral activities.[29]

They also noted that while the spread of human settlements and farming would have negative effects on biodiversity, they recognised that "the presence of humans may create advantages for species with the ability to exploit anthropogenic habitats. Several mammal carnivores, for instance, tend to live at higher densities in humanized habitats than in natural ones."[30] Golden wolves and jackals are prime examples of adaptable opportunists able to vary diet and behaviour to adjust to human expansion.

The domestication and raising of sheep and goats spread to Egypt, the Nile Valley, and the coastal areas of North Africa from West Asia, where there is the earliest evidence of domestication.[31] While it is likely that pastoralists with their stock entered Egypt from the Middle East, some domestication of cattle native to northeast Africa appears to have taken place independently in Egypt.[32] An important site for research into human habitation and dispersal in this period is Nabta Playa, 100km west of Abu Simbel. Here evidence has been found of early Holocene settlements.[33] The area had been dry and unfavourable for settlement prior to the increase in rainfall across much of Africa, but it did have seasonal lakes that filled when it rained and supported vegetation and wildlife. Jórdeczka et al. reported the presence of remains of gazelles, hares, bustards, turtles, ducks, water snails, and what they call the golden jackal (*Canis aureus*),[34] but which we now call the North African golden wolf (*Canis lupaster lupaster*). The mammals, birds, reptiles, and molluscs would all have figured in the diet of the wolf, as they still do. The gazelles and hares are also likely to have been a major source of food for human inhabitants around the seasonal lakes. There were also remains of cattle which, the authors conclude not unreasonably, are unlikely to have survived in the fairly arid conditions without human assistance – something supported by recent DNA evidence of the divergence of African and Eurasian wild cattle species around 25ka and the possibility that the start of domestication of cattle in north-east Africa was about 7,000 years ago.[35] Phillipson says excavations at Dakhleh Oasis in the Egyptian Western Desert found evidence, dated to 9,000–8,500 years ago (7050–6550BCE), that suggested that local wild cattle were already being herded.[36] Domesticated animals and their herders moved into the Nile Delta region between 6,000 and 5,000BCE, as the climate became more arid in the Western Desert and in the Levant to the east, encouraging migration to better watered areas.[37] Domestication of animals for food created a new form of relationship between humans and combined hunters/scavengers like golden wolves and jackals. Competition for carcasses (which humans would invariably have won) was replaced by conflict over protection of livestock and the perception of pastoralists and arable farmers who kept smallstock and poultry that golden wolves were a threat.

Jackals and golden wolves are very adaptable and have found ways of exploiting human activity, whether crops they could eat, livestock they could kill or scavenge, and waste from which they could forage. The expansion of Neolithic cultures and lifestyles would have been advantageous rather than a threat. The African wolf appears to have thrived in a diversity of wilderness and human-inhabited landscapes because of its "habitat plasticity and opportunistic feeding habits."[38] Eddine et al.'s research into the ecological history of the African wolf discovered "consistent evidence supporting our hypothesis that the Northwestern African wolf population experienced a meaningful expansion concurring with a period of rapid population expansion of cattle and other domesticates linked to the advent of agricultural practices during the Neolithic revolution."[39]

This expansion took place around 6,720–3,840BP, with the African wolf expansion "remarkably concurrent with African cattle demographic history with pronounced population expansion approximately 5000 years BP."[40]

The first clearly identified villages in the Nile region date back to 5,200–4,500BCE at Fayum (about 80km south of Cairo) and Merimde Beni-Salama in the Delta.[41] Excavations of waste middens at these sites show the continued presence of wild ungulates such as hartebeest, gazelles, and smaller species of antelopes. Wetterstrom reports that remains of *canis* and *vulpes* (fox) were found at Fayum. The most likely *canis* species to have been present there is the North African golden wolf. By 4,100BCE in Lower Egypt around the Delta, and by 3,800BCE in Upper Egypt further inland along the Nile, farming communities raising crops and livestock were established, creating a harder border with wildlife habitats, though not one that would entirely exclude all wildlife, especially a combined predator/scavenger like the golden wolf that could adapt better than most large predators with more specialisation.[42] At the Qasr el-Sagha Neolithic site in central Egypt west of the Nile Valley, the excavation of sheep, goat, and cattle remains from Neolithic settlements indicated pastoralism, but the discovery alongside them of turtle, crocodile, gazelle, and canis remains, which had been dismembered, suggested the continuation of hunting either for food or, in the case of golden wolves, for pelts and claws.[43]

Domestication of animals and crop cultivation spread across the coastal plains and foothills of the mountain ranges of North Africa, where the golden wolf was widely distributed and abundant. Some estimates put the expansion of pastoralism in North Africa and the Saharan littoral around 6,500BP, when the drying of the Sahara began, though with some areas suitable for nomadic pastoralism.[44] In Morocco, research carried at Kaf Taht el-Ghar indicated that the first cultivation of cereals in the region occurred at about 6,350BCE and that soon after domesticated goats, sheep, pigs, and cattle were raised.[45] There has been less research in the areas of North Africa midway between Morocco and the Nile Delta, but recent work in Libya in north Cyrenaica at the massif of the Jebel Akhdar (Green Mountain) has revealed more about human activity, development of pastoralism, and indications of coexistence of people and indigenous wildlife around 6,800 ± 350BP.[46] The fauna there, evidence of which has been found at sites researched during the 1950s, 1960s, 1980s, and more recently, was dominated by the Barbary sheep (*Ammotragus lervia*). Remains of aurochs, hartebeests, gazelles, hares, golden wolves, and a diversity of bird species were also found.[47]

Pastoralists would have killed or tried to drive off golden wolves and other predators, while land use for cultivation and pastoralism would have progressively restricted food resources and habitats for many prey species of resident predators, though with less success likely for small mammal species such as hares, rats, moles, ground-nesting birds, lizards, snakes, and insects, which make up a large proportion of the golden wolf diet. In some ways, this might have been of advantage for a time for golden wolves and then jackals, as farming and pastoralism spread further south into the Horn of Africa and East Africa. The exclusion, killing, or deprivation of large prey for larger carnivores as human settlements expanded and farming became more widely practised may have benefitted golden wolves and jackals in some parts of their range, as mesopredators like them can thrive when larger predators are removed.[48] Golden wolves (and side-striped or black-backed jackal species to the south and east in Africa) may also have been particularly well placed to start attacking smallstock around settlements as they began to scavenge from the waste from human settlements or temporary encampments. While they developed a wariness of people and the threat they posed, they also adopted strategies to avoid them. They were able to overcome fear to scavenge carcasses or predate on sheep, goats, and poultry around camps and villages at night.

West and Central Africa

From the west-central Sahara, people moved south and west into West Africa, particularly around the Niger River, where there is the first evidence of the cultivation of West African rice from plants gathered from the wild.[49] Yams, taken from wild plants, were cultivated in areas that were not suitable for rice. Over the next 5,000 years, cattle-raising spread and the role of humans as engineers of the environment increasingly affected wildlife, by clearing forests and enclosing land. Archaeological evidence from 9,500–5,000BCE indicates that people were raising animals and growing plants for food in the south-eastern Sahara, the savannas of West Africa, and the Ethiopian highlands.[50] Archaeological finds around Lake Chad show the survival of some nomadic livestock-keeping in highland areas of the central Sahara with sufficient rainfall to support it.[51] Nomadic livestock herding raised few barriers to wildlife, but more settled farming did. The harder the boundaries became, the more adaptable wild species had to be to survive. The survival of golden wolves and side-striped jackals in West and Central Africa demonstrates their more eclectic food acquisition strategies than some predators, like lion and cheetah, which were progressively driven out of settled regions in West and Central Africa.

The Horn and East Africa

Livestock-keeping communities settled in Sudan about 6,000 years ago, spreading from there over the next 1,000–2,000 years to Eritrea, Ethiopia, Somalia, and Kenya. They kept goats, sheep, and cattle.[52] By the start of the third millennium BCE, pastoralist communities were pushing south through Sudan and into northern Kenya, going on to northern Tanzania by the end of the second millennium, encountering and assimilating or pushing into marginal areas the resident hunter-gatherer peoples.[53] This brought the dispersing pastoralists and farmers into contact with black-backed and side-striped jackals of the Horn and East Africa, as well as the African golden wolves (of which they would already have had knowledge).

Archaeological and archaeozoological studies have produced data suggesting that pastoral societies emerged in the Horn of Africa (where golden wolf, side-striped, and black-backed jackal ranges overlap) at the beginning of the second millennium BCE, around 2,000 years later than in the neighbouring regions of the Sahel and parts of north-west Kenya.[54] This may to some extent be related to a series of climatic changes at the end of the Pleistocene and start of the Holocene, with a very dry phase at the end of the Pleistocene, another dry phase with a maximum around 5,500BCE, a wet phase between 5,000 and 3,800BCE, followed by drying with the arid phase that continues until today beginning in 2,500BCE. The paleoclimatic shifts "provoked recompositions of biodiversity and ecological pressures that played out in different ways according to local topography and rainfall," affecting human agricultural development and wildlife habitat.[55] Phillipson believes that pastoralism was being practised on the Sudan-Eritrea border from the sixth millennium CE onwards and that in parts of the Horn of Africa cattle herding may have been taking place before 3,000BCE as herding communities dispersed southwards from Sudan.[56] In northern Ethiopia, excavations at Danei Kawlos rock shelter provided cattle teeth dated to 1741–1540BCE, and to the south at Asa Koma, western Djibouti, cattle, sheep, and goats may have been present from the beginning of the second millennium BCE. Domestic bovine also may present at the junction of the Ethiopian Rift and Afar region. Excavations of faunal remains from these areas around 2,000BCE indicate the presence of animals adapted to a dry environment: side-striped jackal, dorcas gazelle, and dik-dik, with botanical data showing the prevalence of arid steppe vegetation.[57] A later study identified a large number of jackal or golden wolf remains at Asa Koma, second only in number of mammal remains to bovid bones (not differentiated into species).[58]

In northern Kenya, domesticated cattle, sheep, and goats' presence has been indicated by finds at two sites east of Lake Turkana, dated at between 2,800 and 2,200BCE.[59] Robertshaw supports the idea that pastoralists entered northern Kenya in the third millennium BCE, moving on to southern Kenya and Tanzania in the second millennium.[60] The incoming pastoral peoples moved into areas that were inhabited by hunter-gatherers, who may have assimilated pastoralism without totally abandoning hunter-gatherer modes of food acquisition.[61] Excavations of early pastoralist sites in Kenya found that faunal assemblages in and around early settlements and at middens contained overwhelming livestock remains (cattle, sheep, and goats) rather than remains of wild ungulates and that hunting "seems at best to have been a very occasional activity."[62] The keeping of sheep and goats would have created new hunting and scavenging opportunities for East African golden wolves, black-backed, and side-striped jackals. The evidence from the extensive research on human–jackal conflict increasing substantially with the introduction of sheep by humans in South Africa (see Chapter 9) suggests that golden wolves and jackals would have adapted rapidly to a new source of prey and or carcasses (from animals slaughtered by the pastoralists and animals which died from disease or during droughts). This would both have provided a ready source of food and a source of conflict with pastoralists.

The culture of pastoralism spread south into Tanzania as populations expanded and dispersed. Research in north-central and central Tanzania of this period is limited and "remains poorly investigated,"[63] but while pastoral communities spread into the area around Serengeti and then moved south, there is no clear evidence that they passed any of their livestock or their pastoral culture on to the resident hunter-gatherer communities there. The gaps in research include relations between the pastoralists and the resident predators.

Southern Africa

Constructing an evidence-based account of the relationship between the humans and wildlife in Southern Africa is not easy and involves piecing together fragments to try to provide viable interpretations. The historian, David Beach, warns that sources on the development of communities and their relationship to their environment are thin.[64]

The San were the first modern human inhabitants of the region. Lee records that the archaeological record of Dobe in west-central Botswana near the border with Namibia indicates human habitation from at least 20,000BCE, with Middle Stone Age artefacts found there, but with no clear evidence of when the Kung San people arrived there.[65] Lee notes that the Dobe area was replete with wildlife – noting 13 species of large- and medium-sized ungulates and 18 species of carnivores from the banded mongoose up to the lion, including both honey badgers and black-backed jackals, which were both described as common.[66] Around 12,000–10,000BCE, Khoikhoi peoples moved south into the region from near Lake Malawi through eastern Namibia and the Okavango, settling in Namibia and western Botswana. Others migrated from the northern end of Lake Malawi via the Zambezi and Limpopo into eastern Botswana, Zimbabwe, and South Africa.[67] Some San moved eastwards and south in eastern Transvaal, Lesotho, Natal, and the Eastern Cape, the latter often being known as Mountain Bushmen. The San and Khoikhoi both had cultures of cooperative foraging and hunting, including scavenging from carnivore kills. At the stage of the initial dispersion southwards, the San and Khoikhoi were hunter-gatherers with no rearing of livestock or cultivation of food plants. Later contact with early Bantu migrants influenced Khoikhoi adoption of pastoralism, becoming their main form of economic activity, with hunting and foraging as supplementary methods of food acquisition.[68] Walker et al. noted that South Africa's Karoo (later to become the focus of jackal-sheepherder conflict) was peopled by San or their ancestors from 20,000BP, and about 2,000 years ago Khoikhoi pastoralists with flocks of fat-tailed sheep arrived, creating a focus for human–jackal conflict for the first time there.[69]

During the last millennium BCE and the opening centuries CE, southern and south-eastern Africa contained a range of habitats – forest, woodland savanna, dry savanna, montane forest, and arid steppe and desert.[70] Ungulate species were present in large numbers in suitable habitat and included all the current species plus the now-extinct bloubok and quagga, which were found in large numbers in

South Africa's Cape. In grassland, savanna, and semi-arid areas of South Africa, there were huge herds of springbok, black wildebeest, bontebok, and blesbok, which supported large numbers of spotted hyenas, lions, leopards, cheetahs, wild dogs, and jackals – the black-backed being most abundant but the Sundevall side-striped (*Lupulella adustus adustus*) present, particularly in wooded northern areas of the southern African region, including what is now northern Botswana, Namibia, and South Africa and up into Angola, Zambia, Zimbabwe, Mozambique and Malawi.

Egypt – jackals as gods and guides to the afterlife

The development of the immediately pre-Dynastic and then the Egyptian Dynastic (the early period being dated from about 2,925BCE and the Dynastic period as a whole ending in about 343BCE or 332BCE, according to which source you use)[71] civilisations led to the substantial expansion in populations, the size and sophistication of human settlements, the rise of large towns and cities, development of construction and manufacturing techniques, and advances in crop production and animal husbandry. The history of Egypt in this period can be divided into three main periods: the Old Kingdom (2,700–2,200BCE), the Middle Kingdom (2,050–1,800BCE), and the New Kingdom (about 1,550–1,100BCE). And the New Kingdom was followed by the Late New Kingdom, which lasted until the last 40 or so years of the fourth millennium BCE.[72] By 343BCE, the Late New Kingdom Dynasty was crumbling and was hardly in control of the territory over which it and its predecessors had ruled. In 332BCE, Alexander the Great, having destroyed Persian power in the eastern Mediterranean, conquered Egypt and installed his general Ptolemy as ruler. The Ptolemaic Dynasty lasted 300 years until the conquest of Egypt by Rome in 30BCE. The collapse and division of the Roman Empire in the 4th century CE led to the Byzantine rule until the Arab invasion in 639CE. The Arab invaders then moved westward across North Africa occupying Libya and then Tunisia. By the 8th century, they had taken Morocco and moved into Spain.

During the 2,700 years of Dynastic and Ptolemaic rule, the population grew substantially and progressively, and towns, irrigation schemes, and cultivated land created a very hard border with the natural world. This led to the exclusion or progressive extermination of many large herbivores and carnivores. But golden wolves (called jackals in many of the ancient texts and historical accounts of Dynastic Egypt – and where necessary I will continue to use the term "jackals" for them in keeping with usage in historical texts and analysis) clearly survived and, in the form of jackal representations in art, myths, and religion, were clearly numerous, known to most Egyptians across this period, and they became endowed with religious and ritual significance. Images from pre-Dynastic artefacts of the later Neolithic period (3,600–33,300BCE) show spiritual beliefs that included representations of deities in animal forms. By the end of the pre-Dynastic period, jackals, dogs, gazelles, cattle, and rams were appearing in ritual contexts; often these animals would be killed and buried with the corpses of important people.[73] From this period into the Dynastic, the gods of the air, sky, water, and earth were added to with deities drawn from animals such as the jackal, cat, lion, and hawk.

Within the religious system, death and burials were key events. Tombs, coffins, and large sarcophagi would be decorated with images of gods, particularly Osiris, the gold of the underworld, and Anubis (the jackal), associated with burials, the afterlife, and the judgement of the dead.[74] His name is the Greek form of the Egyptian word *anpu*, meaning to decay. He was also known as "The Dog Who Swallows Millions," indicating his association with death and burial.[75] Anubis was seen as the lord of the desert, and the deity Duamutef who oversaw embalming and burial, one of the deities who was son of the god Horus, was depicted as a jackal, and his image adorned canopic jars (into which the internal organs of the deceased would be placed to be buried with the mummified remains) – his jar contained the stomach, oddly appropriate for such an omnivorous animal.[76] Wilkinson believes that the association of jackal deities with death and burial "probably originated in the habit of desert canines scavenging in the shallow graves of early cemeteries and, as was common in Egyptian protective magic, the form of the threat was then utilized in order to provide protection for the dead."[77] Anubis's image would be carved at the entrance to tombs to ward off grave robbers. He was said to command an army of demons and to have flayed the dead deity Seth, who had killed Osiris.[78] Amulets dedicated to Anubis would be placed in tombs. His image probably developed from an older jackal deity, Wepwawet. Both deities were usually represented as muscular jackal-headed men. Input (sometimes Anput) was a female deity, seen as consort to Anubis, often represented as a pregnant or nursing jackal, or as a jackal wielding knives.[79] There were other jackal-headed minor deities, such as Sez and Imit.[80] Beads made from jackals' teeth have been found as grave goods in some tombs.[81]

During the late Dynastic period, the Greek historian Herodotus wrote that in Egypt and Libya, there were "dog-headed men," perhaps referring to images of jackal-headed deities, and he listed jackals among the fauna of the region.[82] When he travelled up the Nile, he started referring to Egyptian wolves – predating the modern classification by nearly 2,500 years. He said they were bigger than foxes but smaller than Greek wolves and that if an Egyptian found one dead, he would bury it[83] – a level of respect not accorded to most other animals. Herodotus would have been familiar with Eurasian jackals and wolves, and it is quite significant that he called the Egyptian ones wolves. Ferguson in his account of the wolf-jackal question in Egypt said that in the settlement in Egypt known as Siout Lykopolis – "the city of the wolf" – the bones of what he called wolf-mummies were found but also said that among these remains were "two kinds of jackals."[84] He said that 19th-century archaeologists had found wolf/jackal skulls in Egypt and believed they were related to wolf because of the size, but they still called them jackals.

Notes

1 R.S. Reid (2012) *Savannas of Our Birth. People, Wildlife and Change in East Africa*, Berkeley: University of California Press, 2012, p. 95.
2 Ibid., p. 63.
3 J. Carruthers & N. Nattrass (2018) History of predator-stock conflict in South Africa, in G.I.H. Kerley, S.L. Wilson & D. Balfour (eds) *Livestock Predation and Its Management in South Africa: A Scientific Assessment*, Port Elizabeth: Centre for African Conservation Ecology, Nelson Mandela University, 30–52, p. 32.
4 A. Roberts (2018) *Evolution the Human Story (Revised edition)*, London: Dorling Kindersley, p. 203.
5 D. Phillipson (2005) *African Archaeology*, Third edition, Cambridge: Cambridge University Press, p. 168.
6 F.J. Alberto et al. (2018) Convergent genomic signatures of domestication in sheep and goats, *Nature Communications*, 9, 813, https://doi.org/10.1038/s41467-018-03206-y accessed 9 May 2019.
7 Roberts, 2018, p. 208.
8 S. Pigott (1950) *Prehistoric India*, Harmondsworth, Middlesex: Pelican, pp. 43–4.
9 Roberts, 2018, pp. 208–9.
10 J.M. Aynard (1972) Animals in Mesopotamia, in A. Houghton Brodrick (ed) *Animals in Archaeology*, London, Barrie and Jenkins, 42–68, p. 43.
11 Ibid.
12 M. Gadgil & R. Guha (1992) *This Fissured Land, an Ecological History of India*, New Delhi: Oxford University Press, p. 63.
13 Ibid., pp. 64–5.
14 Gadgil & Guha, 1992, p, 66.
15 See K. Somerville (2021) *Humans and Hyenas Monster or Misunderstood*, Abingdon, Oxon: Routledge/Earthscan, pp. 23–81; and (2020) *Humans and Lions Conflict, Conservation and Coexistence*, Abingdon, Oxon: Routledge/Earthscan, pp. 29–76.
16 W. Wetterstrom (1993) Foraging and farming in Egypt: The transition from hunting and gathering to horticulture in the Nile Valley, in T. Shaw et al. (eds) *The Archaeology of Africa. Food, Metals and Towns*, London: Routledge, 165–226, p. 167.
17 Phillipson, p. 147.
18 Phillipson, 2005, p. 165.
19 C. Ehret (2016) *The Civilizations of Africa. A History to 1800*, Charlottesville: University of Virginia Press, pp. 33–7.
20 Ibid., p. 149.
21 Wetterstrom, 1993, p. 171.
22 Ibid., pp. 41–2.
23 J. Iliffe (2007) *Africans. The History of a Continent*, Second edition, Cambridge: Cambridge University Press, pp. 12–3.
24 Phillipson, 2005, p. 149.
25 Ibid., 147.
26 Ibid.
27 Ibid., p. 149.
28 Ibid., p. 150.
29 A. Eddine et al. (2020) Demographic expansion of an African opportunistic carnivore during the Neolithic revolution. *Royal Society Biology Letters*, 16, 1–7, pp. 1–2.
30 Ibid.
31 For a more detailed examination of the origins of domesticated cattle, see J.E. Decker et al. (2014) Worldwide patterns of ancestry, divergence, and admixture in domesticated cattle, *PLOS Genetics*, March 27, 2014.
32 For a more detailed examination of the origins of domesticated cattle, see J.E. Decker et al. (2014) Worldwide patterns of ancestry, divergence, and admixture in domesticated cattle, *PLOS Genetics*, March 27, 2014.

33 M. Jórdeczka et al. (2015) Here comes the rain again… The early Holocene El Adam occupation of the Western Desert, Nabta Playa, Egypt: Site E-08-2, *Azania: Archaeological Research in Africa*, 50(1), 3–26, p. 3.
34 Ibid., p. 8.
35 Ibid.
36 Phillipson, 2005, pp. 150–1.
37 Wetterstrom, 1993, p. 167.
38 Eddine, 2020, pp. 1–2.
39 Ibid., p. 5.
40 Ibid., p. 6.
41 Ibid., p. 201.
42 K. Somerville (2021) *Humans and Hyenas Monster or Misunderstood?* Abingdon, Oxon: Routledge/Earthscan, p. 55.
43 Ibid., p. 208.
44 Roberts, 2018, p. 212.
45 De Faucamberge, 2016, p. 145.
46 Ibid., p. 146.
47 Ibid.
48 L.R. Prugh et al. (2009) The Rise of the Mesopredator, *BioScience*, 59, 9, 779–91, p. 779.
49 B.W. Andha (1993) Identifying early farming traditions of west Africa, in Shaw et al., 241–54, p. 245.
50 C. Ehret (2016) *The Civilisations of Africa. A History to 1800*, Charlottesville: University of Virginia Press, pp. 25–33.
51 Ibid., p. 13.
52 Reid, 2012, p. 96.
53 Robertshaw, 1993, p. 358.
54 J. Lesur (2014) The advent of herding in the Horn of Africa: New data from Ethiopia, Djibouti and Somaliland, *Quaternary International*, 343, 148–58, p. 148.
55 Ibid., p. 149.
56 Phillipson, 2005, p. 205.
57 Lesur, 2014, p. 155.
58 L. Coudert et al. (2018) New archaeozoological results from Asa Koma (Djibouti): Contributing to the understanding of faunal exploitation during the 3rd millennium BC in the Horn of Africa, *Quaternary International*, 471, 219–28, p. 223.
59 Lesur, 2014, p. 150.
60 P. Robertshaw (1993) The beginnings of food production in southwestern Kenya, in Shaw et al., 358–71, p. 358.
61 Ibid., p. 359.
62 P. Robertshaw (1990) *Early Pastoralists of South-Western Kenya*, Nairobi: British Institute in East Africa, p. 1.
63 Phillipson, 2005, p. 212.
64 D.N. Beach (1980) *The Shona and Zimbabwe 900-1850*, Gweru, Zimbabwe: Mambo Press, p. 53.
65 R.B. Lee (1979) *The Dobe !Kung*, Cambridge: Cambridge University Press, p. 17.
66 Ibid., p. 24.
67 A. Barnard (1992) *Hunters and Herders of Southern Africa. A Comparative Ethnography of the Khoisan Peoples*, Cambridge: Cambridge University Press, 1992, p. 34.
68 Ibid., pp. 30–2.
69 C. Walker et al. (2018) Drivers and trajectories of social and ecological change in the Karoo, South Africa, *African Journal of Range & Forage Science*, 35(3), 157–77, p. 161.
70 Ehret, 1998, pp. 212–3.
71 Phillipson, 2005, p. 190 and *Ancient Egypt, 3c Dynasties*, https://www.ushistory.org/civ/3c.asp accessed 10 December 2021.
72 *Ancient Egypt, 3c Dynasties*, https://www.ushistory.org/civ/3c.asp accessed 10 December 2021.

73 R.W. Wilkinson (2017) *The Complete Gods and Goddesses of Ancient Egypt*, London: Thames and Hudson, p. 12.

74 Ibid., p. 76.

75 J.J. Mark (2016) Anubis, https://www.ancient.eu/Anubis/ accessed 30 November 2020.

76 Ibid., p. 88.

77 Ibid., p. 187. See also L. Oakes & L. Ghalin (2018) *Ancient Egypt*, London: Lorenz Books, p. 281.

78 Ancient History Encyclopedia (no date) Anubis, https://www.ancient.eu/Anubis/#:~:text=Anubis%20is%20the%20Egyptian%20god,whom%20he%20is%20often%20confused Accessed 9 December 2020.

79 Ibid.

80 W. Cline (1948) Notes on Cultural Innovations in Dynastic Egypt, *Southwestern Journal of Anthropology*, 4, 1, 1–30, p. 6.

81 T. Rehrenn (2013) 5,000 years old Egyptian iron beads made from hammered meteoritic iron other excavations, *Journal of Archaeological Science*, 40, 4785–92, p. 4786.

82 Herodotus, trans. R. Waterfield (1988) *The Histories*, Oxford: Oxford World's Classics, Book Four 191–3, p. 298–9.

83 Ibid., p. 122.

84 W.V. Ferguson (1981) The systematic position of Canis aureus lupaster (*Carnivora: Canidae*) and the occurrence of Canis lupusin North Africa, Egypt and Sinai, *Mammalia*, 45(4), 459–65, p. 465.

4

JACKALS AND HUMANS IN AFRICA IN THE PRE-COLONIAL ERA

Trying to tease out evidence of the presence of jackals and golden wolves in Africa and assess the level of conflict with humans are not easy in this period. The three African species which are my focus, with black-backed jackals at the forefront in this narrative, are patchily mentioned in early accounts of the fauna of the region. They were, as demonstrated in the previous chapter, important in the belief systems of the Egyptians, an indication of an abundance, of human awareness of their role as scavengers (including suspicions of grave-robbing and consumption of human remains), but elsewhere on the continent, they were less culturally important and often ignored as unimportant scavengers rather than being charismatic species that were respected, feared, or hated.[1] Often, animals not important for people are not included in folktales, belief systems, or recorded in images. Similarly, outsiders visiting Africa commented on elephants, lions, rhino, leopards, cheetah, gorillas, and charismatic mammals, not jackals. Monsarrat and Kerley noted that European traders, settlers, missionaries, and naturalists of the 16th–20th centuries in southern Africa "provide valuable information on the past composition of mammal fauna, but the taxonomic biases in these records remain to be investigated and understood"[2] and may distort the actual faunal record by concentrating on species for reasons of charisma, nuisance value, commercial value, or sporting challenge. The following narrative is, because of the scarcity of records, patchy and stitched together from the limited material available.

In their study of the historical faunal record of the Cape Floristic Region of south-western South Africa, Monsarrat and Kerley examined 780 historical records of mammals – which at the time of the first European occupation at Cape Town included a great diversity of herbivores and carnivores, the latter including black-backed jackals, caracals, lions, leopards, wild dogs, and spotted hyenas. There were 31 species of large terrestrial mammals contained in the records, with

DOI: 10.4324/9781003199793-5

the first dating back to 1407 and the Portuguese explorer Vasco da Gama, who saw herds of elephants at Mosselbaai.[3] The presence of buffalo, lion, and elephant was over-reported in comparison with their numbers, and

> the dataset is strongly biased against 15 species (aardvark, aardwolf, African wild cat, blue duiker, bushpig, Cape fox, Cape grysbok, Cape porcupine, common duiker, honey badger, klipspringer, mountain reedbuck, small spotted cat, steenbok and vervet monkey) … Black-backed jackals were measured with an observational abundance of 17, compared with the calculated historical abundance of 10,703.[4]

Boshoff and Kerley also noted that black-backed jackals were unrealistically under-documented in historical accounts of the fauna of South Africa,[5] something redolent of my research using published sources and accounts of journeys by naturalists, hunters, traders, and other travellers in sub-Saharan Africa from the first European arrival in the last years of the 15th century.

Sudan and the Horn of Africa

In the last millennium BCE and the Common Era up to the present, there is evidence of diverse faunal presence and human occupation of much of southern Sudan and neighbouring areas of Ethiopia and Kenya. Archaeological investigations have shown a mixture of pastoralism and continuing hunter-gatherer activities, with middens at old permanent settlements and at seasonal cattle or fishing camps containing large numbers of fish bones, the bones of wild ungulates and of domestic livestock, along with evidence of the presence of carnivores and scavengers, which would have included North African golden wolves, Equatorial and Kaffa side-striped jackals, and East African black-backed jackals, present in the region today.

Archaeological discoveries in Lakes Province, South Sudan, stretching over the time period noted above, show that human settlements were in areas of savanna/woodland, with evidence of large numbers of trees having been felled to clear land for grazing.[6] Less permanent camps were found where the terrain changed to open plains, which merged with seasonal floodplains and swamps of the Sudd area of the Nile. Cattle were moved across these areas, being gathered in camps situated near permanent rivers during the dry season to take advantage of remaining grazing and access to water.[7] Excavations of middens and ash mounds showed evidence of the disposal of waste, including livestock, fish, and wild ungulate bones, including kongoni, kob, bushbuck, and other small- to medium-sized antelope.[8] One site contained canid remains that could have been from jackals or golden wolves but were not identified as a specific species.[9]

The 19th-century diplomat Mansfield Parkyns conducted several expeditions to explore Ethiopia and recorded the use of jackal skins as part of the apparel of warriors in the Galla region. He wrote that in addition to leopard skins being worn,

> I might name those of the jackal from Simeon [the Simien region] ... Some persons have, however, a prejudice against the skin of the red jackal, as it is supposed that, should a lance piercing the skin inflict a wound on its wearer, and by ill luck a single hair of the jackal enter the wound, the patient is sure to die.[10]

The red jackal may refer to the Ethiopia wolf, but the generalised description of jackal could refer to the golden wolf, the side-striped jackal, or the black-backed jackal, all found there then and now. Parkyns said he wore a jackal skin over his shoulders at times. He noted that in the Ado region community ceremonies in which oxen or sheep are slaughtered benefit jackals and hyenas, as the slaughtered animals are not consumed but left for predators,[11] something that parallels the interesting coexistence achieved by some Ethiopian communities with carnivores that clear up animal carcasses and other human-created refuse and are often treated as efficient street cleaners, ridding towns and villages of potentially harmful waste and diseased carcasses.[12] Parkyns later notes that the jackal, like the hyena, "is very common in the populous districts of Abyssinia, where he finds plenty of offal."[13]

Parkyns exhibited no contempt or hatred for the jackal and, for a period during his stay in Ethiopia, kept one as a pet, which formed a friendship with his dog.[14] He did also suggest that jackals would make excellent quarry for fox-hunting on horseback, "were not the country too difficult to ride over."[15] He said there were different species of jackal in Ethiopia, though he may include the Ethiopian wolf as a jackal, and obviously, as most zoologists did recently, refers to the golden wolf as a jackal. He wrote of the species: "There are three or four sorts of jackals in Abyssinia. The common grey one is found everywhere; another is larger and of a bright chestnut and grey colour, while that of Simeon [Simien] is yellow."[16]

Naturalists and travellers in the Horn of Africa in the 19th century commented on the substantial numbers of jackals – usually without identifying the species. W.G. Blanford, a corresponding member of the Zoological Society of London (ZSL), wrote in February 1868 that at Annesley Bay, in Eritrea, he had seen jackals that he said were quite different from the Indian species, but he regretted he had been unable to get a specimen.[17] An expedition to Ethiopia by naturalists and geographers connected with the ZSL reported seeing a jackal they labelled as *Canis anthus* – presumably referring to what was then called the North African golden jackal but giving it the Latin name for the West African golden jackal rather than *Canis aureus* or *lupaster*.[18] To the south, in Somaliland, a British naval officer, Frederick Forbes, reported that at the port of Berbera, jackals seemed common around the town. Walking on the beach on the outskirts of the town, he said,

> Having taken a short turn among the sand hills I ascended ye high land to the right of the wells. It consisted of loose blocks & boulders of aluminous

stone quite sonorous & as hard as flint. There were some caves in ye face of ye precipice near ye top where ye jackals or hyenas appear to frequent.[19]

In the interior of Somaliland, jackals and golden wolves were common. The hunter C.J. Mellis recalled that on hunting trips, he always heard as the sun went down "the sound of the wailing cry of the jackal."[20] A well-known hunter and specimen collector for museums, R. Hawker, shot more than a dozen black-backed jackals during a hunting expedition to Somaliland in 1898 and said that they were "everywhere very common" there.[21]

East Africa

Over millennia, pastoral communities in East Africa developed considerable skill in hunting predators, initially to protect their cattle and communities, and in some this evolved into culturally important rituals. The hunting of lions, such as by the Maasai, Nandi, and Barabaig, became part of the rites of passage to adulthood. The killing of leopards was also seen as a brave act that protected the community. This never extended to jackals. Dr Amy Dickman, who founded the Ruaha Carnivore Project in Tanzania and has studied Maasai and Barabaig conflict with carnivores, told the author that huge kudos was attached to killing lions, but there was no respect to be gained for killing jackals, which were generally ignored, though in modern times, they might fall victim to pastoralists poisoning carcasses of livestock killed by lions, leopards, or hyenas.[22]

In East Africa, especially the savannas and open woodlands of central-southern Kenya and northern Tanzania, human populations were in a state of flux in the second millennium of the CE period. In the 17th–19th centuries, there were extensive movements of pastoralist communities. It is hard to establish exact timelines, but Robertshaw suggests that prior to the 17th century, which saw the arrival and establishment of dominance by Maa-speakers (Maasai and Samburu), groups of Southern Nilotic-speaking peoples such as the Datoga and southern Kalenjin occupied much of the land suitable for pastoralism in central and southern Kenya and northern Tanzania. They mixed cultivation in suitable areas with livestock husbandry.[23] Early Maa-speaking communities moved into the area around 1600.[24] They were the ancestors of the Omotik and Okiek Ndorobo (the latter word being a pejorative one applied by the Maasai to these communities, meaning "without cattle"). They displaced many of the existing inhabitants but were then displaced by the Maasai/Samburu influx. The Maasai and Samburu took over grassland areas from Laikipia down into Tanzania. Maasai dominance was established in the 18th and 19th centuries in many areas around Laikipia, and then south towards the Mara, and into Tanzania's steppe.[25] The dominance of pastoralism at this time was based on the fertile grasslands and open woodland areas, which were the habitat of the huge herds of wildebeest, zebra, gazelles, buffalo, and other ungulates, which supported large populations of hyenas, lions, wild dogs, leopards and cheetahs, and East African golden wolves and black-backed and side-striped jackals.

To the west of the Maasai, the Nandi and other Kalenjin-speaking groups settled on the highlands of the Rift Valley. They combined livestock husbandry with crop production, supplemented with hunting when necessary. The Nandi developed male age-sets and through manhood rituals established them as warrior groups tasked with protecting cattle against raiders (chiefly Maasai) but also from attack by predators. To the south-west, on what is now the western border of the Serengeti National Park, communities such as the Ishenyi and the Ikoma lived alongside the considerable concentrations of game animals and the carnivores preying on them. Hunting was for meat, hides, protection of stock, and ritual purposes.

The development of trade on the Indian Ocean littoral, controlled by Arab, Afro-Shirazi, or Swahili traders, had an effect on the hunting practices of communities across the region, as demand grew for ivory, animal skins, teeth, and claws; there is no indication of a large trade in jackal skins or claws. The growth in trade in wildlife products did have an effect on the ability of East African communities to kill hunt, as it brought firearms into the region. The use of firearms had a huge and lasting effect on wildlife. They were often antiquated muzzle-loaders but increased the ability of communities to hunt for food, skins, and ivory, to protect themselves and raid for slaves. By 1888, around 100,000 guns a year were being traded to people through Zanzibar alone.[26] Game hunting was at the heart of early European penetration of East Africa – with ivory the most sought-after commodity and the trade as one of the drivers of commercial and then political interests in the region on the part of Britain and Germany.[27]

Jackals and golden wolves were not hunted much for sport or to protect livestock in East Africa but were common across Kenya and Tanzania. In Tanzania, the East African golden wolf was limited in its distribution, to the northern, drier areas, which it shared with the East African black-backed jackal, with the side-striped in woodland and thick bush rather than the open grasslands and open woodlands. The side-striped was present in the foothills of Kilimanjaro and around the Moshi area, where it was reported to the ZSL by H.H. Johnson as very common and found around villages, where it foraged for food among refuse put out by the villagers.[28] The explorer Richard Burton reported the areas of east-central Tanzania (near Morogoro), through which he passed journeying from the coast to Lake Tanganyika, had abundant wildlife including what he called the "beautiful silver jackal,"[29] which could have been the black-backed or the side-striped. Few of the other explorers, traders, and hunters journeying through Kenya or Tanzania mentioned the jackal.

Southern Africa

In southern Africa before the arrival of Bantu migrants from the north, San and Khoikhoi communities peopled the region, having themselves dispersed from East and Central Africa over several millennia. At the stage of the initial dispersions southwards, the San and Khoikhoi were hunter-gatherers with no livestock or cultivation of food plants. Later interactions with the Bantu migrants led to the Khoikhoi adoption of pastoralism, with hunting and foraging as supplementary methods of food acquisition. Many of the hunter-gatherer and early pastoral communities of southern Africa were pushed into arid regions or were assimilated when Bantu migrants occupied the land around 300–200BCE, basing their livelihoods on pastoralism and cultivation.[30] Bantu migrants brought iron tools, along with their agricultural methods, at a time when the Khoikhoi and San were still using stone tools and weapons.[31] The San remained totally dependent on hunting and foraging for wild foods, with no assimilation of pastoralism, though they did trade with farming communities.[32] The Bantu migrants – Tswana speakers in the case of Botswana and the ancestors of the Nguni, Xhosa, and Sotho peoples in South Africa – settled in the better-watered north-eastern and eastern areas of the countries. The pastoralist communities, or crop cultivators who kept a few goats, sheep, cattle, or chickens, as Carruthers and Nattrass believe, would often have "suffered predation on their livestock from dangerous wild animals,"[33] and these would have included jackals preying on young or weak sheep, goats, and on poultry.

As agriculture became more developed and settlements expanded, land was increasingly cleared of forest for cultivation or grazing – the forging of iron tools enabled more efficient forest and bush clearance and establishment of larger settlements in south-east Africa by the 3rd century CE.[34] The migrant communities grew in size and developed hierarchical societies ruled by chiefs or kings – developing eventually into large states like the Mapungubwe kingdom of northern South Africa and eastern Botswana, the Monomotapa of Zimbabwe and eventually the Zulu Kingdom, its Ndebele offshoot, the Xhosa chieftaincies of the Eastern Cape, and the Tswana chieftaincies of South Africa and Botswana. They hunted to supplement food production and for sport. But a diversity of wildlife species survived across southern Africa, in substantial numbers in some areas and with no evidence of extinctions. It was only the arrival of European settlers from the mid-17th century onwards that led to the extermination of game in much of the Cape, Natal, Free State, and Transvaal (as the regions became known under Dutch then British/Afrikaner rule) and the start of the concerted persecution of the jackal.

Angola, Botswana, and Namibia

In Angola, Portuguese colonisation started in the late 15th century. Over the next four centuries, their presence grew, mainly due to the slave trade, which provided labour for Portuguese plantations in Brazil and on islands like São Tomé and Príncipe. Plantation agriculture was developed in north and central Angola by the Portuguese, where rainfall and soils were better suited for growing coffee, cotton, and other cash crops. In the south and south-east, population density was low, Portuguese settlement was minimal, and the initial effects on wildlife were consequently limited. The areas adjacent to Namibia along the Kunene River were rich in wildlife, especially the coastal areas and inland from the coast north of the Kunene, in what is now Iona National Park. To the east, the arid savanna and dry woodland supported a similar range of antelope as the nearby areas of Namibia and northern Botswana, with large predator populations. Early accounts by explorers and settlers gave insights into the wildlife they encountered. In 1578, the Portuguese trader Duarte Lopes travelled to the Kingdom of Kongo on the border of present-day Angola and Congo. He noted that the area around Ambriz, north of Luanda, was inhabited by "wolves" which had a great fondness of the fruits of oil palm trees – he gave no further clues to their species, with the possibility that they could be side-striped or black-backed jackals or brown hyenas.[35] One of the first detailed accounts of Angola's wildlife came from Joachim John Monteiro, a Portuguese colonial official and naturalist. He worked in Angola from 1858 to 1876 and was a keen collector of natural history specimens. Monteiro noted that in Cuanza Norte, east of Luanda, cattle were predominant with the result that wildlife of all sorts was scarce, but he did report the presence of jackals, hyenas, and leopards.

The black-backed jackal and the side-striped jackal were resident in Angola throughout this period – the black-backed mainly to the south and east in open woodland savanna and arid regions, including coastal desert and dunes, and the side-striped in central and northern Angola, especially in wooded areas or savanna-woodland mosaics. In the region around the capital, Luanda, and inland to the Dembos forest, the Mbundu community has a variety of folktales involving animals, but jackals are rarely mentioned. Courlander in his compendious collection of African folktales lists only one involving a jackal. That story says that a jackal used to mix with domestic dogs. One day the jackal goes with a dog into the bush but sends the dog back to its home to get fire, so they can burn the long grass and catch the locusts hiding there for them to eat. But the dog finds food in the village and does not return to the jackal. The howling of the jackal is supposed to be the animal saying, "I am surprised, I, jackal of Ngonga; dog, whom I sent for fire, when he found mush [food], he was seduced; he stayed for good."[36]

San people and Khoikhoi peoples populated many of the arid regions of Botswana and neighbouring areas of north-west South Africa and eastern Namibia. Human population densities were low, and the effects of hunting by the human inhabitants on wildlife numbers were negligible. Tswana migration into and settlement in the region from as early as 400CE[37] (in more fertile east and north-east) pushed the San into more arid areas or assimilated them into the pastoral communities that settled the land, using them as herders or trackers for hunting.[38] The Tswana were pastoralists and manufactured iron implements and weapons. Several Iron Age cultures developed in the region they inhabited from around 900BCE. One was Toutswe, centred on Toutswemogala Hill settlement about 50km north of Palapye in central Botswana. Cattle were the centre of the Tswana economies and culture. The Tswana also traded in livestock, hides, and wildlife products from hunting with neighbouring peoples along the Limpopo to the Indian Ocean. Ivory and rhino horn were traded by the Tswana, and possibly also jackal-skin karosses in exchange for beads, clothes, and manufactured goods. The Tswana also traded with the San for skins and ivory.

Namibia was home to San and Khoikhoi communities several millennia ago, with more Khoikhoi moving into what is now Namibia as they were pushed out of South Africa by incoming Bantu farmers. Bantu peoples, ancestors of the Ovambo, Herero, and Himba, moved into northern Namibia from Angola along the Kunene and Cubango Rivers. The first evidence of domestic livestock in northern Angola dates to 1,800BP.[39] The areas of northern Angola and less arid districts of Damaraland could accommodate pastoralists, despite having a diverse wildlife population, which faunal remains show included routes of large herds of springbok, ostrich, zebra, gemsbok, kudu, duiker, steenbok, lion, leopard, cheetah, wild dog, and jackal.[40] Even before the opportunity to scavenge from human settlements, kill young livestock, and feed from carcasses of livestock killed by larger predators or by drought or disease, black-backed and side-striped jackals in the wooded areas of northern Namibia and the Caprivi-Zambezi region would have had abundant prey and foraging opportunities. They would have come into little conflict with the San, and conflict may have developed with Khoikhoi pastoralists who kept sheep and goats.

Excavations near Katima Mulilo in Namibia's Zambezi (Caprivi) region indicated that people living there as hunter-gatherers consumed a great variety of wild fauna as part of their diet. Phillipson says that in common with communities to the south in the Linyanti, Chobe, and central Kalahari districts of Botswana, the range of prey killed for food or skins would have been large. Faunal remains from hunter-gatherer settlements south of Zambezi included "440 tortoises, 295 rodents, 227 springhares, 161 birds, 21 pints of termites and ants, 68 steenbok, 65 duiker, and a mixed bag of 134 animals belonging to 11 larger species, including fox, jackal, wildebeest, hartebeest, eland, kudu, and giraffe."[41] Along the Namibian coast lived the Topnaar Khoikhoi community. They were an offshoot of the Nama people, who lived by small-scale pastoralism, hunting and harvesting wild cucumbers.[42] When a Dutch East India Company ship stopped at Sandwich Harbour in 1670, the crew met a group of Topnaar armed with bows and assegais, who attacked the Dutch.[43] The men carried sticks with jackal tails attached, used as fly-whisks. Over 150 years later, a French surveying ship landed at Walvis Bay and encountered members of the Topnaar community. The crew described the people as wearing skin cloaks and jackal-skin caps,[44] indicating the killing of jackals for their skins.

Across this whole pre-colonial period, black-backed jackals were common across most of Namibia, including the arid pre-Namib, Kaokoveld, and Damaraland. Game was plentiful, with large herds of gemsbok, springbok, zebra, giraffe, wildebeest, hartebeest, kudu, and other antelopes. In the 19th century, density of the human population – San, Nama, Topnaar, Orlam Herero, Damara, Ovambo, plus Boer farmers who had trekked north – was low, and the main forms of livelihood were pastoralism, hunting, and some cultivation in Ovamboland and the Caprivi. The Nama, Damara, and other peoples of Khoikhoi descent mixed livestock husbandry with hunting and foraging. The Nama had been pushed north by Dutch settlement in the Cape in the 18th century. By the time of their settlement in Namibia, many of them had horses and guns, which increased their ability to conduct raids for cattle against neighbouring communities and protect their stock against predators.[45] Charles Andersson accompanied the British geographer Francis Galton on a Royal Geographical Society expedition to Namibia in 1850. Throughout his journey, he kept a diary of his travels, regularly mentioning his hunting exploits and the wildlife he encountered. Along the coast near Walvis Bay, he tells of the frequenting of the beaches by jackals and hyenas, who take advantage of the carcasses of fish or marine mammals (especially seals) found there and the waterfowl that fed along the sand.[46]

Travelling along the Swakop River, Andersson met a Danish hunter, Hans Larsen, who told him that stock killing was such a problem that many farmers set up gun traps to kill predators. Jackals and hyenas were particularly prone to being killed as they scavenged the remains of livestock killed by lions, leopards, or cheetahs. Further north, at the scene of a massacre of Damara people by the Nama, Andersson reported finding the bones of those killed, "but we were unable to ascertain the exact number of people killed, as the jackals and hyaenas had carried away and demolished many parts of the skeletons."[47] In the south-east of the country, on the border with South Africa (adjacent to what is now the Kgalagadi Transfrontier Park), Andersson noted that along the Nossob River that game had been decimated and, "[w]ith the exception of hyaenas and jackals, beasts of any size were scarce."[48] While camping in southern Namibia, he expressed anger at the depredations of jackals in and around his camp, vowing vengeance after jackals ate leather straps used to secure their belongings to the oxen. One entry reported that,

> the following morning, the first object that met my half-sleepy gaze was a jackal, busily engaged examining our baggage. Having no gun within reach, I threw a handful of sand at the impudent fellow, on which he saluted me with a mocking laugh, and slowly retreated ... The brute had, indeed, devoured one of the 'rims', with which we secured the packs on the oxen.[49]

Andersson frequently mentioned jackals but seemed to be in some confusion about the species and deciding what was a jackal and what was a fox. In talking about dogs kept by local people, he wrote that they were like "the black-backed fox of Southern Africa, the jackal as he is falsely called, *canis mesomelas*."[50] His frequent references and his anger at their raiding of his camp suggested that jackals had adapted to the human presence and developed ways of benefitting from it through guile and intelligence as foragers. Francis Galton, who led the expedition, also noted the almost total lack of fear shown by jackals when they came into the camps: "jackals are always seen and are always amusing; their impudence is intolerable; they know that you do not want to shoot them, and will often sit in front of your screen and stare you in the face."[51] In an area with a substantial elephant and game population, which Andersson calls Lake Gamut, he says jackals were seen in large groups. One of his hunting dogs killed a jackal and brought it back into the camp, where some of the local men working for him retrieved the jackal from the dog and skinned it, as they valued its fur. He later reports seeing a village chief wearing a strikingly attractive jackal fur kaross over his shoulders.[52]

The relationship between people and wildlife in Namibia began to change in the 1880s, when Germany laid claim to the region, with the first German outpost established at Luderitz in 1882. This enabled German chancellor Bismarck to lay claim to what became known as German South-West Africa at the Berlin Conference of 1884, at which the major European powers divided up Africa between them. The influx of German administrators, traders, settlers, and soldiers completely altered the nature of the region, with commercial farming, the search for valuable minerals, and hunting for profit and to protect livestock taking its toll on wildlife in many areas.

South Africa

By the end of the BCE and start of the Common Era, San and Khoikhoi communities had settled across much of South Africa.[53] Most San and Khoikhoi communities relied on the hunter-gatherer mode of production, but there is evidence that some Khoikhoi had assimilated or developed pastoralism as they moved south from northern Botswana, through contact with pastoralists such as the Tswana. Evidence suggests that some pastoralism might have been practised in South Africa as early as 2,500BP.[54] At this stage, the hunting by the San and Khoikhoi had a minimal effect on biodiversity, including on predators like jackals. Hunting was purely for food or skins to make bedding or clothes. Excavations along the west coast of South Africa have shown extensive consumption of shellfish by coastal communities, as evidenced by large shell middens at some ancient settlements that have been excavated.[55] Some of the middens were found to contain mammal remains, including those of black-backed jackals. This could indicate hunting of the jackals, most likely for their luxuriant skins and furs,[56] or that jackals regularly scavenged among the middens and some died there. At another site near Saldanha Bay, middens were found that included Cape fur seal remains, which made up 93% of mammal bones, and small numbers of black-backed jackal bones in or near middens.[57] This indicates likely consumption of seals by the Khoisan communities along the coast – a dietary choice they shared with the black-backed jackals. Given the substantial numbers of fur seals present now – and likely to have been even more numerous around 2,000–3,000BP – there is unlikely to have been competition between humans and jackals for seals, and jackals may have opportunistically scavenged the carcasses of seals killed by humans. Early San rock art found at various places across South Africa depicted people and animals. Jeffreys noted one showing jackals chasing adult gazelles.[58] Jolly reported that an old San woman he met talked of rock paintings being important to the San to represent their world and the animals with which they shared it. He also recorded accounts of San ritual dances in which dancers wore jackal skins "complete with head and ears."[59]

As the Khoikhoi spread across southern Africa, some groups brought sheep with them – which were present in the Western Cape from 1,860 years ago, with cattle and dogs from 1,300 years ago.[60] The arrival of sheep turned a new page in human–jackal interactions, with the predation issue now developing as jackals would have taken advantage of new sources of food – whether as prey or carcasses to scavenge.[61] The arrival of dogs initiated the possibility both of conflict, including through the use of dogs to guard livestock, and of the transmission of diseases between the species.

South Africa's human population was in a state of flux during the opening millennia of the Common Era, with migration into the region and dispersal of San and Khoikhoi communities by incoming Bantu communities, who settled in the better-watered north-east and east. The forging of iron tools enabled more efficient hunting and facilitated the clearance of woodland, expansion of agriculture, and establishment of larger settlements in south-east Africa by the end of the 2nd century CE.[62] By 500CE, pastoralism was expanding, and in the Zoutpansberg area of northern Transvaal, communities were building homesteads of circular huts around central cattle pens to protect against predators.[63]

By 700CE, Bantu-speaking communities had established settled communities in south-west Zimbabwe/north-eastern Botswana/northern South Africa in what became known as the Leopard's Kopje culture. They combined pastoralism and cultivation with iron-working.[64] By 1075, the Leopard's Kopje communities had grown into a kingdom based on Mapungubwe on the border of Botswana, Zimbabwe, and South Africa. The Leopard's Kopje people hunted as a supplement to livestock and arable farming, but there is no convincing evidence that they seriously depleted wildlife or permanently excluded it from farming areas within their territory. Further south in what is now the KwaZulu-Natal Province, there is evidence of Iron Age Bantu settlement. The excavation site, whose finds are dated at 993–1210CE, included remains of black-backed jackals and honey badgers, perhaps used by herbalists and traditional healers.[65]

Returning to the western half of South Africa, which was not as suitable for settlement by Bantu migrants cultivating crops, the San and Khoikhoi communities were left largely undisturbed there until Bantu groups moved south through Angola and the arrival of European settlers with the Dutch establishment of a provisioning station at Cape Town in 1652. The fauna of the Western Cape was far richer than today with an extensive range of sub-Saharan African ungulates and carnivores. The same is true for the eastern and northern areas of South Africa. There were massive herds of springbok, blesbok, bontebok, blaubok (bluebuck), black wildebeest and eland, as well as kudu reedbuck, a range of smaller antelopes like steenbok and dik-dik, warthogs, zebra, and quagga. The blaubok and quagga became extinct due to European hunting, while the bontebok, blesbok, and black wildebeest almost shared their fate. The carnivore population included the range of carnivores currently found there but in greater numbers and with a distribution across the whole of the country. Black-backed jackals were common and widely distributed, while side-striped jackals were limited to the wetter, wooded areas of north-eastern and eastern South Africa. They had abundant sources of prey among the smaller antelopes, young of larger species, rodents, birds, reptiles, insects, and fruits, as well as carcasses from kills by larger carnivores. Springbok would have been a significant source of prey for jackals, particularly newborn and young animals and old, sick, or injured adults. Green, in his historical account of the Karoo, its wildlife and human settlement, refers to the massive herds of springbok that migrated across the grassland, semi-desert, and desert fringes of southern Africa before European settlement. The herds were decimated by European settlers for meat and to remove competition for grazing with sheep and cattle.[66]

The arrival of the Europeans

In 1487, the Portuguese explorer Bartolomeu Dias rounded the Cape, anchoring at Mosselbaai on the south coast. Ten years later, Vasco da Gama sailed into the Indian Ocean, opening the way for Portuguese trading expeditions. Commerce developed with East Asia, and ports were established on the Mozambican coast, trading for gold, ivory, and skins with the peoples of Mozambique, Malawi, and Zimbabwe. By the end of the 16th century, Dutch, English, and French ships had rounded the Cape to trade in Asia. By 1590, Dutch and English ships were stopping at Table Bay en route to India.[67] The Khoikhoi started trading beef and mutton with the ships in return for tobacco, copper, and iron.

In 1649, a Dutch expedition wintered in Table Bay, and the Dutch East India Company (VOC) decided to occupy the area to reprovision ships sailing to Asia. In 1652, Jan van Riebeeck arrived as commander of the expedition to establish a settlement.[68] The company bought commodities, such as spices, ivory, and animal skins, from around the world and sold them not only in Europe but also to the newly opened trading ports in Japan. The early settlers encountered Khoikhoi pastoralists in the Cape and regularly bought sheep and cattle. In his extensive accounts of life at the Cape settlement, van Riebeeck makes few references to jackals. But in December 1652, van Riebeeck reported the first case of stock predation, writing that seven or eight "wild beasts" killed a sheep. On 12 January 1653, he is more specific saying that a lion killed a sheep and was driven off with musket fire; another lion was seen near the carcass, and another lion and several jackals were sighted the same day.[69] On 27 January 1653, van Riebeeck recorded that a "wolf" had killed a sheep and taken away part of the carcass.[70] The wolf could be a brown or spotted hyena, a jackal, or even a wild dog – the identification is not clear. In July 1661, he recorded again that a "wolf" had been shot by freemen (as opposed to company employees) settlers, and a reward was paid to them.[71]

For decades after the first settlement was established, predators (lions, leopards, cheetahs, wild dogs, jackals, spotted and brown hyenas) were common in the Western Cape. They were gradually driven out as the settlement grew, and settlers expanded their livestock herds, in the process killing substantial numbers of resident ungulates for meat and hides and reducing competition for grazing.[72] They declared war on predators and did all in their power to kill or drive them from farming districts, though not as successfully with jackals as with larger predators.[73] The early depredation by predators had led the Dutch to establish bounties for killing them – van Riebeeck said that in June 1656, he set the rewards at "a lion 6; a tiger or wolf 4; and a leopard 3 reals."[74] The confusing use of tiger and wolf does not make it clear if jackals were included in the bounties.

While most settlers stayed near Cape Town under Company rule, some sought to evade its jurisdiction and went on hunting expeditions as far east as the Great Fish River, where they encountered Xhosa communities.[75] By 1707, there were over 700 company employees at the Cape, a settler community of 2,000, and a large population of servants, local pastoralists who traded with the Dutch, and slaves captured during conflicts with indigenous groups or imported from Benin and Angola. As early as 1710, the company established outposts along the Atlantic coast past Saldanha Bay, inland as far north-east as Graaf-Reinet, and in the east to the Fish River beyond Algoa Bay (Port Elizabeth).[76] The free burghers, as the white settlers were known, were harsh and racist in their treatment of indigenous people but rebellious towards the Company – starting the tradition of the trekboers, the families who moved to new land to avoid regulation (by the Dutch and then the British). They lived by stock farming, cultivation, and extensive hunting for meat, hides, ivory, rhino horn, lion skins, claws, etc. to sell. As they moved east and north, the trekboers relied on killing game for meat and hides, to preserve the livestock which were their basic resource for establishing farms and pulling wagons. From the mid-17th to the mid-19th centuries, they wiped out the huge herds of grazers in the Cape and Karoo. Large predators were killed in huge numbers, but black-backed jackals remained common in the area, adapting to the hunting/foraging opportunities offered by human settlements, their livestock, and carcasses of animals killed for food by the settlers and local indigenous communities.

European settlement, the development of pastoralism, and cultivation of grains, fruits, and vegetables to supply the Dutch trading fleets and the growing number of settlers in the Cape region became the dominant factor in substantial changes in land use and the destruction of the San and Khoikhoi ways of life in the Cape and into Namaqualand. In 1779, the explorer-soldier Robert Gordon noted that few traditional Khoikhoi kraals were to be seen in Namaqualand, with most Khoikhoi forced into the employment of the Dutch colonists or driven away.[77] European occupation and farming practices, expansion of livestock herds, and hunting for meat and hides also led to the progressive depletion (and in the cases of the quagga and the blaubok, extinction) of the huge ungulate herds and the predators that preyed on them. "The impact of settlement on the biodiversity of these regions was enormous: on wildlife (e.g. the extinction of the quagga in the late 1870s) and eventual local extinction of livestock predators, such as wild dog, lion and brown hyaena,"[78] leaving the leopard and caracal, and the adaptable jackal, as apex predators. Jackals, with their plasticity of diet, enabling them to consume a wide range of foods, survived, adapting their behaviour to counteract concerted, long-term persecution as stock-killers. It should be noted here that stock predation by jackals did not derive solely from changes in natural prey and the increase in livestock farming followed by European settlement. Jackals would have preyed on the smallstock and young animals of Khoikhoi herders for many hundreds of years before European settlement and, as Beinart argues, "must already have been attuned to the availability of domesticated species," but the balance changed with European settlement as more land was given over to livestock husbandry and natural prey, especially as large- and medium-sized ungulates, which would have provided both prey and carcasses for jackals, disappeared.[79]

During the 18th century, livestock became a mainstay of farming communities. Cattle and sheep were bought or stolen from the Khoikhoi (and later the Xhosa of the Eastern Cape) or imported, especially wool-bearing sheep to supplement the indigenous breeds which produced lower quality fleeces. For European settlers, the Khoikhoi and, in the east, the Xhosa and Nguni, "meat, milk products, fat and skins were vital in the subsistence and material culture of rural communities, black and white";[80] thus, the fierce hatred of and desire to exterminate predators like the jackal. In the 1720s, the colony had more than 250,000 sheep and fewer than 100,000 cattle, but by 1770, "there were more than a million sheep and around 250,000 cattle, and the numbers continued to grow thereafter, to over 1¾ million sheep and over 300,000 cattle by the 1820s."[81] As settlements spread northwards and eastwards from the Western Cape, Boer and some English-speaking farmers found that sheep-farming could prosper in the drier areas of the Cape, including the Karoo and neighbouring arid areas not suitable for cattle or arable farming.[82]

African beliefs, superstitions, and folktales about jackals

This section is based on historical, archival, and oral history and European travellers' sources about jackals. One interesting general point is that apart from southern African folktales and beliefs from the San, Khoikhoi, Xosa, Zulu, Sotho, and Swazi peoples, jackals are very under-represented in the depiction of animals in sub-Saharan oral histories, belief systems, and folktales. Courlander's extensive collection of beliefs and folktales has very few references to jackals.[83] Most references are almost always about the cunning of the jackal, its intelligence, and hunger for food. Across the continent,

[l]ike the fox in European folklore, the jackal is often represented in African folk tales as a trickster. Its ability to adapt to changing circumstances and its legendary stealth and cunning have inspired stories about the wily creature that dodges traps and avoids hunters year in year out. The jackal is reputed to be able to obliterate its tracks, feign death and rid itself of fleas by immersing itself in water.[84]

The San communities have a deep and long-lasting connection with the animals around them – as sources of food, as beings with which they coexist in a world where all things are imbued with spiritual values, and as sentient individuals with particular features, some of which may render them taboo.[85] Black-backed jackals are considered part of what is termed the *paaxo* category of carnivorous animals, which include dangerous animals like the lion, leopard, and hyena, even though jackals are clearly not regarded by the San as threats to life.[86] There are many stories derived from San and Khoikhoi folklore which have jackals and hyenas outwitting lions and killing them – using cunning to beat the lion, even though at times stories show both animals to be led into trouble by greed.[87] Barnard records that the jackal and the hare appear as great tricksters in the folktales of the /Xam, !King, Nharo, and many other San mythologies and Khoikhoi stories.[88] Wittenberg has pointed out,

> In Khoi orature, satire is associated with the transgressive trickster figure of the jackal; he is attractive, roguish and able to outwit the powerful, in particular predators such as lions. In /Xam [San] story telling on the other hand, the jackal does not function as a likable trickster but has negative traits such as cunning, cowardice and selfishness.[89]

Some stories represent the observations of these peoples living with jackals and others, showing that they recognised relationships between jackals and large predators like lions and leopards, whose kills the jackals would scavenge, but which would sometimes kill jackals. Bleek and Lloyd retold a San story about a jackal begging for meat from a leopard kill and annoying the leopard so much that it kills the jackal.[90] This story demonstrates the San view of the relationship between the jackal and the leopard, for whom the San have great respect, while the jackal was seen as cunning but cowardly – and that if the meat of a jackal was eaten, they would never give the heart to a child as it would render it a coward.[91] But they valued the skin and fur of the jackal that would be made into a kaross. Bleek and Lloyd noted that the San were aware of what could be called the dark side of their relationship with humans, that jackals would consume human remains left out in the bush.[92]

Many Khoisan stories were taken up by the Afrikaner writer G.R. von Wielligh in his children's stories in Afrikaans. The jackal appears prominently in the stories and continually outwits the greedy, stupid hyena. One story has the jackal cheating the hyena of a hartebeest he has killed and is taking home to his mother. Over several days, the jackal gets the hartebeest, an eland, and a small antelope by tricking the hyena and his mother.[93] Other stories involve the hyena being tricked into challenging the lion to a fight and then being persuaded to fight a human, who shoots at him and hits him with a sabre.[94] As with the story above about the hyena injuring its back legs, Wielligh recounted one in which the jackal persuades the hyena to raid a house to steal meat. In the process, a door falls on the hyena leaving it with shorter back legs.[95] One San tale involves a jackal who goes hunting with the lion and promises to give the lion's family most of the meat while only keeping a little. The lion then finds his family starving and goes to attack the jackal. The jackal tricks the lion by offering him soup made from the animal killed by the lion. The soup contains a red hot rock, which the lion swallows. The lion dies, and the jackals retains the rest of the carcass and is rid of the dangerous lion.[96] The Khoikhoi have a similar version, but according to Wittenberg, their one is more graphic and tells of the lion trying to pass out the hot stone with his faeces and dying in the process – he noted that this version was gathered by Leonhard Schultze and published in *Aus Namaland und Kalahari* and that his accounts of Khoikhoi tales are less tidied up and Europeanised than those of Bleek and Lloyd, the latter also making greater reference than Schultze to the "treachery" of jackals, seemingly transferring San images that convey this to Khoikhoi oral traditions.[97]

Wittenberg noted that "in Khoi orature, the jackal is clearly a figure of identification for Khoi story-tellers and listeners, while the lion is the symbolically defeated and discredited enemy at whose misfortune one laughs."[98] He suggests that this tale has within in it the Khoikhoi use of animals to represent human life, well-armed Dutch settlers represented by the lions, and the Khoikhoi by the clever jackal.[99] Wittenberg points out,

> The liberating comic potential of satire depends on such a referentiality in which the historic oppressor and the barely fictionalised target become merged … it draws attention to its fantastical, absurd and at times grotesque aspects, an overt form of fictionalisation that shields its speakers from retribution … the Khoi story-teller can also shield himself behind the proxy figure of the crafty jackal,[100]

while in fact promoting a culture of mocking the oppressor and praising those who resisted or undermined the Boers. One story from Khoikhoi oral history relates how a jackal sold a horse to the Boers. The "impish jackal" outwitted a greedy but dim-witted Boer who "retaliates with characteristic and unrelenting violence."[101] The jackal told the Boer he had to pay well for the horse as in three days, it "will shit money." When the Boer realised he had been tricked, he tried to kill the jackal, but the trickster jackal persuaded him it was another jackal, so the Boer rides off to shoot an entirely innocent jackal.[102]

Animals are important in the spiritual beliefs and cultural traditions of other South African communities. Vusamazulu Credo Mutwa, the Zulu historian and collector of traditional stories, in his compendious work *Indaba, My Children*,[103] recounted their spiritual beliefs and creation stories, taking in long retold stories of other language communities that were part of the Bantu migration from West and Central Africa. It included the role of animals in the spiritual beliefs and oral histories. One belief held by the Zulus, as recounted by Mutwa, is that it was the jackal – called *mpungushe* in Zulu – that stole fire from the village of the gods to warm the cave of the first man and woman on Earth.[104] The story starts with the Sun arguing with his wife, the Earth, and pretending to be ill. The Sun doesn't shine, and all the creatures on Earth are cold and short of food. The lion survives by killing weaker animals. One day the lion chases a thin jackal into a cave occupied by a cold and hungry man, called Kintu, and his wife. Kintu kills the lion and eats its meat, giving some to the jackal, which amazes him by talking. The man laments that they have no way of getting warm because the sun is not shining, and they have no fire.

The jackal says he will go to the village of the gods and get fire; in return for this, the man must let the jackal spend a night with his wife. The man intends to trick the jackal and thinks of him as "a four-footed, wretched, mangy, stinking little beast."[105] The jackal goes to his cave, where he fetches his pipe, because the jackal had discovered the secret of marijuana. After a dangerous climb up through the mountains, the "brave and cunning jackal" reached the home of the gods.[106] He met a dragon, messenger of the gods, and made friends with him by singing to him. The dragon agreed to carry the jackal across the remaining mountains to the fertile land of the gods. The guardian of the gods, a giant called Ngozi, seized the jackal, but the jackal let him smoke his marijuana pipe, and the giant fell asleep. The jackal entered a large hut and found fire sticks and a stone. He stole them and returned to Kintu and his wife. They used the fire sticks to light a great fire. But Kintu did not want the jackal to sleep with his wife, so he summoned a great bird who punished those who wronged the gods and told the bird that the jackal stole fire from the gods. The great bird killed the jackal.

In another story, Credo Mutwa tells of a witch doctor wearing a jackal skin headdress and loincloth,[107] a reference to the frequent use in southern Africa of the skin of the black-backed jackal to make headdresses or capes known as karosses. Jackals were credited in these traditions as being cunning and clever, with a man called Vumba in one story being referred to "as more clever than the fabled jackal that stole the skin of the Sun Elephant."[108] Another story said that after a battle, in which Nguni and Mambo warriors killed their enemies, the bodies of the dead were buried with their legs sticking out of the ground; "[t]hus they mark the spot, so to speak, for hyaenas and jackals to find them and drag them out," highlighting the role of jackals and other carnivores in clearing up the bodies of the dead and scavenging from battlefields.[109]

If the jackal was presented in San, Khoikhoi, Zulu, and Xhosa stories as full of cunning and tricks, the Afrikaners used the term "jackal" to mean liar.[110] There were stories from Afrikaner farmers that jackals deliberately brushed their tails along the ground to obliterate their tracks and had tricks by which they rendered themselves invisible to humans.[111] Tales of trickery, greed, and cunning concerning jackals can also be found in Xhosa, Sotho, and Swazi folklore, which have been recounted by Savory in his collection of stories – they mirror those of the San and Khoikhoi, showing jackals outwitting other animals or gullible humans to get food.[112] Courlander reported the existence of similar stories in Ethiopia, Tigre, and Somalia, all showing the jackal as cunning and not to be trusted.[113]

Notes

1 S. Monsarrat & G.I.H. Kerley (2018) Charismatic species of the past: Biases in reporting of large mammals in historical written sources, *Biological Conservation*, 68–75, p. 68.
2 Ibid., pp. 68–9.
3 Ibid., p. 69.
4 Ibid., pp. 70–2.
5 A.F. Boshoff & G.I.H. Kerley (2010) Historical mammal distribution data: How reliable are written records? *South African Journal of Science*, 106, ½, 1–8, p. 1.
6 P. Robertshaw & A. Siiriäinen (1985) Excavations in Lakes Province, Southern Sudan, *Azania: Archaeological Research in Africa*, 20, 1, 89–161, p. 89.
7 Ibid., p. 91.
8 Ibid., p. 102.
9 Ibid.
10 M. Parkyns (1853) Life in Abyssinia: Being notes collected during Three Years' Residence and Travels in that Country Volume 2, London: J. Murray, published by HardPress, Kindle Edition, 2017, loc 154.
11 Ibid., loc 1021.
12 K. Somerville (2021) *Humans and Hyenas Monster or Misunderstood*, Abingdon, Oxon: Routledge, pp. 98–9.
13 Ibid., loc 1819.
14 Ibid., loc 3612.
15 M. Parkyns (1868) *Life in Abyssinia: Being Notes Collected during Three Years' Residence and Travels in that Country 2nd edition*, London: J. Murray. Kindle Edition via Hard Press 2017, loc 743.
16 Parkyns, 1853, loc 3639.
17 *Proceedings of the Zoological Society of London* (henceforth referred to as *Proceedings*), 27 February 1868, p. 157.
18 *Proceedings*, 11 February 1869, p. 113.
19 R. Bridges (1986) The visit of Frederick Forbes to the Somali Coast in 1833, *International Journal of African Historical Studies*, 19, 4, 679–91, p. 687.
20 J. Mellis (1895) *Lion Hunting in Somaliland*, Reprinted in 1991 by St Martin's Press, New York, p. 6.
21 *Proceedings*, 15 November 1898, Mammals of Somaliland, p. 765.
22 Personal communication with Dr Amy Dickman, 4 May 2021.
23 P. Robertshaw (1990) *Early Pastoralists of South-Western Kenya*, Nairobi: British Institute in Eastern Africa, p. 17.
24 Ibid., p. 19.
25 Robertshaw, 1990, p. 19.
26 R.S. Reid (2012) *Savannas of our Birth. People, Wildlife and Change in East Africa*, Berkeley: University of California Press, p. 105.
27 K. Somerville (2019) *Ivory. Power and Poaching in Africa*, Updated edition, London: Hurst and Company, pp. 29–35 and 60–79.
28 *Proceedings*, 3 March 1885, p. 220.
29 R. Burton (1860) *The Lake Regions of Central Africa: A Picture of Exploration Volume 1*, New York: Harper Brothers (Reprinted by Hard Press 2017) Kindle Edition, loc 3412.
30 J. Carruthers & N. Nattrass (2018) History of predator-stock conflict in South Africa, in G.I.H. Kerley, S.L. Wilson & D. Balfour (eds) *Livestock Predation and its Management in South Africa: A Scientific Assessment*, Port Elizabeth: Centre for African Conservation Ecology, Nelson Mandela University, 30–52, p. 33.
31 Ehret, 2016, 165.
32 Ibid., pp. 170–2.
33 Carruthers & Nattrass, 2018, p. 33.

34 J. Ki-Zerbo & D. Tamsir Niane (1997) *Africa from the Twelfth to the Sixteenth Century. General History of Africa IV,* London: James Currey/UNESCO, p. 209.
35 D. Lopes & F. Pigafetta (translated by Maragerite Hutchinson) (1881) *A Report of the Kingdom of Congo: And of the Surrounding Countries, 1591,* London: John Murray (reprinted by Leopold Classic Library), p. 52.
36 Courlander, 1996, p. 304.
37 M. Van Der Ryst (2004) Rocks of potency: Engravings and cupules from the Dovedale Ward, Southern Tuli Block, Botswana [with Comment], *South African Archaeological Bulletin* , 59, 179, pp. 1–11, pp. 1–2.
38 Personal communication from San trackers working for photographic safari operators in the central Kalahari.
39 A.B. Smith (1995) Excavations at Geduld and the appearance of early domestic stock in Namibia, *South African Archaeological Bulletin,* 50, 161, 3–20, p. 3.
40 Ibid., pp. 3–4.
41 L. Phillipson (1976) Survey of the Stone Age archaeology of the Upper Zambezi Valley: II. Excavations at Kandanda, *AZANIA: Journal of the British Institute in Eastern Africa,* 11, 1, 49–81, p. 51.
42 R. Vigne, "The Hard Road to Colonization: The Topnaar (Aonin) of Namibia, 1670-1878." *Journal of Colonialism and Colonial History,* 1, 2, 2000. MUSE, doi:10.1353/cch.2000.0007. No page numbers.
43 Sydow, 1973, p. 73.
44 Ibid., p. 74.
45 G. Owen-Smith (2010) *An Arod Eden. A Personal Account of Conservation in the Kaokoveld,* Jeppestown, SA: Jonathan Ball, p. 77.
46 C. John Andersson (1857) *Lake Ngami: Or Explorations and Discoveries during Four Years' Wanderings in the Wilds of South Western Africa.* London: Hurst and Blackett, reproduced as Kindle Edition, Hard Press 2017, loc 309.
47 Andersson, 1957, loc 1716.
48 Ibid., loc 4962.
49 Ibid., loc 5102–9.
50 Ibid., loc 3863.
51 F. Galton (1889) *Francis. Francis Galton's Narrative of His Exploration of Namibia in 1851,* Swakopmund: Keith Irwin. Kindle Edition, loc 4639.
52 Ibid., loc 5877.
53 L. Thompson (2001) *A History of South Africa,* New Haven, CT: Yale University Press, p. 6.
54 Ibid., p. 12.
55 A. Jerardino (2010) Large shell middens in Lamberts Bay, South Africa: A case of hunter gatherer resource intensification, *Journal of Archaeological Science,* 37, 2291–302, p. 2291.
56 Ibid., p. 2300.
57 J. Orton (2009) Rescue excavations at Diaz Street Midden, Saldanha Bay, South Africa, *Azania: Archaeological Research in Africa,* 44, 1, 107–20, pp. 116–7.
58 M.D.W. Jeffreys (1962) Bushman Art, *South African Archaeological Bulletin,* 68, p. 241.
59 P. Jolly (1986) A first generation descendant of the Transkei San, *South African Archaeological Bulletin,* 1986, 41, 6–9, p. 8.
60 R.G. Klein & K. Cruz-Uribe (1989) faunal evidence for prehistoric Herder-Forager activities at Kasteelberg, Western Cape Province, South Africa, *South African Archaeological Bulletin,* 44, 150, pp. 82–97, p. 95.
61 D. Balfour & G. Kerley (2018) Introduction – The need for, and value of, a scientific assessment of Livestock predation in South Africa, in G.I.H. Kerley, S.L. Wilson & D. Balfour (eds) *Livestock Predation and Its Management in South Africa: A Scientific Assessment,* Port Elizabeth: Centre for African Conservation Ecology, Nelson Mandela University, 15–29, p. 15.

62 Ki-Zerbo & Niane, p. 209.
63 Iliffe, 2007, pp. 100–01.
64 Hrbek, 1992, p. 318.
65 A. le Roux & S. Badenhorst (2016) Iron Age fauna from Sibudu Cave in KwaZulu-Natal, South Africa, *Azania: Archaeological Research in Africa*, 51(3), 307–26, pp. 311 and 317.
66 L.G. Green (1955) *Karoo*, Cape Town: Howard Timmins, pp. 24, 35–36.
67 Elphick & Malherbe, 1979, loc 652.
68 C.R. Boxer (1965) *The Dutch Seaborne Empire 1600-1800*, London: Hutchinson, pp. 22–7.
69 J. van Riebeeck (1897) *Precis of the Archives of the Cape of Good Hope, December 1651-December 1653, Riebeeck's Journal, Volume I*, Cape Town: W.A. Richards and Sons (Reprinted Lightning Source UK, no date), pp. 55 and 59.
70 Ibid., p. 64.
71 J. van Riebeeck (1897) *Precis of the Archives of the Cape of Good Hope, December 1651-December 1653, Riebeeck's Journal, Volume III*, Cape Town: W.A. Richards and Sons (Reprinted Lightning Source UK, no date), p. 282.
72 Beinart, 2003, p. 196.
73 Somerville, 2021, pp. 104–5.
74 J. van Riebeeck (1897) *Precis of the Archives of the Cape of Good Hope, December 1651-December 1653, Riebeeck's Journal, Volume II*, Cape Town: W.A. Richards and Sons (Reprinted by SN Books World, India, no date), p. 23.
75 N. Mostert (1992) *Frontiers. The Epic of South Africa's Creation and the Tragedy of the Xhosa People*, London: Jonathan Cape, p. 153.
76 Thompson, 2001, p. 35.
77 Walker et al., 2018, p. 163.
78 Ibid.
79 W. Beinart (2003) *The Rise of Conservation in South Africa. Settlers, Livestock, and the Environment 1770-1950*, Oxford: Oxford University Press, p. 203.
80 Beinart, 2003, p. 7.
81 R. Ross (1979) The cape of good hope and the world economy, 1652–1835, in R. Elphick & H. Giliomee (eds) *The Shaping of South African Society 1652-1840*. Cape Town: Wesleyan University Press, Kindle edition, 6215–7174 loc 6369–85.
82 J. Carruthers & N. Nattrass (2018) History of predator-stock conflict in South Africa, in G.I.H. Kerley, S.L. Wilson & D. Balfour (eds) *Livestock Predation and Its Management in South Africa: A Scientific Assessment*, Port Elizabeth: Centre for African Conservation Ecology, Nelson Mandela University, 30–52, p. 36.
83 H. Courlander (1996) *A Treasury of African Folklore*, New York: Marlowe and Co.
84 B. Grahl (2016) Why is the Jackal called a trickster? Gondwana Collection, Namibia, https://www.gondwana-collection.com/blog/jackal/#:~:text=Like%20the%20 fox%20in%20European,hunters%20year%20in%20year%20out, accessed 3 March 2021.
85 K. Sugawara (2001) Cognitive space concerning habitual thought and practice toward animals among the Central San (|Gui and ||Gana): deictic/indirect cognition and prospective/retrospective intention, *African Study Monographs*, 21, 61–98, p. 61.
86 Ibid., pp. 65–6.
87 W.H.I. Bleek & L.C. Lloyd, *Bushman Folklore*, Kindle edition, loc 349; see also, Folklore and Symbolism http://www.heartofsnow.net/hyena/hyenalore.php, accessed 24 February 2020, and http://www.sacred-texts.com/afr/sbf/sbf16.htm, accessed 24 February 2020.
88 A. Barnard (1992) *Hunters and Herders of Southern Africa. A Comparative Ethnography of the Khoisan peoples*, Cambridge: Cambridge University Press, p. 258.
89 H. Wittenberg (2014a) The Boer and the Jackal: Satire and resistance in Khoi orature, *Multilingual Margins*, 1, 1, 75–89, p. 76.

90 Bleek & Lloyd, loc 2311.
91 Ibid.
92 Ibid., loc 2874.
93 G.R. von Wielligh (2011) *Animal Tales 1*, Pretoria: Protea Book House, pp. 55–8.
94 Ibid., p. 77.
95 G.R. von Wielligh (2011) *Animal Tales 3*, Pretoria: Protea Book House, pp. 38–9.
96 South Africa: Jackal Fools Lion Again, Mythology and Folklore UN-Textbook: South Africa: Jackal Fools Lion Again (mythfolklore.blogspot.com) accessed 3 March 2021.
97 Wittenberg, 2014a, p. 79.
98 Ibid.
99 Cited by Wittenberg, 2014a, pp. 79–80.
100 Wittenberg, 2014b, p. 606.
101 Ibid.
102 Ibid., pp. 85–6.
103 V. Credo Mutwa (1972) *Indaba, My Children*, New York: Viking Press.
104 V. Credo Mutwa (1996) *Zulu Shaman Dreams, Prophecies and Mysteries*, Rochester, VT: Destiny Books, p. 78.
105 Ibid., p. 85.
106 Ibid., p. 86.
107 Credo Mutwa, 1972, p. 232.
108 Credo Mutwa, 1972, p. 235.
109 Ibid., p. 396.
110 F. Brownlee (1985) Dove and jackal – Collected from the Xhosa, in S. Gray (ed) *The Penguin Book of South African Stories*, Harmondsworth: Penguin, p. 14.
111 Beinart, 2003, p. 206.
112 P. Savory (2014) *The Best of African Folklore*, Cape Town: Struik/Penguin Random House, Kindle Edition.
113 Courlander, 1996, pp. 561 and 573–4.

5

THE JACKALS OF EURASIA

The areas suitable for the first development by people of cultivation and pastoralism were the fertile and well-watered areas of West Asia near the Tigris and Euphrates Rivers, from where farming spread west to the Mediterranean and Europe and east into Asia. These areas supported a diversity of wildlife, with the likelihood that the golden jackal was abundant and widely distributed, from the Balkans, Greece, eastern Europe, and Turkey across the Levant, Palestine, the Arabian peninsula, Central and West Asia, and into South and Southeast Asia.[1] Jackals were present in the Balkans, Greece, Anatolia, and the Caucasus during the Neolithic, overlapping in their range there and in Arabia, Iraq, Iran, Afghanistan, Pakistan, India, and some other parts of Asia with grey wolves; though, as will be examined, there are questions about whether the presence of grey wolves limited or even excluded jackal populations.

The International Union for the Conservation of Nature (IUCN) in 2013 estimated the global golden jackal population at over 130,000, with 54,350–55,271 or about 43% of that total occurring in Europe.[2] A later IUCN estimate projection in 2016 put the number in Europe at around 70,000,[3] but a 2018 estimate suggested the number might have exceeded 117,000,[4] which seems rather high and remains to be verified.

Europe

Fossil evidence suggests the golden jackal has been "absent from most of Europe since the Pleistocene, with a range restricted to parts of the Adriatic, Mediterranean and Black Sea coasts until the nineteenth century."[5] Sommer and Benecke's study of canids in Europe during the late Pleistocene and early Holocene recorded that only four golden jackal fossils had been found in Europe in that period and dated to the Neolithic (7,000–3,200BCE) and in areas within the

DOI: 10.4324/9781003199793-6

current range of European golden jackals (*Canis aureus mormoreoticus*).[6] This indicated that jackals lived along the coasts of Croatia and Greece around 7,000–6,500BP.[7] The earliest jackal fossils in the Caucasus dated to the Bronze Age (3,300–1,200BCE).[8] The European range of the jackal was limited in the BCE and the first two millennia of the CE, only expanding in the late 1900s CE. Today, jackals are expanding west, north, and east from their Balkan and southeast European ranges; this expansion is gaining increasing media and public attention.[9]

It is not clear when golden jackals dispersed into Europe from the east, but it has been suggested that they appeared in the Holocene having migrated from the Caucasus "via a land bridge between the Balkans and Anatolia during the Würm" – the Würm being the last glacial period, ending at the start of the Holocene.[10] They reached as far as Croatia, and the range stayed stable in the Balkans and south-eastern Europe, not extending further west or north.[11] But data is scarce.

Spassov and Acosta-Pankov examined and rejected the possibility that jackals could have arrived in Europe at the end of the Pleistocene via the northern Black Sea coast or across the Bosporus. They believed migration across the Bosporus was possible, as it did not split apart until 9,300–7,600BP, but that there was a complete absence of fossil evidence for a Pleistocene dispersal, and it must have been in the Holocene.[12] Hatlauf et al. also noted the sparse data on jackal origins in Europe but concluded that "the first verified golden jackal records date from around the year 1384 in south-eastern Europe, near Sofia (Bulgaria)."[13] Spassov and Acosta-Pankov also said the first record of jackals in Europe dated to the Middle Ages, and the "earliest reliable historical data are from the end of the 14 century (the vicinity of Sofia), from Turkish chronicles, during the siege of the town," and other accounts suggest they were in southern Ukraine along the Black Sea from the 16th to the 18th centuries.[14] Records of presence in the Pannonian/Carpathian region centred on Romania date from the 19th century.[15] In the 1800s, there are references to their presence in Hungary, where they are referred to as "reed wolf" because of their liking for wetland areas, and it is possible that individuals may have dispersed from Hungary to Austria.[16] Jackals are recorded as having a

> preference for agricultural areas and wetlands with adequate cover in lower elevations. Intensively cultivated areas without cover are not suitable, although human activity often increases food availability. While subpopulations comprise fewer than 1,000 adults, the species is common and numerous where food and cover are abundant.[17]

From the 19th century through until around the 1930s, jackals were mainly distributed in the Balkans, with stable local populations in Northern Thrace (Bulgaria), Eastern Thrace (Greece), Western Thrace (European Turkey), Dalmatia, and the Peloponnese, with expansions to the west/northwest during favourable periods, on the Bulgarian Black Sea coast, west and north Bulgaria, and possibly northern Serbia.[18] The Greek population was mainly found between Marmara and the Aegean, and in Chalkidiki. Along the Adriatic, there were small populations from Albania north through Montenegro into Croatia. In Romania, the jackal was recorded as a periodic visitor but, by 1929, was said to be present in Wallachia and the Dobrudja district.[19] There was a progressive expansion of the European jackal range in the late 19th and early 20th centuries, "in what is arguably the most dramatic recent expansion among the native predators on the continent,"[20] as the Balkan and south-east European populations merged with jackals migrating from the Caucasus. Since 1945, the jackal population has been spreading across western, Central, and northern Europe.[21]

The European golden jackal population declined substantially in the first half of the 20th century, as a result of sport hunting and persecution as suspected stock predators. Lapini wrote that at the time of the jackal decline, wolves, which dominated jackals, were nearly exterminated or severely reduced in numbers over much of their Balkan, Central, and northern European range.[22] Wolves had retreated to more mountainous areas with thick snow cover, and jackals then had the opportunity to increase in lowland areas. And a change began to take place in the 1950s–60s, perhaps related to these factors. In the 1960s, the jackals received official protection in Bulgaria, enabling the first big expansion of European jackals. At the same time, the reduction in the number of wolves in the Balkans may have, given the possibility that a large wolf presence deters jackals, enabled a substantial increase in Croatia's jackal population, with the first wave of expansion in the 1950s, when vagrants also reached north-west Slovenia.[23] Further range and population expansion occurred in the 1980s and 2000s. The largest populations in the Balkans, Central, and south-east Europe are now found in Bulgaria (72%), Hungary (13%), Serbia (9%), Romania (4%), and Greece (2–3%). Jackals are resident in most of Bulgaria, with the highest densities in the south-east, north-east, and north-central.[24] Reproductive populations are established in "Central, East, and North Europe with vagrant animals or small groups found in parts of Western Europe".[25]

The precise reasons for expansion are still being studied. As Krofel wrote:

> Factors affecting the jackal expansion across Europe are not completely understood and several hypotheses have been proposed. Recent studies have suggested that jackal expansion was triggered by intensive persecution and resulting decline of the European apex predator, the gray wolf ... the historic changes in the distribution and abundance of the two canid species ... seem to support the mesopredator release hypothesis. Recent data suggest that also nowadays top-down suppression may be weakened where wolves are intensively persecuted by humans or occur at reduced densities. Additional factors that have been suggested to influence jackal expansion in Europe include abundant and easily accessible anthropogenic food resources, changes in land use and wildlife management, and climate change.[26]

In Central Europe, following the initial dispersals from Bulgaria, the jackal colonised Hungary (where it had disappeared completely by the 1940s), growing from a few vagrant animals in the 1970s to a viable breeding population from 1991 to 1992 and now over 7,200 in southern Hungary alone according to Deinet et al.[27] In Romania, jackals are believed to have dispersed from Bulgaria around 1929, while the long-standing but small Greek population survived in scattered pockets, with numbers decreasing in many areas, though some increase probably on the island of Samos.[28] Giannatos, in his 2004 action plan for jackals, put the decline down to loss of habitat due to changes in land use, which resulted mainly in reduced cover and possibly a reduced food base, too. He noted that "jackals seem to do well in moderately modified agrosystems with non-invasive human activities. Barriers for jackal expansion and population recovery seem to be the mountains with extensive high forests or unbroken scrub, heavy snowy winters, with a long period of thick snow cover, and irregular food supply, large intensively cultivated areas without cover, urbanisation, and established wolf populations."[29] In addition, the Greek Ministry of Agriculture carried out poisoning campaigns against carnivores up to 1981. "According to the Ministry's statistics, on the average 1000 jackals, 740 wolves and 50,000–74,000 foxes were killed annually in Greece, by hunters and organised carnivore poisoning campaigns. The jackal was considered vermin up to 1990 and until 1981 a bounty was paid for each animal killed."[30] The result of anthropogenic effects, wolf presence, and culling is that the Greek population is in fragments across areas of the country with suitable habitat but is nowhere thriving, except on Samos. They became extinct on Corfu in the 1990s, having been very common 30 years before.[31] In 2000, there were estimated to be 1,000 adult jackals and 153 sub-adults in Greece, which is worrying given the killing of nearly 400 jackals a year in Greece by hunters and a high road-death toll.[32]

The European range has moved gradually north and west, with individuals or groups reaching Czech Republic, Slovakia, Ukraine, Poland, Latvia, Lithuania Estonia, Finland, and Norway.[33] In the west, jackals have dispersed into Switzerland, Austria, and Germany and have now been sighted in France, Lichtenstein, Belgium, the Netherlands, and Denmark. In Austria, it was confirmed in 2007 that jackals were breeding there; this has also been recently reported from Italy.[34] Hatlauf et al. noted in 2021 that golden jackals had been reported in 33 European countries, in 11 of which they were recorded for the first time during the past 10 years – Switzerland (recorded on camera traps in 2011and 2012), Estonia and Latvia (from about 2013), Poland (2015), Denmark (2015), Lithuania (2015), the Netherlands (2016), France (2017), Liechtenstein (2018), Finland (2019), and Norway (2021).[35] A lone jackal was photographed in Norway, north of the Arctic Circle in March 2021, and the Norwegian Environmental Protection Agency in Lakselv said, "It took some time to confirm that it was a golden jackal. Now we are also trying to get DNA from it."[36] The golden jackal is currently the most successful carnivore species of the continent with a continuously expanding population and distributional area.[37] The expansion of the range is both "significant and fast" in terms of the conservation status, population, and distribution.[38]

Central Europe and the Balkans

Bulgaria has the largest jackal population in Europe, and jackals expanded into the Balkans and Central Europe from there, colonising parts of Europe which had not previously had jackal populations either in the recent or in the distant past. Range expansion in Bulgaria probably started in the early 1970s, following the institution of some legal protection in 1962, increases in populations of small game species, and a decline in wolf numbers.[39] The increase was despite a major cull which killed between 5,999 and 10,790 jackals from 1976 to 1981.[40] Giannatos believes the jackal population experienced a 33-fold increase between the early 1960s and the mid-1980s, as land-use changes provided habitat and food sources increased, including animal carcasses from state livestock and game farms.[41] Estimates in 2010 put the population at 36,000.[42] By 2012, the range had increased with high population densities in south-eastern Bulgaria from the Kamchia River basin to the Turkish border to the south, from the Black Sea to Stara Zagora in the west, in north-eastern Bulgaria, and the central Danubian Plain of northern Bulgaria.[43] The thinly wooded areas of the Eastern Rhodope Mountains, which experience low winter snow cover, have a substantial population of jackals.

Amid the expansion of numbers and range, there was a high level of sport hunting and culling – in 1983, 5,538 killed; 1999, 7,422; and 2010, a massive 26,570 jackals.[44] The continued population growth demonstrates the high reproductive rate despite human persecution. A study in the Western Upper Thracian Valley of Bulgaria showed that despite accusations of livestock and excessive roe deer predation by jackals, small rodents were the main source of food at 42.9% of intake, followed by brown hare (20.1%) and fruits or other plant food (19.7%). Markov and Lanszki reported no livestock or large ungulate remains in jackal scats they examined.[45] In central Bulgaria, Raichev et al. found that in winter there was a far greater dependence on domestic animals in some areas, either predated or carcasses of animals already dead (30.2%), but wild ungulates were the main source of food (47.9%),[46] with rodents presumably in short supply in winter. The evidence of either predation or scavenging from carcasses of domestic animals is the source of human–jackal conflict in Bulgaria.[47] Stoyanov found that in general, the jackal diet consisted, in terms of the proportion of scats containing the following remains, mainly of mammals (87.5%, of which 41.7% was rodents and the rest wild ungulates or domestic animals, either predated or scavenged), birds (36.1%), vegetable matter (30.6%, mainly fruits, seeds, acorns, and crops), fish (12.5%), insects (9.7%) reptiles (1.4%), and crustations (2.8%).[48]

In Romania, jackals were recorded regularly in the 19th century. Tóth, in his survey of published sources on jackals in Romania, reported they were numerous in Bihor County, on the western border with Hungary.[49] Reports of hunting of jackals indicated their continued presence into the 20th century, though with no clear evidence of numbers or exact range.[50] A sighting of what the observer called "small reddish reed wolves" was made near Dumitra, Bistriţa-Năsăud

County, in 1942, and in the same year, a male jackal was shot by Baron Ferenc Daniel at Etéd, Harghita County.[51] Thirty years later, the number and range of jackals in Romania had expanded with signs of them found at the Humor Monastery's hunting district; two jackals were hunted near Voronet between 1971 and 1975, and they were recorded as present in Buzau County, near Bucharest.[52] In Romania, the jackal has been hunted continuously as a killer of livestock but still increased in numbers and expanded their Romanian range from this period to the present, though without reliable estimates of numbers. The population size can only be guessed at using official hunting figures – in 2008, 1,061 jackals were killed; by 2012–13, this had risen to 2,502.[53] Papp et al. estimated in 2013 that despite the high hunting claims, the Romanian population was about 6,400 spread across 28 counties.[54] In a 2018 report, the same authors said that the population might have more than doubled to 14,273, with jackals present in 39 out of 41 counties.[55]

The expansion of numbers could result from a number of factors – poor animal waste disposal by farmers and food businesses, elimination of wolves in many areas, and an increase in protected areas where hunting was prohibited.[56] Many livestock or poultry farmers accuse jackals killing livestock; this has particularly been the case around the from Danube Delta Biosphere Reserve. Hunters believe jackals are reducing the numbers of game species.[57] Marinov et al. sent out questionnaires to over 150 farmers in the Delta and recorded that 88% of those questioned reported damage to livestock by jackals; 390 attacks were on goats and 355 on sheep. But 44% reported that some of the attacks attributed to the jackals could have been by domestic/feral dogs or foxes. Most farmers believed that "the jackal is a voracious predator that decimated the livestock and should be controlled."[58]

Up to the last half of the 20th century, Hungary was considered to be on the northern edge of the jackal range; now it sits between the large population of Bulgaria and the dispersals of the past 40–50 years into Central, northern, and western Europe.[59] Hungarian records of jackals date back to the early 1800s, some of them in areas bordering Croatia in the south-west and Ukraine in the north-east. Tóth noted, "Until the 1920s, the jackal was observed in Hungary only along major rivers and between the Danube and Tisza, and records are almost missing in the 1920–1945 period, whereas in the last 50 years surveyed, the number of observations increased"; in 2007, the population was put at 1,500.[60] Lapini put it at 2,763 in 2008.[61] Jackals are included in the Hungarian Game Management Database and are culled by gamekeepers on hunting land, by livestock farmers and sports hunters.

The study of jackal and red fox diet in Hungary conducted by Lanszki et al. indicated that from 814 scats examined, the major jackal prey were small mammals like voles and mice, which made up 70–90% of consumed biomass, with consumption of weasels, foxes, or other small carnivores rare and brown hare present in a surprisingly small amount (0–10% presence in the scats). Wild ungulate remains were present in 0–43% of scats, with young wild boar were found in scats in early spring, when boar give birth. Domestic animal remains (mainly from cattle and pig carcasses) were present in 0–29% of scats.[62] Lanszki et al.'s studies of jackal diet have not shown that they consume game animals to any great extent and do not, as some hunters contend, pose a threat to game hunting.[63]

In the early 20th century, the jackal population across the Balkans had "declined dramatically, which led to several local extinctions and its distribution became highly fragmented,"[64] but much of the Balkans has been recolonised. Szabo et al. reported in 2009 that jackals had been observed again in Macedonia and Albania, and there was a stable population on the Dalmatian coast of Croatia.[65] Populations are also well established in Serbia, Bosnia-Herzegovina, Kosovo, Montenegro, North Macedonia, and Slovenia. Šálek et al. carried out density studies of Balkan jackal populations and found in 2014 that the greatest density was in lowlands around the Danube in Serbia, with an average density of 1.1 groups/10km^2.[66] They found that across the Balkans, Bulgaria, and Romania, jackals increased in localities with shrubs to provide cover and mixed land use but decreased where there was extensive arable land. Districts with "woody shrubs and herbaceous vegetation … are structurally highly diverse and provide … and serve for protection against human hunting pressure" while offering diverse opportunities for foraging for berries and other fruits and hunting small mammals.[67] Hunting of jackals occurs across the Balkans. Banea researched hunting records and found that 623 were killed in Croatia in 2012 and 1403 in Serbia in 2008 – in addition to 3,267 in Hungary (2015), 5,371 in Romania (2016), and 20,179 in Bulgaria (2015).[68]

The dispersal of jackals in the Balkan and Black Sea regions in the last 70 years took them into Slovenia, from where they were to move into Italy, Austria, and then further north and west. The first reported presence in Slovenia was in 1952.[69] Passov and Acosta-Pankov said that genetic analysis had shown that the Slovenian population had originated in Bulgaria and had moved north and west through Romania and Serbia – from Slovenia some had dispersed into Italy, along with jackals from Dalmatia.[70] Potočnik has examined the expansion of jackals into Slovenia and noted that there has been an increase in the jackal presence and that, unusually, it took place at the same time as the wolf population was increasing, somewhat in conflict with the generally accepted idea that the presence of wolves may deter jackal colonisation.[71]

Italy, Austria, and into Central and northern Europe

In Italy, jackals have been present since 1985 in the Pozzuolo del Friuli area, near the border with Slovenia, occurring as vagrant individuals, originating from Slovenia and the Istrian peninsula.[72] By 2010, they had expanded into the Julian Prealps, and a single jackal was spotted near Venice. Verified sightings confirmed they had been seen in the Bolzano-Bozen Province and South Tyrol. It was thought that breeding pairs had established themselves in some of these areas.[73] Fannin wrote in 2018 that jackals had been present in south-eastern Friuli

Venezia Giulia since 1996 and were blamed by farmers for attacks on sheep.[74] Examination of carcasses and use of camera traps showed that jackals had preyed on sick animals and lambs, killing them with bites to the throat.[75] Two jackals were found poisoned in the region, which was believed by Pesaro et al. to have several breeding packs.[76] By late 2021, jackals had spread from their initial locations in the north-east into central Italy, being recorded on a camera trap near Florence, and they are said to be moving further south.[77] The *Rewuilding Appenines* web page on Facebook reported on 4 February 2022 that jackals had moved into the Central Apennines, and one had been photographed in Lazio.[78]

In Austria, they are reported to have been "sporadically present" from the late 1980s, having come from Hungary, though some may have dispersed via Slovenia.[79] The first validated specimen in Austria was a dead one found in Styria in 1987, with vagrants or individuals shot or found dead in the provinces of Carinthia, Styria, Burgenland, and Upper and Lower Austria in the following few years.[80] Sightings and jackals killed on the roads continued to be reported into the 2000s, with 17 confirmed and 3 unconfirmed records by 2007. In that year, the first confirmed breeding of jackals in Austria occurred by the Hungarian border near Lake Neusiedl.[81] Jackals have also spread to Switzerland where they were rare or completely absent in most areas until very recently. They have been sighted in the cantons of Waadt, Bern, and Freiburg.[82] In the Czech Republic, jackals are present, but numbers and distribution aren't clear. The first confirmed sighting was in 2006, when a dead jackal was found in southern Moravia near the Austrian and Slovak borders.[83] In Slovakia, the jackal was added to the national mammal checklist in 1947, but the first recent confirmation of presence was in 1989, based on a jackal killed by a hunter in the south-east. Between 1995 and 2012, another five jackals were killed by hunters, confirming continued presence.[84] Even though numbers are believed to be low, there is an official jackal hunting season which runs from 1 August to the end of February.[85] Hunting statistics show that numbers are increasing, as the annual tally of animals killed has gone from 3 in 2009 to 125 on average between 2012 and 2015.[86]

In Ukraine, jackals were recorded from the end of the 1990s and have become established since, notably in the Tuzlivski Lymany National Nature Park on the Black Sea, near the border with Romania and Moldova.[87] There are five jackal groups in the park, which have been breeding for a number of years, leading to a population of about 150. A population of 70, possibly part of the Tuzlivski population or animals dispersing from there, has been reported in the Odessa region to the north-east.[88] There have been reports of vagrants dispersing into northern Ukraine. Jackals are hunted in the country, and between 1998 and 2018, 318 jackals were killed by hunters in southern Ukraine.[89] The jackal there is thought to have originated in Romania but with some animals moving into Ukraine from the Krasnodar, Don, and Donetsk regions of southern Russia.[90] Jackals migrating from southern Russia are found in small numbers in the formerly Ukrainian region of the Crimea. From Ukraine, jackals have dispersed into Belarus and Poland[91] and from there into the Baltic states and then on to

Finland and Norway. Jackals were first reported in Belarus in 2011, but there is a lack of data on their numbers or range there.[92]

The Baltics and Finland

The first evidence of jackal presence and reproduction in Boreal/Baltic region was recorded in Estonia in February 2013, when a young female was shot in Western Estonia. The species was identified by physical characteristics and DNA analysis. Another jackal was shot in Estonia in the same year. Since then, jackals appear to have expanded their range, with confirmed presence of territorial groups in Salevere and Läänemaa (Western Estonia).[93] In 2016–17, jackals were distributed along almost all the western coast of Estonia, with at least ten reproductive pairs.[94] When first recorded, jackals were regarded as alien/vagrant species in all Baltic states. They are now treated as a naturally occurring, if new, species and included on the lists of game species in all three Baltic countries. In Estonia, the number hunted in 2016 and 2017 was 32 and 26, respectively.[95] In Latvia, jackal reproduction was confirmed in 2013 and 2014 in inland habitats in the south, but no more evidence of reproductions was recorded afterwards. In Lithuania, there have been two confirmed reports of jackal presence without any known reproduction. In February 2015, a jackal was confirmed have been shot in Lithuania, though no exact location was published.[96] In December 2013, a female jackal was recorded as having been shot by a hunter near Jelgava close to Lielupe River in central Latvia.[97] The first confirmed sighting of a jackal in Finland occurred in 2019. It was presumed to have dispersed naturally from the Baltic states via Russia. The sighting set off a fierce debate over whether the jackal should be considered an invasive alien species, albeit one that had migrated naturally and not been introduced by humans. The *Helsinki Times* reported a passionate debate concerning the future of the jackal:

> "It should be killed on sight," stated Mikko Kärnä (Centre), a member of the Finnish Parliament's Agriculture and Forestry Committee. "The committee notes in its statement that the species should be regarded as an invasive alien species, meaning that it is not protected ... The Finnish Association for Nature Conservation (SLL) has contrastively demanded that the golden jackal not be treated as an invasive alien species. "The golden jackal is an alien species that has spread naturally, and thereby it is protected under nature conservation laws. Killing the jackal would constitute a nature conservation offence..." reminded Tapani Veistola, a special advisor at SLL. Jenni Pitko (Greens) pointed out that species ... are moving north as a consequence of climate change ... The Ministry of Agriculture and Forestry has similarly argued that the fact that the specimen entered the country independently makes it a non-invasive species. "The golden jackal sighted in Rautavaara is a young specimen that, according to experts, left its pack to look for its own new territory .... Killing it is prohibited," it stressed in a press release.[98]

The jackal sighting in Norway, north of the Arctic Circle, came two years after the Finnish one.

Why have jackals started expanding northwards into the Baltic region and westward into Germany, France, Belgium, and the Netherlands? Ranc has suggested that there are a number of environmental reasons – he put snow cover at the most important, with 37.2% in terms of the strength of influence, with wolf presence at 20.8% – the probability of jackal presence or colonisation was highest in areas with the shortest period of snow cover and which had a mosaic of forest and agricultural land that provided cover and foraging opportunities.[99] The probability of jackal presence was increased where there wasn't a permanent or substantial wolf presence.

The sparse, though recently increasing, wolf population and declining snow cover may explain why jackals have migrated westward into France, where they were recorded in 2017 and have been seen periodically since.[100] Camera trap footage showed them in the Chablais region, near the border with Switzerland.[101] In Germany, a male jackal was first reported around the Brandenburg region in 1996, but it was shot in 1998.[102] But the Wild Dog Foundation reported on its Facebook site in 2021 that four jackals had been seen – separately not as a group – in Baden-Württemberg, and genetic tests on scats found there had revealed that there was a family group of a mated pair and two pups.[103] Jackals have also been caught on camera traps in the Black Forest near Freiburg. At first the Black Forest jackals were only caught singly on cameras, but then a pair was seen together, and it is believed that there is a mated pair. Scat samples confirm that there is a male and a female in the forest, with the likelihood that they could breed.[104]

Dispersing, probably vagrant rather than settled, jackals have been caught on camera and video in the Netherlands – though it is not entirely clear if it is one animal photographed repeatedly or there are several. Dutch conservation scientists have been quoted as saying there is good habitat for jackals in the Netherlands, and there is an established wolf population there, with possibly 10–20 wolves in residence including several breeding pairs,[105] but that this is not sufficient to act as a major deterrent to jackals if they are dispersing into the country.[106] Wennink and Lelieveld estimated that there was significant potential for jackal colonisation, though they would avoid heavily built-up areas with large roads and areas where wolves were present. They concluded that the core areas suitable for jackals

> consisted of high quality areas of at least 6 km^2 consecutive area without any roads ... [and] lower quality area ... of at least 12 km^2 of consecutive area. The results showed that core areas can support more than 100 family groups, while the highly suitable areas can support an additional 50 family groups. The remaining suitable areas can support up to around 1200 more family groups. In total the Netherlands could support around 1450 golden jackal family groups. Reestablishment of wolves could decrease these numbers to around 800 golden jackal family groups.[107]

This seems highly optimistic, but it demonstrates the belief that jackals will continue to expand in numbers and range in western Europe, something demonstrated by sightings in Denmark in 2015 and 2017 – in 2015, a dead jackal was found by a roadside having been killed by a car.[108]

Problems of expansion in Europe – economic and legal

In their original range in the Balkans and south-eastern Europe, and now in their expanded range, farmers and hunters are concerned that the increase in jackal numbers and their expanding range will increase livestock predation and killing of deer, hare, rabbits, and other game species that are hunted, providing both income from hunting tourism and food in rural areas. In Romania, for example, hunters kill jackals and want their numbers reduced because of a perceived, but not yet proven, reduction in numbers of game animals.[109] In many of the countries in which they are now present in substantial numbers, forms of population control – whether through quotas for recreational hunting or targets for culling – are used to limit numbers. Jackals are often also killed during the culling of foxes using traps, poison, or by shooting.[110] Deinet reported that in countries such as Slovenia and Hungary, hunters have a very negative attitude towards jackals, blaming them for declines in roe deer, and these countries and several others are already or have announced the intention to include jackals in hunting regulations to allow legal shooting to limit numbers and the effects on the game industries.[111] Bulgaria, with the largest jackal population in Europe, already has regulations in place to allow recreational hunting and culling of jackals.

Many countries in Central and northern Europe do not include jackals in their national lists of resident species, and the legal position regarding population management is not clear.[112] Reports of livestock predation and the damaging of the game industry, combined with the recent arrival of jackals in many countries, have led to a confused situation within the European Union (EU) region, for example, with no clarity about which conservation/environmental regulations apply. Hatlauf et al. outlined the questions that still need answers – questions such as:

> Is the culling of golden jackals possible within the applicable legislation, and what are the most efficient methods? Should the golden jackal be listed as a game species, a protected species, or as neither of those? Should it be hunted year-round or only during a specific hunting season? Will there be damage to livestock? How can potential damage be proven? Should there be damage mitigation and/or compensation for losses in animal husbandry? How can livestock be protected from golden jackals? Is golden jackal monitoring necessary? Does a clear legal framework, combined with damage mitigation and compensation schemes, contribute to a better human–predator coexistence?[113]

There is no clear framework across the EU, but it is more defined in some countries. In Austria, the jackal is a protected species but can be hunted as game species in the province of Salzburg, but in Germany and the Czech Republic, it is not included among game species and has no protection under law. In Slovakia, it can be hunted between 1 August and 1 February each year.[114] In Italy, despite fears of farmers about stock predation, the golden jackal is a specially protected species and has been since at least 1992. In Bosnia, Bulgaria, Croatia, Hungary, Kosovo, Latvia, Lithuania, Montenegro, Romania, Slovenia, and Ukraine, they can be hunted under existing game laws.[115]

Russia, the Caucasus, Turkey, and Central Asia

This section will start with a brief review of the status of jackals within the territory of the former Soviet Union, as set out in huge detail by Heptner and Naumov.[116]

Jackals were found in southern Russia north of the Caucasus around Novorossiysk and Krasnodar, east to the Soviet Central Asian republics, and south throughout the Caucasus. In the latter region, the range of the jackal is in the plains and foothills. In the western half of this part of Russia, just north of the Caucasus, jackals inhabit the entire Black Sea coast from Novorossiysk to Batumi but are rarely found above 500m. In the 19th century, jackals were also found along the eastern shore of the Azov Sea, along the lower Don, the Volga Delta, and the Don steppes but disappeared from there mainly due to increasing human populations and changes in agricultural methods, especially during and after the collectivisation of agriculture.[117]

In the Caucasus, jackals were found in the plains, river valleys, and foothills in regions that are part of Russia, Georgia, Armenia, and Azerbaijan – in the last years of the Soviet Union, jackals were said to be abundant in Georgia and southern Azerbaijan but less so in Armenia. Their main food items were small rodents, hares, pheasants, partridges, ducks, and other waterfowl, but they also hunted or scavenged the carcasses of lambs, sheep, goats, and the young of cattle and domesticated buffalo. The drying and shrinking of the Caspian Sea and the loss of large wetland areas with reed beds led to a shrinking of the jackal range by the 1960s.[118] Their high population and wide range in the Caucasus prior to World War II (WWII) is demonstrated by the annual harvesting of jackal pelts, which averaged 17,000 a year across the Caucasian Soviet Republics in the 1930s – about 9,400 from Azerbaijan, more than 2,300 from Georgia, and the rest from Caucasian regions of southern Russia and from Armenia.[119] As well as being hunted for their skins, jackals were "considered the most harmful destroyers of game" and also destructive of introduced species such as nutria and muskrat, which were trapped for their fur.[120] Jackals are still very numerous in independent Georgia, where they overlap to a surprising degree with the wolf population. There is some separation of main populations, and while both species are found in eastern and western Georgia, the jackal is more common in lowlands, particularly in the western Georgia/Colchis lowlands, while the wolf is particularly common in the mountains and throughout eastern Georgia.[121]

Shakarashvili et al. believe jackals have been present there since the Early Bronze Age (from 3,500BCE).[122]

Jackals are believed to have been present in Turkey at least as far back as the Bronze Age, but there are surprisingly few references the author has been able to trace in the opening 15 centuries of the Common Era. In the middle of the 17th century, a description of a victory parade through Istanbul celebrating the conquests of Murad IV (Sultan from 1623–40CE) was recorded as including 10 lions, 5 leopards, 12 tigers, and a number of jackals, wolves, and hyenas walking tamely alongside soldiers and imperial officials. In 1877, a survey of the mammals of Asia Minor (Turkey) presented to the Zoological Society of London (ZSL) by Charles Danford and Edward Alston noted that the jackal (*Canis aureus*) was common across Turkey and particularly abundant in southern regions.[123] Another survey for the ZSL noted their abundance in Turkey, particularly in coastal regions and around Adana in south-east Turkey.[124] No estimates are given of numbers and no detail, which is also surprising, of the threat they might have posed to livestock or hunting of them to remove this threat or for sport. A study published in 2016 said the jackal was found across Turkey, with the exception of a small area along the Black Sea coast. In several areas it coexisted with both wolves and striped hyenas.[125] It was particularly found in areas below 800m in height, in river valleys, and along coastal plains. Ambar and Bilgin suggested that while being widely distributed, they had been exterminated by poisoning in some livestock-raising districts along the Mediterranean coast.[126] One unusual source of food for the jackals living on the eastern Mediterranean coast at Akyatan is the green turtle. A study showed that 75% of turtle nests there failed, usually after they were raided by jackals and foxes. Jackals were observed going from nest to nest along the beach.[127]

The most extensive details of golden jackals in Central Asia were published by Heptner and Naumov in their study of the mammals of the Soviet Union in 1967. They recorded that jackals were found widely distributed across Kazakhstan, Kyrgyz Republic, Tajikistan, Turkmenistan, and Uzbekistan. In Turkmenistan, they were found in river valleys and the foothills of mountains but not higher in mountainous areas but could tolerate arid conditions and were present in small numbers on the edge of Central Asian deserts.[128] The same distribution was found in the other republics – in Tajikistan's Pyandzh Valley, 133–61 jackals were present in one district with a density of about 0.2 jackals per hectare.[129] During the Soviet period, jackals were trapped or shot for their pelts, and on average, Uzbekistan produced 6,200 a year – indicating that there must have been a very large population to sustain that annual offtake.[130] In Kazakhstan, the offtake was much lower and declined over time, falling from 250 skins annually in the late 1920s to 104 just after WWII.[131]

Palestine, the Levant, and the Arabian peninsula

A source, though a rather anecdotal and far from unimpeachable one, for the presence and human–jackal interactions in Palestine and the Levant during the BCE period is the Bible.[132] References to jackals are numerous and far from positive. Curiously, some translations refer to jackals, while others refer in the same verses to dragons. Thus, in Psalms 44:19, the verse reads, "Though you have crushed us in the haunt of jackals, and covered us with the shadow of death," but in some it says haunt of dragons. Jackals are used to signify the desert and desolation or to refer to jackals as consumers of human corpses (as in Psalms 63:10). Interestingly, there are no references to jackals in the Koran.[133]

Something from the region that showed a long-standing human–predator conflict and the trapping or hunting of jackals, wolves, and other predators was the discovery by archaeologists of ancient stone carnivore traps in the Negev and Judaean Deserts of Israel and neighbouring regions stretching down to Yemen, where they are called wolf traps but are clearly used to trap any large predator, including golden jackals.[134] Avner et al. discovered a number of these traps in the 'Ein Gedi Oasis in the Judaean Desert and the 'Uvda Valley in the southern Negev Desert. They were built of unworked stones and were baited with meat. The trap was operated with a rope and trap door, which would close on the animal when it took the meat."[135] The exact age of the traps is unclear, but artefacts from 4,500 to 3,500BCE and Roman lamp fragments (c30BCE–70AD) have been found in the traps.[136] There is also evidence of newer traps having been built by Bedouin pastoralists in the last few centuries. Conflict between pastoralists and predators like jackals, wolves, and leopards has continued across the centuries. Records of jackals in this region are few and far between, and there is little to report until we jump forward to the 19th century. In 1866, the Rev. H.B. Tristram reported to the ZSL on the mammals of Palestine, noting that while wolves were hated by herders, shepherds in particular, because of predation on livestock, the far more numerous jackals were far less destructive, even though the jackal "swarms in incredible numbers in every part of the country."[137]

There has been far more contemporary study of the region's jackals, particularly in Israel. Borkowski et al. noted that with the extermination of lions during the 19th century and then by the 20th-century bears, cheetahs, and leopards (apart from a very small number that may still survive in the Judean Hills and other remote areas), jackals, striped hyenas, and wolves became the apex predators in Israel, Palestine, and surrounding areas of the Levant and Jordan.[138] The grey wolf numbers have declined dramatically as a result of fragmentation and shrinking of habitat, but jackals have survived better and in places thrive on the waste produced by Israel's livestock and poultry sector, in competition with hyenas. Despite a poisoning campaign against jackals and other carnivores in 1963–64, to deal with the threat of rabies, jackals have maintained a stable population and in some areas expanded their numbers, because of their adaptability to the presence of humans and plasticity in diet, with a high density around human settlements. Density can be as high as 2.5 per km^2, and in some areas around the Israeli-occupied Golan Heights, the illegal dumping of turkey, chicken, and cattle carcasses by farmers has enabled numbers to reach 11 per km^2 – there was also predation by jackals of calves reported there.[139] Most attacks on calves, 75.5%, took place within two days of birth.[140]

A study of jackals in Park Britannia in central Israel indicated that they had a wide diet, but scats showed the following occurrence in the food intake – ungulates (39.4% frequency of occurrence), 80% of which were domestic animals, assumed to be mostly consumed as carrion; fruits (31.3%); birds (30%); small mammals (23.5%); and invertebrates (21.2%), while garbage was found in only 9.1% of the scats.[141] No definitive figure has been given for local or national jackal populations, but the size can be guessed at by the fact that around Park Britannia, 50–65 jackals were killed on the roads annually in 2002–05.[142] In many areas of high jackal density, they forage at sites where carcasses and other human waste are dumped, and this enables them to live in stable groups, larger than those where food is less plentiful. Such clumped resources mean that groups of 10–20 jackals come together and have only small areas that they defend, tolerating other jackals at feeding sites.[143]

In neighbouring Lebanon, the Syrian sub-species of the golden jackal (*Canis aureus syriacus*) is found particularly in low-lying coastal areas, and its range stretches into Syria and Palestine. It coexists in some areas with the Persian and Indian wolf (*Canis lupus pallipes*).[144] Like their Israeli neighbours, the jackals of Lebanon live in close proximity to human settlements, whose waste is an important source of food. To the south, in Jordan, the jackal has declined over the last 120 years, mainly due to habitat loss.[145] Persecution by humans, related to suspected livestock predation and superstitious dislike of the animal, has also been blamed for declines in numbers.[146] Superstition and the belief that jackals and hyenas are grave robbers have led to persecution on the Arabian peninsula.[147] In the waterless desert regions of Arabia, jackals are absent, but they are to be found in areas with water and livestock, such as Al Jubah and Wadi as Sirhan where there are farms and water is available throughout the year.[148] Numbers are unlikely to be high, and one must be aware that desert conditions, widespread persecution, or recreational hunting of wildlife has rendered the fauna of Saudi Arabia one of the poorest in the world.[149]

Iraq and Iran

Both the Syrian and Persian sub-species of golden jackal are found in Iraq.[150] Not surprisingly, decades of conflict have rendered detailed surveys or study of particular populations impossible. The jackal has been there for millennia, and in the early 20th century, British naturalists noted its presence in date palm plantations along the banks of the Shatt al-Arab River in southern Iraq and along the Tigris and Euphrates Rivers.[151] Cheesman reported that in 1917 jackals were raiding fields used to grow vegetables for the British troops garrisoned in Iraq, destroying crops of melons and vegetables. As a result, the troops were encouraged to shoot them – 60 being shot in a couple of weeks.[152]

Iran has a long-standing population of jackals. There is a lack of historical records of their range, numbers, and interactions with people, but one study noted 245 referenced records of jackals in Iran between 1865 and 2017.[153] The Persian jackal (*Canis aureus aureus*) is present, but not always in large numbers, across Iran and has long coexisted with humans and with other carnivores, including wolves, but also come into conflict with them. Álvares's study in the Kerman region of south-east Iran found that people interviewed in 12 rural villages considered jackals to be responsible for damage to crops and the killing of poultry and livestock. Although they used dogs to guard livestock and fences to try to keep out predators, they still said they lost animals to jackals and resorted to poisoning, shooting, and trapping.[154] The study concluded that despite conflict with humans and persecution by them, jackals in Iran are widespread and live close to humans, because of the feeding opportunities they provide.[155] The jackal coexists, sometimes with conflict, with leopards, the rare Asiatic cheetah, wolves, striped hyenas, and bears.[156] Using anthropogenic food sources, including predation on livestock, jackals survived on agricultural land but also in mountainous areas, with major jackal populations in the arid regions of the Alborz, Zagros, and Kerman mountain ranges.[157] They avoided large urban areas and or areas with large wolf or other large carnivore populations.[158]

Mohammadi et al. reported conflict between wolves and jackals in western Iran's Hamadan Province.[159] There, habitat transformation and agricultural land uses have significantly reduced wild prey available for jackals and wolves. This has pushed them to rely increasingly on food that can be found around human habitations, particularly poorly disposed of livestock, poultry, and other animal carcasses.[160] This brings wolves and jackals into conflict with each other over shrinking food sources, with wolves under particular pressure through shrinking habitat and growing conflict with people over livestock predation.[161] Direct conflict was discovered with a wolf killing a jackal at Nashar village, within an area where wildlife was protected. The remains of a recently killed jackal were found near a poultry farm, around which people dumped chicken carcasses and other animal waste, on which both wolves and jackals have been observed feeding.[162] The jackal had been clearly killed by a wolf but not consumed, suggesting carnivore guild conflict rather than predation for food.

South Asia and Southeast Asia – from Pakistan and India to Indochina

The range of the jackal sub-species in South Asia spans Pakistan, India, Bhutan, Nepal, Bangladesh, Sri Lanka, and into Myanmar, Thailand, Laos, and probably also Cambodia and Vietnam. This includes a diversity of ecosystems of the types found in the South Asian, Malayan, and Indochinese regions.[163] Jackals inhabit forests, deserts, mangroves, and swamp areas but not dense rainforests and high mountains.[164]

Charting the journey across time of Indian, Sri Lankan, and Indochinese sub-species of the golden jackal is not easy. Sources are few and far between for the early history of the region, those which mention wildlife even more scarce. Eminent historians of India have noted the particular problems of reconstructing its history. Majumder wrote in his 11-volume history of the Indian people, "prior to the thirteenth century A.D., we possess no historical text of any kind, much less a detailed narrative as we possess in the case of Greece, Rome or China."[165] Sharma supports this view, noting that in comparison with the wealth of sources on the Roman Empire, the evidence of Indian societal and economic development is weak with an unhelpful dearth of written sources to support archaeological evidence.[166] The result, as Keay said, is that historians have to reconstruct the past from "reluctant materials," involving piecing together an account from "random inscriptions, titbits of oral tradition, literary compositions and religious texts" to add to what can be ascertained from archaeological research.[167] In trying to give a picture of the effects of human economic and societal development on wildlife and the environment, and specifically jackals, I have concentrated on the rise and fall of major civilisations and their probable – I write probable because of the paucity of documentary or archaeological evidence to go on – effects on jackals and, in the later chapters, honey badgers.

Research into early Indian history suggests that the move from early humans' hunter-gatherer modes of subsistence took place initially in areas suitable for the development of agriculture-based societies. In very arid areas, such as the Thar Desert (bordering north-western India and Pakistan), the dense humid forests of the north-east, the Eastern and Western Ghats of central-southern India, and other mountainous or thickly forested regions, hunter-gatherer modes survived for millennia.[168] Shifting slash-and-burn-type agriculture developed first and settled agriculture from 10,000 to 4,000 years ago, with settlement and population expansion taking off around 4,000BP.[169] The earliest evidence from the region of domestication of animals comes from Mehrgarh in Baluchistan, Pakistan, in 7,000BCE (with the growing of wheat, barley, pulses, and the keeping of sheep, goats, and cattle),[170] though some put it around 8,000 years ago.[171] The development of cultivation and livestock husbandry did not mean the end of hunting. Jukar et al. believe that humans in forested areas hunted small forest-dwelling prey. But in more open grassland areas and with improvements in weapons, by the early Holocene, evidence from butchery sites suggests a growing preference for the abundant wild bovids and cervids.[172]

Estimates of the start of rice cultivation vary between 7,000BP[173] and 3,000BP.[174] Farmers and pastoralists spread across India over several thousand years as livestock husbandry, based particularly on zebu cattle, and the cultivation of rice, gram, and beans enabled populations to expand, and this involved extensive clearing of forests.[175] The clearing of land and need to protect livestock will have led to attempts to exclude wildlife from cultivated fields and deter or kill predators. But well into the Common Era, wildlife remained diverse, widely distributed, with wild prey (particularly small mammals, birds, reptiles, and insects) and other resources providing food for jackals. As with the jackals and golden wolves in Africa, Europe, and West Asia, the jackal sub-species found across South Asia and Indochina appear to have adapted to life around human settlements, feeding opportunistically on waste produced by humans and on the carcasses of livestock that died from disease or other natural causes. It is notable that the ideal jackal habitats, in areas with open woodland, tall grassland, scrub, and hills with valleys that provide den sites, were often in areas where pastoralism and cultivation developed.

There were, dating back many thousands of years to the pre-Vedic period of Indian history prior to 5,000BCE, many spiritual connections between people and their environment and the wildlife it contained. Spiritual connections and the beliefs derived from them arose through the greater embedding of people within the natural environment and the way that communities fed themselves and interacted with wildlife. Spiritual and religious beliefs are one way of charting the relationship of people in India with wildlife over time. Rao wrote that in the Vedic period of Indian history and culture (2,000BCE–600BCE), references to wildlife in the Upanishads and other historical or religious works divide animals into types, with jackals, one presumes honey badgers and other carnivores, in the first grouping of quadrupeds and birds that hunt other animals.[176] As Singh et al. have identified, in this period "communities consider themselves connected with their biophysical environment in a web of spiritual relationship. These rural communities consider specific plants, animals, or even rivers and mountains as their ancestors and protect them."[177] The authors pointed out, "One of the important traditions of nature reverence is to conserve those patches of forest that have been dedicated to a god or goddess or ancestral spirits as 'sacred groves.'"[178] The conservation of areas of forest, riverine habitat, and mountains perceived by communities to be sacred or areas containing shrines or temples, by no means limited to India, can perform an important environmental role in preventing deforestation, protecting water sources, protecting wildlife species, and maintaining harvestable natural resources for local communities.[179] India is estimated to still have between 100,000 and 150,000 sacred groves and other natural sites,[180] although some have estimated the number of protected, community sacred sites lower at 4,215, covering an area of 39,063 hectares distributed across India.[181] Whatever the precise definition and number, as Gokhale et al. contend, "The 'sacred groves' are in fact the 'reserve forests' of the local tribes/communities who maintain/conserve these patches of woodlands in a religious manner. They act as natural gene pool reservations and serve as an example of habitat preservation through community participation."[182] They can serve as refuges for persecuted wildlife species.

The expansion of the human population, growth of crop cultivation, and pastoralism and clearing of forests meant humans would have both coexisted with and come into conflict with the diversity of predators indigenous to South Asia, including not only jackals but also tigers, leopards, cheetahs, wolves, wild dogs (dholes), striped hyenas, and bears. The rise of the Indus Valley civilisations in India around 3,300–1,300BCE, equivalent to the Bronze Age cultures of Egypt and Mesopotamia, quickened the pace of population growth and sophistication of food production. More advanced agricultural and societal forms spread with the peoples of this culture and reached Gujarat from the north.[183] This was true of many of the cultures and dynasties that developed in northern and western India, such as the Harappan (c2,600–1,500BCE), Yadavas, the Mauryan, the Kshatrap dynasty, the Gupta Empire, and then the Pratiharas (which succeeded one another and ruled parts of the region from the second millennium BCE through to the 9th century CE).[184] The presence of jackals and other lesser predators is not as well recorded, unlike lions or tigers in regional histories of the time or in artistic or religious images. But carvings from Bharhut in Madhya Pradesh from the 2nd century BCE depict jackals as beasts of ill omen and show them in graveyards and on corpse-strewn battlefields[185] – a not inaccurate portrayal but one that would have helped instil prejudices against them as scavengers, grave robbers, and consumers of human remains.

Indian history from these early dynastic periods is hard to piece together and relies heavily on tracing people and events, not to mention animal life at the time, from sections of literary and religious works.[186] Pigott records that evidence from the Harappa culture from the mid-third millennium BCE onwards, in the form of domestic and wild animal remains found at Harappan sites, indicates the presence of jackals and wolves close to large human settlements, suggesting a very fast adaptation by them to the opportunities presented by human activity and the establishment of towns. The faunal remains included barasingha, sambur and hog deer, small mammals, reptiles, and birds, as well as domesticated oxen, sheep, goats, and even horses.[187] Harappan art shows wild mammals, including tigers, rhinoceros, and water buffalo, which are also likely to have been common in many areas of Harappa. Wild fauna would have provided prey and carcasses (of larger animals beyond the power of jackals to overcome) and the remains of domestic stock after slaughter or which died from disease or age. But the evidence suggests little other than the presence of jackals near villages and towns.

From around 1,500–600BCE, Aryan communities from the north dominated much of the region once occupied by the Harappans, with the influx from the north lasting from around 1,500–1,200BCE. These communities gave rise to the Vedic culture, the foundation of modern Hinduism, and came to be dominant in Punjab and parts of the Gangetic Plain.[188] The religious, quasi-historical works that emerged from this culture – the *Brahmanas, Mahabharata,* and the *Ramayana* – comprise both a literary heritage and a means of trying to piece together an otherwise largely unrecorded period of history,[189] working in spiritual beliefs and concepts of the natural world, with Hindu deities having incarnations or avatars in the form of various animals from fish to elephants.[190] The goddess Durga, represented as not only a protective mother but also the goddess of war, strength, and protection,[191] was created by the gods Brahma, Vishnu, and Shiva to kill the buffalo demon Mahishasura. Wilkins notes that in some of the accounts of Durga, a jackal is depicted as her representative and as protecting Krishna, the eighth incarnation of Vishnu, from the king Kansa, who wanted to kill him.[192]

Other early accounts of animals in Indian oral history and cultural systems come from the *Pañĉatantra* (Sanskrit: पञ्चतन्त्र, also called the *Five Treatises*), a collection of interrelated animal fables in Sanskrit, dating from about 200–300CE, based on older oral traditions.[193] Jackals occur regularly in a number of fables and tales in the *Panćatantra*. The names of the two jackals who are referred to most often, Karataka and Damanaka, are described by the translator Chandra Rajan in his foreword as "Wary and Wily."[194] This sort of representation in oral history and literature establishes in Indian culture the reputation of the jackal for its cunning. In one story, "The Trial of Dimnah," the jackal is tried for the crime of treachery at the bidding of "the Queen Mother, the lion's mother, and a leopard minister is the chief witness in the court. Dimnah, the villain is tried, sentenced and executed."[195] Treachery, double-dealing, cunning, and meanness are qualities ascribed to jackals in many stories and fables included in the collection.[196] In book 1, a story depicts the jackal as an "ambitious, callously opportunistic politician out for what he can get," and in stories it tells how the jackal espouses the view that in using cunning and treachery, the end justifies the means.[197] In the stories involving the jackals and Tawny, the lion king, the jackals are depicted as devious servants trying to trick or make a fool of their lord.[198]

Another of the tales, which suggests cunning, trickery, but eventual death because of the failure of deception, is of the blue or indigo jackal. The tale relates how a jackal called Fierce Howl, who lived in a cave on the outskirts of a big town, had to enter the town searching for food, where he was set upon by feral dogs, which bit and snapped at him. To save himself, Fierce Howl

> rushed into the house of an artisan, where he tumbled into a huge vat of indigo dye. The dogs ran away in the direction they had come from. The poor jackal whose allotted span of life had not yet run its full course, clambered out of the vat with much difficulty and ran back to the forest on the outskirts of the city, where a crowd of animals of various kinds, who were roaming around in the vicinity, took one look at him dyed a brilliant indigo and fled, their eyes widening and quivering in terror; and they ran crying out, "Ayo, ayo … what weirdly-coloured animal is this that has come into our midst?"[199]

Seeing the animals fleeing, Fierce Yowl the jackal understood that they were afraid of him, and he shouted to them,

> Hey there, hey, you wild creatures; why are you all fleeing in such terror? The sovereign of the gods, noticing that the wild creatures of this forest have no sovereign of their own, anointed me, Fierce Yowl, to rule over you as your lord. Come back, come and live happily in the safety of the cage of my paws, strong as thunderbolts.[200]

This worked, and all the wild animals – lions, tigers, leopards, monkeys, gazelles, hares, jackals, and the rest – came forward, bowed, and paid homage to Fierce Yowl, who made the lion his chief minister, the tiger chamberlain; the leopard he put in charge of the royal betel nuts made the elephant the royal doorkeeper and the monkey, bearer of the royal umbrella. But he banished all the other jackals. The indigo jackal then ruled, and the lion hunted his food for him. But one day he heard jackals howling, and "his eyes filled with tears of joy; and he too began howling in a shrill, high-pitched tone. The lion and other members of the court hearing it gasped in surprise and exclaimed, "Good Heavens! This is but a jackal … Listen; we have been taken for a ride by this jackal; let's kill the scurvy fellow." Fierce Yowl heard this and tried to escape but was caught by a tiger and torn to bits."[201] Throughout the reciting of all of the tales, the two main jackal characters, Wary and Wily, are sowing dissension and trying to find ways of getting more food by persuading the lion to kill other animals.[202] One verse from the collection sums up the view of jackals as clever but utterly self-serving:

> The shrewd person who accomplishes both the ruin of another and his own good, keeps his counsel, gives no hint of his aims, as Sly, the jackal did, deep in the woods.[203]

The jackal's cunning and its supposed willingness to help its friends by using its wits are demonstrated in a more positive tale, of the Crab and the Crane, in which the jackal tells to its friend, the crow, who is threatened by a cobra. The jackal tells the crow to steal some jewellery from a palace. When people chase the crow back to its nest, they encounter the cobra and kill it, thereby saving the crow.[204] A more positive image that shows not only intelligence but also loyalty to a friend, one might surmise a result of people seeing the presence, without conflict, of jackals and crows at carcasses. But this better image of a jackal as having some loyalty to a friend is in contrast the majority of the stories in the *Pañćatantra*, which depict the jackal as entirely self-serving, greedy for meat, and ready to betray anyone for their own advantage – but they are depicted as clever, with a certain courage and resourcefulness, unlike Kipling's much later depiction of the jackal Tabaqui in *The Jungle Book*, as a low, snivelling, treacherous wretch – "Tabaqui, the Dish-licker—and the wolves of India despise Tabaqui because of the truth he runs about making mischief, and telling recollections, and consuming rags and pieces of leather-based-based from the village rubbish-heaps."[205]

Donald in his interesting account of the Indian jackal in the *Bombay Journal of the Natural History Society* rather derides the popular stereotype of the jackal that Kipling seized on. He says that the image of the jackal as "dirty, skulking scavengers" is well known to most people in India, who see it as the harijan of the animal kingdom ("harijan" being a disparaging word for the caste of poor Indians called Dalits but known as Untouchables in pejorative parlance).[206] Donald comes to their defence and talks of the four tame jackals which he had at various times, which accompanied him and his dogs on walks in the hills. He says the jackals became very close to the dogs and would defend them if threatened by other dogs, even attacking much larger dogs, and also showed themselves to be more intelligent and better problem solvers than domestic dogs, when it came to stealing food from the table.[207]

One of the most famous folktales involving a jackal, from southern India, is that of the tiger, the Brahmin, and the jackal. This tale represents the cunning of the jackal, the naivety of the Brahmin, and the stupidity of the tiger.[208] It was collected and published in 1868 by Mary Frere, a British woman living in India, but is older than that. The basic story is that a hungry tiger is stuck in a cage. The tiger pleads with a Brahmin to let him out to get a drink of water and some food. Eventually, despite the obvious danger, the Brahmin releases the tiger and becomes his prisoner and likely meal. But the Brahmin persuades the tiger to seek six opinions on whether he should be eaten. A tree, a bullock, a camel, an eagle, and a crocodile all say the tiger should eat him. The sixth, a jackal, feigns stupidity and asks the tiger to show what had happened to lead to the Brahmin releasing the tiger and then being threatened with becoming the tiger's meal. The jackal tricks the tiger into getting back into the cage and so saves the Brahmin. The jackal's parting words to the tiger are, "Oh, you wicked and ungrateful Tiger! … Proceed on your journey, friend Brahman. Your road lies that way, and mine this." Another depiction of the cunning of the jackal, but this is a more unusual one, in that it shows the jackal as an instrument of justice rather than self-interest.

Returning to the historical detail of the spread and turnover of successive empires and princely states, the incoming pastoral communities spread south and south-east and across the plains along the Ganges, pushing out hunter-gatherer peoples and expanding the reach of their spiritual beliefs and deities. The first half of the final millennium BCE was one of expansion to the east and south into Gujarat and south-central India, with kingdoms being created and, by 600–500BCE, a new wave of urbanisation developing.[209] There was the incursion into and occupation of parts of north-western India and Punjab by Persian armies in the mid-6th century BCE, bringing it into the Achaemenid Empire of Persia. In north-western India and the Indus Valley, Achaemenid rule was ended by Alexander the Great's conquest of Persia in 331–330BCE and his progress through what is now Afghanistan into Pakistan and then India, the latter around 326BCE, when he reached the Indus Valley.[210] The Greek occupation was short-lived as Alexander's troops mutinied and refused to push further eastward when they came up against the large army of the Nanda kingdom of eastern India.[211]

The Nanda kingdom, centred on Bihar, was overthrown by the founder of the Mauryan dynasty, Chandragupta Maurya, whose new state allied itself with Magadha rulers to the east but then superseded them as the most powerful rulers in India. The succession of Mauryan rulers (323–185BCE) created the largest empire in Indian history, stretching from the Himalayas in the north down towards what is now Kerala and from Gujarat in the west to Assam in the east. The Mauryans were avid hunters and established hunting grounds across their lands. In the Saurashtra region, they established hunting grounds in and around the Gir Forest in the Kathiawar peninsula, which today survive in the form of the Gir Forest Reserve, the last refuge of the Asiatic lion and an important protected area for other predators. The forest areas of the peninsula were well populated with a diversity of not only antelopes and wild boars, but also jackals, leopards, and hyenas. The Mauryan rulers hunted for sport and also allowed farmers and hunting communities to kill wildlife that invaded farmland or killed stock.[212]

After the decline of the Mauryan Empire, other kingdoms appeared over several centuries, though no single, dominant empire. The history of this period is not well documented, and Keay wrote that "India's history plummets again to a murky obscurity," until the rise of the Gupta rulers in 320CE.[213] The Guptas (4th–6th century CE) established another huge empire. A series of regional kingdoms coexisted with the Guptas – the Vakata in the Deccan, the Kamarupa in Assam, and the Pallava in southern India. In the 8–10th centuries, the Pratiharas enjoyed ascendancy in northern India in the 8th–10th centuries CE. The Muslim Delhi Sultanate was a regionally powerful state that evolved before Mughal dominance in the 16th century.[214] Its rule was ended by the arrival from Central Asia of the armies of Babur, founder of the Mughal Empire.

The Mughal Empire was established in 1526 in the plain of the Ganges River by a Muslim chieftain called Babur, who had been driven from his homeland near Samarkand by powerful Uzbek invaders. He led his army and other followers south from Central Asia and took control over parts of Afghanistan, modern-day Pakistan, and India, laying the foundations for an empire that eventually reached as far east as Assam and western Bangladesh and as far south in India as the Deccan Plateau.[215] The empire was to last, though in its final years in a much weakened state, until the 1857 Indian rebellion against British occupation, but was at its most powerful between 1526 and 1720.[216] Mughal rule saw an expansion of agriculture, the taxing of agricultural produce and other economic activities by the Mughals, and the reduction in wildlife habitat and so the ranges of wildlife species, especially large predators and herbivores that were steadily excluded from arable land, much of which was created by clearing forests.[217]

Rao notes in his study of the history of Indian knowledge of fauna that under the Mughals, memoirs written by nobles, officials, and learned men indicate an keen interest in the natural world and the animals species with which they coexisted, chronicling the presence of jackals, hyenas, wolves, wild dogs, lions, tigers, cheetahs, lynx, and caracals.[218] The agricultural and other land-use changes under the Mughals, and rulers in South Asia outside the Mughal orbit, were occurring just as British penetration and control increased. The British East India Company, at first in competition with Dutch, Portuguese, and French, was at the forefront of European exploitation and colonisation. In 1634, Company traders gained access to trading stations in Bengal, through an agreement with the Mughals.[219] Trade was the focus of the Company's activities, but its interests turned gradually to control of territory and hegemony over Indian rulers. This occurred amid competition with France, Portugal, and Holland for control of commerce with Asia and as the Mughal Empire declined in power, leading to its eventual disintegration into competing states with their own locally powerful rulers.[220] The power of the Company increased following British victories during the Seven Years' War, during which the declining Mughal Empire, allied to the French, was defeated by the British and their allies. Mughal power and that of the Maratha state, another powerful polity, was brought to an end by a series of wars won by the Company. Thereafter, the British, through the Company, became the overlords of the Mughals, laying the foundation for British colonial rule until Indian independence in 1947.[221]

There is little recorded in the first 1,800 years of the CE period of the numbers and ranges of jackals in India. In the late 19th century, it was thought that the limit of the eastern range of the jackal in Asia was the Bay of Bengal, and, according to Lydekker in his *A Geographical History of Mammals*, "the jackal does not range east of Burma."[222] But R.C. Wroughton, in his survey of mammals of India, Burma, and Sri Lanka in 1914, wrote that the jackal was widely distributed and "fairly plentiful" in Burma and was well known around Mandalay, Mt. Popa, and Mingum as well as up onto the Shan Plateau, where the local people had once hunted them.[223]

In the 18th and early 19th centuries, north and central India had a diverse population of wildlife, including large, medium, and small ungulates that supported substantial numbers of tigers, lions (their range was steadily shrinking in the face of human expansion and, particularly, hunting by Indian rulers and by the ever-growing number of British soldiers, merchants, and administrators), leopards, and the jackals and hyenas that benefitted by scavenging from their kills. But much of the wild prey and the carnivores was exterminated or had disappeared from many areas by the end of the 19th century, "strong human and wild animal interface, human–wildlife conflict" being the major causes.[224] After the 1857 rebellion against British rule, self-servingly described as the Indian Mutiny by the British, the British government took over the government of India from the East Indian Company and forestry acts followed in 1865 and 1878, with forestry officers appointed to regulate logging and ensure the long-term survival of forest, more out of economic self-interest than interest in environmental conservation.[225] While the growth in the human population and agricultural expansion, along with sport hunting, forced lions, tigers, wolves, and dholes from substantial parts of their range, jackals with their ability to survive in close proximity to humans and to scavenge waste from human settlements, from the carcasses of livestock that died from disease, age, or road collisions, as well as scavenging carcasses of dead wildlife, enabled them to survive in areas that would not support larger carnivores. Their ability to be almost invisible at times helped them avoid persecution that was meted out to more visible predators and scavengers.

Hunting was a major cause of wildlife loss in British-controlled India. While the colonial authorities restricted the relatively small-scale hunting by remaining hunter-gatherer communities, they positively encouraged hunting by British officials and soldiers, and their devolution of local power to the remaining Indian rulers allowed them to control hunting in their areas and to continue with extensive sports hunting by the rulers and elites.[226] There is little recorded in the late 18th and then 19th centuries by British colonial officials, hunters, and early conservationists about the presence, ranges, and numbers of jackals in British-ruled South Asia. Major Sykes exhibited skins of a variety of Indian mammals to a meeting of the ZSL on 12 July 1831. Sykes said that golden jackals, called kholah in the Mahratta region of India (south of the Narmada River and north of Karnataka), were common, and the males had a strong "almost unbearable" scent.[227] In his compendious survey of the mammals of India, Burma, and Ceylon, published over several years in the *Bombay Journal of the Natural History Society*, Wroughton said that his contributors had found jackals across the majority of India, and he particularly noted Kumaon, Bengal, Cutch, the Central provinces, Kathiawar, Sikkim, and Rajputana, as well as being well distributed in both Burma and Ceylon.[228] Wroughton believed that the Indian, South Indian, and Ceylonese (Sri Lankan) jackals were three different species rather than, as is clear now, sub-species of *Canis aureus*,[229] with just one Indian sub-species and one Sri Lankan. Blanford noted that the jackal was common throughout India and even extended as high as 4,000ft in the foothills of the Himalayas.[230] Col A.E. Ward, in a survey of the wildlife of Kashmir, wrote that they were often found above 5,000ft in Kashmir and even as high as 7,500ft. He noted that they were very destructive of wild and domestic sheep.[231] Their reputation for predating livestock meant that in some areas they were vermin, with a price on their heads. The Maharajah of Patiala in Punjab paid a bounty of four annas for the production of proof of the killing of a jackal.[232] In the 1950s, following independence, the jackal (along with the leopard, snow leopard, and wild cat) was registered as vermin, and people were encouraged to kill them.[233]

Daniel Johnson wrote about hunting in India in 1882 and recounted a story about a Rajah in Bahar (possibly Bihar) who was given as present a pair of

> large Persian grey-hounds which he took out on a sporting excursion with a party of gentlemen … He slipped them after a jackal, and rode off himself in the direction of the animal, hallooing the dogs, who mistaking the object intended for them, attacked the Rajah's horse, and obliged him to ride into a neighbouring river,

and he later refers to local Indian rulers hunting jackals with hounds.[234] Another account of hunting jackals with packs of dogs comes from Green Price who recounts a jackal being hunted and torn to pieces by a pack of hounds owned by a British hunter, despite showing amazing speed and stamina in trying to escape the hounds. During his accounts Johnson refers several times to the cry of the jackal at night and of the presence of jackals deterring deer he wanted to hunt.[235] He also refers to being attacked by a rabid jackal, which then attacked cattle and a herder, the latter killing the jackal; Johnson's hunting dogs once killed a rabid jackal, were bitten in the process, and had to be killed.[236]

By 1938, it was reported that jackals had become very scarce in some areas because of persecution and habitat loss and were rarely seen in Mysore state, the disappearance coinciding and suspected to be linked with a serious outbreak of mange among dholes. It also followed the deaths of large numbers of cattle from rinderpest, with jackals observed feeding from the carcasses of the diseased cattle.[237] By 1968, a survey of wildlife in South and West India was able to report that jackals had recovered somewhat in numbers and were common in some protected areas of Mysore, notably the Venugopal Wildlife Park.[238] The same survey reported that jackals were common in Madras state's Mudumalai Wildlife Sanctuary and were to be found at Point Calimere Sanctuary in Madras.[239] They were also found, along with tigers, leopards, and sloth bears, plus a range of potential ungulate prey species, at Kodaikanal Hills in Madras.[240] Throughout this time, the fable continued to be widely believed that jackals could lead tigers and, in their surviving range in the Gir Forest, lions to kills. Eardley-Wilmot commented that jackals were often seen in close proximity to tigers, with no obvious aggression by the tiger. He added that they scavenged tiger kills and also those of leopards, which seemed more often to drive off jackals.[241] Hill, in his notes on jackals, tigers, and lions, rejected the idea that the jackal was a "provider" for tigers or lions.[242]

From the end of British rule to the present, the jackal has survived well, described by Menon as being found throughout India in a wide variety of habitats, except for high mountain regions. They are common in many areas and are not seen as being of concern in conservation terms. It hunts rodents, reptiles, birds, and some small antelopes or their offspring. It has "an undeserved reputation as a scavenger."[243] Jhala seconded this view, reporting that

> [d]ue to its adaptive nature and omnivorous diet, the golden jackal occurs across all habitats in India except the higher Himalayas, Sundarban mangroves and extreme parts of the Thar desert. The jackal often occurs in suburban areas, across rural India, and is encountered as one of the most frequent road killed mammal. The jackal subsists primarily on meat obtained through scavenging on livestock carcasses, kills of other predators and from garbage dumps. The jackal is also an effective predator primarily targeting young of ungulates, ground birds, reptiles and insects.[244]

Srivathsa et al. highlighted areas of conflict between people in modern India and jackals, noting a very immediate and growing reason for human–carnivore conflict – "India harbours 23% of the world's carnivore species that share space with 1.3 billion people in approximately 2.3% of the global land area ... Carnivore occupancy ranged from 12% for dholes to 86% for jackals."[245] Interestingly they recorded, "Depredation incidents were recorded from 68 sites for wolf, nine sites for dhole and 44 sites for fox. There were no records of depredation by hyena, but incidents attributed to jackal were reported in more than 95% of the sites."[246] Jackals were mainly blamed for attacks on poultry.

Jackals in India have proved adept at moving into areas where human activity produces food sources – whether livestock and poultry or waste dumps – and into areas made into uninhabited wasteland by human economic activity. One good example of this is the colonisation of areas of grassland and scrub around airports.

In such areas jackals can commute to human settlements and the airports themselves to forage human waste and hunt rodents and other small prey. They seem to cause little trouble, but do gather in large numbers around dumps, according to eye witnesses.[247] The Zoological Survey of India plans to radio collar jackals near the airport to monitor their numbers and activities. They appear to have moved into the airport environs from habitat around Rajarhat, New Town, and other expanding residential areas around the city. They are also being monitored at Guwahati airport.

In Pakistan, despite persecution by pastoralist communities and small-scale farmers who keep a few livestock alongside their crop cultivation, golden jackals are still found across many areas of the country. They haven't been intensively studied, but studies of human–wildlife conflict have often named jackals as one of the predators blamed for livestock losses, particularly sheep and goats. The study by Iftikhar Darr et al. of leopard predation around the boundaries of Machiara National Park in the Azad Jammu and Kashmir Himalayan area of Pakistan found that jackals ranked after leopards and black bears as suspected livestock killers – being blamed for six sheep and goat deaths over a 3-year period (compared with 309 blamed on leopards).[248] A paper based on questionnaires and discussions with farmers in the Dkirkot region of Azad Jammu and Kashmir brought the response that jackals were blamed for 80% of poultry predation.[249] The most serious threat to jackals, after habitat loss and human persecution, was being killed on the roads. Akram looked at wildlife mortality on roads on the Potwar Plateau in Pakistan's Punjab Province. Between 2 March and 12 February 2014, 131 carcasses were found that resulted from being hit by vehicles – 49 were jackals, the animal most often found.[250]

Jackals are also widespread in Bangladesh and Nepal. In the latter they are frequently found in the very common and large sugar cane fields – which offer cover and food in the form of small mammals, birds, insects, and the sugar cane itself.[251] Jackals were persecuted as pests by the farmers, who would drive the jackals from the fields and kill them.[252] They are also important as scavengers, clearing up the carcasses of roadkill, livestock that have died of disease of remains of animals slaughtered for food. In Nepal, they are present particularly in the foothills and small areas of plain or plateau. In hillier areas, they may be found in river valleys but not at high altitudes or on steep hill or mountainsides.[253] In the high forests and more open, tall grassland areas of Nepal, golden jackals are fairly common. Byrne reported that in the region that became the 150,000acre Royal Shuklaphanta Wildlife Reserve (aka White Grass Plains Wildlife Reserve), "[i]ts pristine forests are home to five species of deer – sambar, swamp deer, axis deer, hog deer, and barking deer – and to tiger, leopard, wild boar, mugger crocodile, jackal, wild dog, hyena, leopard cat, jungle cat, and pangolin, the scaly anteater."[254] Byrne noted that hunters in India and Nepal were generally after trophies and rarely shot "any of the smaller animals, such as monkeys, jackals, badgers, or civet cats."[255] A report on the status of jackals in Nepal in 2011 described them as a versatile species occurring in more open country and often near human settlements, from which they scavenged food, such as livestock and poultry carcasses.[256]

Cambodia, Laos, and Vietnam

The Indochinese sub-species of golden jackal (*Canis aureus cruesemanni*) is found in eastern India, Myanmar, Thailand, Laos, Cambodia, and Vietnam. They are one of the least-studied sub-species. They prefer wetland, mangroves, evergreen forests, agricultural land, and even semi-urban areas. Kamler found in Cambodia that they were often found in low densities and had large territories as a result.[257] They seemed to avoid areas populated by leopards and dholes. Scat analysis showed that

> the main food items consumed by jackals were processional termites (Hospitalitermes spp.; 26% biomass consumed), followed by wild pig ... 20%, muntjac.20%, and civets (17%) ... jackals were not random in their consumption of ungulates because muntjac were selectively consumed over larger ungulate species.[258]

They were common but not especially numerous in Cambodia, though that may be a result of the paucity of recent studies. In Vietnam, their status was uncertain because of a lack of studies and no confirmed sightings since 2004.[259]

Notes

1 N. Huisman (2020) First golden jackal depredation confirmed in the Netherlands, *European Wilderness Society* https://wilderness-society.org/first-golden-jackal-depredation-confirmed-in-the-netherlands/ accessed 25 January 2021.
2 S. Deinet et al. (2013) *Wildlife Comeback in Europe: The Recovery of Selected Mammal and Bird Species. Final Report to Rewilding Europe*, London: ZSL, BirdLife International and the European Bird Census Council, Wildlife-Comeback-in-Europe-the-recovery-of-selected-mammal-and-bird-species.pdf accessed 30 December 2021, p. 91.
3 IUCN Red List (No date) Canis aureus, Golden Jackal, https://www.iucnredlist.org/species/pdf/46194820 accessed 16 February 2022.
4 M.A.E. Rossberg (2018) Golden jackal takes on Europe, https://wilderness-society.org/golden-jackal-takes-on-europe/ accessed 16 February 2022.
5 H. Tsunoda (2017) Food niche segregation between sympatric golden jackals and red foxes in central Bulgaria, *Journal of Zoology*, 303, 64–71, p. 64.
6 R. Sommer & N. Benecke (2005) Late-Pleistocene and early Holocene history of the canid fauna of Europe (Canidae). *Mammalian Biology*, 70, 227–41, p. 234.
7 R. Rutkowski (2015) A European concern? Genetic structure and expansion of golden jackals (Canis aureus) in Europe and the Caucasus. *PLoS ONE*, 10(11), 1–22, p. 3.
8 M. Shakarashvili et al. (2020) Population genetic structure and dispersal patterns of grey wolfs (Canis lupus) and golden jackals (Canis aureus) in Georgia, the Caucasus, *Journal of Zoology*, 312, 227–38, p. 227.
9 T. Wyatt (2021) Golden jackal spreading across Europe as climate warms, *Independent (UK)*, 13/12/21, Golden jackal spreading across Europe as climate warms | The Independent, accessed 17 December 2021.
10 Deinet et al., 2013, p. 8.
11 B. Krystuffek, D. Murariu & C. Kurtoner (1997) Present distribution of the Golden Jackal Canis aureus in the Balkans and adjacent regions, *Mammal Review*, 27, 109–14, pp. 112–3.

12 N. Spassov & I. Acosta-Pankov (2019) Dispersal history of the golden jackal (*Canis aureus moreoticus* Geoffroy, 1835) in Europe and possible causes of its recent population explosion. *Biodiversity Data Journal*, 7, 1–22, pp. 4–5.

13 J. Hatlauf et al. (2021) New rules or old concepts? The golden jackal (Canis aureus) and its legal status in Central Europe, *European Journal of Wildlife Research*, 67,25, 1–15, p. 1.

14 Spassov and QAcosta-Pankov, 1997, pp. 4–5.

15 A. Trouwborst, M. Krofel & J.D.C. Linnell (2015) Legal implications of range expansions in a terrestrial carnivore: the case of the golden jackal (Canis aureus), in Europe, *Biodiversity Conservation*, 24, 2593–610, p. 2595.

16 Hatlauf et al., 2021, p. 1.

17 Deinet et al., 2013, p. 8.

18 Spassov & Acosta-Pankov, 2019, p. 5.

19 Ibid., p. 7.

20 M. Krofel (2018) Golden jackal expansion across Europe: causes and consequences, in G. Giannatos et al. (eds) *Proceedings of the 2nd International Jackal Symposium, Marathon Bay, Attiki Greece 2018*. Hellenic Zoological Archives. 9, November 2018, p. 105.

21 Ibid.

22 L. Lapini (2011) The Golden Jackal in Europe, Lapini 2011Golden Jackal in Europe accessed 17 December 2020.

23 Ibid.

24 Deinet et al., 2013, p. 91.

25 Krofel, 2018, p. 105.

26 Ibid.

27 Deinet et al., 2013, p. 91.

28 Ibid.

29 G. Giannatos (2004) *Conservation Action Plan for the Golden Jackal Canis aureus L. in Greece*. Athens: WWF Greece, p. 1.

30 Ibid., p. 18.

31 I. Gasteratos & Z. Fondoulakou (2018) The presence and the extinction of the golden jackal from the island of Corfu, north-western Greece, in Giannatos et al., p. 86.

32 Giannatos, 2004, pp. 5–6.

33 J. Arnold et al. (2012) Current status and distribution of golden jackals *Canis aureus* in Europe, *Mammal Review*, 42,1, 1–11, p. 1.

34 Ibid.

35 Hatlauf et al., 2021, p. 1.

36 O.C. Banea (2021) The Jackal from Rome crossed the Arctic Circle, 70° North. Welcome to Norway, https://gojage.blogspot.com/2021/?m=0 accessed 29 March 2022.

37 M. Heltai et al. (2018) Golden jackal population dynamics in certain study areas of Hungary – examples for hectic population change and invasion, in G. Giannatos et al., p. 50.

38 C.M. Gherman & A.D. Mihalca (2017) A synoptic overview of golden jackal parasites reveals high diversity of species, *Parasites and Vectors*, 10(419), 1–40, p. 1.

39 Krystuffek, Murariu & Kurtoner, 1997, p. 111.

40 Ibid.

41 Giannatos, 2004, p. 5.

42 G. Markov & J. Lanszki (2012) Diet composition of the golden jackal, Canis aureus in an agricultural environment Authors, *Folia Zoologica*, 61(1), 44–8, p. 44.

43 Arnold et al., 2012, p. 5.

44 Markov & Lanszki, 2012, p. 4.

45 Ibid., p. 46.

46 E.G. Raichev et al. (2013) The reliance of the golden jackal (Canis aureus) on anthropogenic foods in winter in central Bulgaria, *Mammal Study*, 38, 19–27, p. 19.

47 Ibid., p. 26.

48 S. Stoyanov (2012) Golden jackal (*Canis Aureu*) in Bulgaria. Current status, distribution, demography and diet, *International symposium on hunting, 'Modern aspects of sustainable management of game population'*, Zemun-Belgrade, Serbia, 22–4, June, 2012, 48–55, p. 53.

49 T. Tóth (2009) Records of the golden jackal (Canis aureus Linnaeus, 1758) in Hungary from 1800th until 2007, based on a literature survey, *North-Western Journal of Zoology*, 5(2), 386–405, pp. 394–5.

50 Ibid., p. 396.

51 Tóth, 2012, p. 398.

52 Spassov & Acosta-Pankov, 2019, p. 7.

53 Papp et al., 2013.

54 Ibid.

55 C.R. Papp, R. Papp & O.C. Banea (2018) Population dynamics and current status of the golden jackal in Romania, in Giannatos et al., pp. 146–7.

56 M. Marinov et al. (2018) Research regarding the damages caused by the golden jackal in Danube Delta Biosphere Reserve and in surrounding areas (Romania), in Giannatos et al., p. 154.

57 Ibid.

58 Ibid.

59 T. Tóth (2009) Records of the golden jackal (Canis aureus Linnaeus, 1758) in Hungary from 1800th until 2007, based on a literature survey, *North-Western Journal of Zoology*, 5(2), 386–405.

60 Ibid, p. 386.

61 Lapini, 2011.

62 M.H. Lanszki & L. Szabo (2006) Feeding habits and trophic niche overlap between sympatric golden jackal (Canis aureus) and red fox (Vulpes vulpes) in the Pannonian ecoregion (Hungary), *Canadian Journal of Zoology*, 84, 1647–56, pp. 1650–1.

63 J. Lanszki et al. (2015) Diet composition of the golden jackal in an area of intensive big game management. *Annales Zoologici Fennici*, 52, 243–55, p. 243.

64 M. Šálek et al. (2014) Population densities and habitat use of the golden jackal (Canis aureus) in farmlands across the Balkan Peninsula, *European Journal of Wildlife Research*, 60, 193–200, p. 194.

65 La´szlo Szabo et al. (2009) Expansion range of the golden jackal in Hungary between 1997 and 2006, *Mammalia*, 73, 307–11, p. 307.

66 Šálek et al., 2014, p. 194, p. 195.

67 Ibid., p. 197.

68 O.C. Banea (2018) Red fox and golden jackal hunting bag differences in countries from central and south-eastern Europe. Population trend and management aspects, in Giannatos et al., pp. 118–9.

69 Arnold, 2012, p. 8.

70 Spassov & Acosta-Pankov, 2019, pp. 7–9.

71 H. Potočnik (2018) Potential versus actual "habitat interference" between expanding golden jackals and wolves in Slovenia, in Giannatos et al., p. 114.

72 Ibid., p. 7.

73 Ibid.

74 Y. Fannin (2018) Golden jackal (Canis aureus moreoticus Geoffroy, 1835) predatory behaviour and carcass consumption of livestock in north-east Italy, in Giannatos et al., 93–4, p. 93.

75 Ibid.

76 S. Pesaro et al. (2018) Thanatological and necroscopic remarks about two suspected poisoned golden jackals (Canis aureus moreoticus) from north-eastern Italy, in Giannatos et al., pp. 95–6.

77 T. Wyatt (2021) Golden jackal spreading across Europe as climate warms, *Independent (UK)*, 13/12/21, Golden jackal spreading across Europe as climate warms | The Independent, accessed 17 December 2021.

78 Rewilding Appenines on FB 4/2/2022, accessed 4 February 2022.
79 Lapini, 2011.
80 Arnold, 2012, pp. 3–4.
81 Ibid.
82 Papp et al., 2013.
83 Arnold, 2012, p. 6.
84 M. Slamka, P. Kaštier & M. Schwarz (2017) The golden jackal in Slovakia, *Canid Biology & Conservation*, 20(9), 38–41, p. 38.
85 N. Guimãraes, J. Buêko a & P. Urban (2018) The evolution of the presence of the golden jackal in Slovakia, Giannatos et al., p. 144.
86 Slamka, Kaštier & Schwarz, 2017, p. 38.
87 I. Rusev (2020) The golden jackal (Canis aureus) in the Tuzlivski Lymany National Nature Park, *Theriologia Ukrainica*, 20, 46–57, p. 46.
88 Lapini, 2011.
89 M. Rozhenko & V. Kormyzhenkog (2018) Some aspects of the golden jackal invasion in the south of Ukraine. Giannatos et al. p. 19.
90 A.M. Volokh (2018) Distribution of the golden jackal in Ukraine and its trophy value, in Giannatos et al., p. 131.
91 Ibid.
92 D. Yurchenko et al. (2018) Identification of golden jackal in Belarus with the help of mitochondrial genetic markers, in Giannatos et al., p. 85.
93 Jackal Ecology (no date) Golden Jackal Dispersal Movements in Europe, https// goldenjackal.eu accessed 18 December 2020.
94 P. Männil & M. Mustasaar (2018) Jackal′s expansion towards north: Can they survive in boreal ecosystem? in Giannatos 2018, p. 110.
95 Ibid.
96 Jackal Ecology (no date) Golden Jackal Dispersal Movements In Europe, https// goldenjackal.eu accessed 18 December 2020.
97 Ibid.
98 *Helsinki Times* (2019) Finland's first sighting of golden jackal sets off political debate, 29 July 2019, https://www.helsinkitimes.fi/finland/finland-news/domestic/16594-finland-s-first-sighting-of-golden-jackal-sets-off-political-debate.html, accessed 21 February 2022.
99 N. Ranc (2018) The golden jackal in Europe: Where to go next? in Giannatos et al., pp. 108–9.
100 N. Huisman (2017) First golden jackals arrived in France, *European Wilderness Society*, https://wilderness-society.org/first-golden-jackals-arrived-france/ accessed 26 January 2022.
101 Ferus (no date) Le chascal dorć est arrivćen France, https://www.ferus.fr/actualite/le-chacal-dore-est-arrive-en-france accessed 16 February 2022.
102 Arnold et al., 2012, p. 6.
103 Facebook Wild Dog Foundation 21/11/2021 NEWGROUND.
104 Erstes Goldschakalpärchen Deutschlands: Liebesspiele im Schwarzwald – Schwarzwald aktuell (schwarzwald-aktuell.eu) accessed 16 February 2022.
105 Personal communication with Nick Huisman, 6 March 2022.
106 N. Huisman (2021) Unique video of golden jackal in The Netherlands, Unique video of golden jackal in The Netherlands (wilderness-society.org) accessed 13 May 2021.
107 J. Wennink & G. Lelieveld (2018) Habitat suitability analysis on the golden jackal for the Netherlands, in Giannotos et al., pp. 121–2.
108 Jackal Ecology (no date) Golden Jackal Dispersal Movements in Europe, https// goldenjackal.eu accessed 18 December 2020.
109 C.-R. Papp, O. C-tin Banea & A. Iuliana Szekely-Sitea (2013) Applied ecology and management aspects related to the golden jackal specific ecological system in

Romania, *Acta Musei Maramorosiensis IX*, Sighetu Marmației, Ianuarie 2014, no page numbers.
110 Deinet et al., 2013, p. 91.
111 Ibid.
112 Hatlauf et al., 2021, p. 1.
113 Ibid., p. 2.
114 Ibid.
115 Trouwborst et al., 2015, p. 2597.
116 V.G. Heptner & N.P. Naumov (1967) *Mammals of the Soviet Union, Volume II, Part 1a*, Moscow: Vysshaya Shkola Publishers, Mammals of the Soviet Union (archive.org) accessed 28 June 2021.
117 Ibid., p. 134.
118 Ibid., pp. 135–6 and 141.
119 Ibid., p. 141.
120 Ibid., p. 147.
121 M. Shakarashvili et al. (2020) Population genetic structure and dispersal patterns of grey wolfs (Canis lupus) and golden jackals (Canis aureus) in Georgia, the Caucasus, *Journal of Zoology*, 312, 227–38, p. 227.
122 Ibid.
123 *Proceedings of the Zoological Society of London* (henceforth referred to as *Proceedings*), 20 March 1877, p. 273.
124 Ibid., 3 February 1880, p. 53.
125 M.W. Chynoweth et al. (2016) Human–wildlife conflict as a barrier to large carnivore management and conservation in Turkey, *Turkish Journal of Zoology*, 40, 1–12, pp. 2–3.
126 H. Ambar & C. Can Bilgin (2013) First record of a melanistic golden jackal (Canis aureus, Canidae) from Turkey, *Mqammalia*, 772,2, 219–22, p. 219.
127 L. Brown & D.W. Macdonald (1995) predation on green turtle (*Chelonia mydas*) by wild canids at Akyatan beach, Turkey, *Biological Conservation*, 71, 55–60, p. 57.
128 Heptner & Naumov, 1967, pp. 142–3.
129 Ibid.
130 Ibid.
131 Ibid., pp. 153–4.
132 *Holy Bible, Authorised King James Version*, London: Collins, 1991.
133 J. Charteris-Black (2004) *Corpus Approaches to Critical Metaphor Analysis*, London: Palgrave-Macmillan, recorded, p. 238.
134 U. Avner et al. (2011) Carnivore traps in the Negev and Judaean deserts (Israel): function, location and chronology, *Prédateurs dans tous leurs états. Évolution, Biodiversité, Interactions, Mythes, Symboles. XXXIe rencontres internationales d'archéologie et d'histoire d'Antibes*, 253–68, p. 255.
135 Ibid., pp. 255–7.
136 Ibid., p. 259.
137 *Proceedings*, 13 February 1866, p. 91.
138 J. Borkowski, A. Zalewski & R. Manor (2011) Diet Composition of Golden Jackals in Israel, *Annales Zoologici Fennici* 48(2), 108–18, p. 109. See also M.B. Qumsiyeh (1996) *Mammals of the Holy Land*, Lubbock: Texas tech University Press.
139 Borkowski, Zalewski & Manor, 2011, p. 109.
140 Y. Yom-Tov, S. Ashkenazi & O. Viner (1995) Cattle predation by the golden jackal Canis aureus in the Golan Heights, Israel, *Biological Conservation*, 73, 19–22, p. 21.
141 Borkowski, Zalewski & Manor, 2011, p. 108.
142 Ibid., pp. 110–1.
143 N. Nattrass et al. (2017) Understanding the black-backed jackal, *CSSR Working Paper*, No. 399, Institute for Communities and Wildlife in Africa, http://cssr.uct.ac.za/pub/wp/399, accessed 4 June 2021, p. 10.

144 R.E. Lewis & J.H. Lewis (1968) A review of Lebanese mammals. Carnivora, Pinnipedia, Hyracoidea and Artiodactyla, *Journal of Zoology*, 154, 517–31, p. 517.

145 M.B. Qumsiyeh, Z.S. Amr & D.M. Shafei (1993) Status and conservation of carnivores in Jordan, *Mammalia*, 57, 1, 55–62, p. 56.

146 F. Bunaian (2001) The Carnivores of the Northeastern Badia, Jordan, *Turkish Journal of Zoology*, 25, 19–25, p. 24.

147 D.J. Osborn & I. Helmy (1980) The Contemporary Land Mammals of Egypt (including Sinai). Field Museum of Natural History, digitized version, https://archive.org/details/contemporaryland05osbo/page/422 accessed 24 July 2019, pp. 428–9.

148 A.A. Green (1986) Status of large mammals of northern Saudi Arabia, *Mammalia*, 50(4), no page numbers.

149 I. Khorozyan et al. (2014) Patterns of co-existence between humans and mammals in Yemen: some species thrive while others are nearly extinct, *Biodiversity Conservation*, 23, 1995–2013, p. 1996.

150 J.E. Castelló (2018) *Canids of the World*, Princeton, NJ: Princeton University Press/Princeton Field Guides, pp. 133–7.

151 R.E. Cheesman (1920) Reports on the Mammals of Mesopotamia, *Bombay Journal of the Natural History Society*, XXVII, 1920–22, 323–46, p. 325.

152 Ibid., p. 334.

153 F. Álvares (2018) Golden jackals in Iran: distribution, population genetics and ecology, in G. Giannatos et al., 34–6, p. 34.

154 Ibid.

155 Ibid., p. 36.

156 G. Hosein Yusefi et al. (2021) Habitat use and population genetics of golden jackals in Iran: Insights from a generalist species in a highly heterogeneous landscape, *Journal of Zoological Systematics and Evolutionary Research*, 59, 1503–15, p. 1504.

157 Ibid., p. 1507.

158 Ibid., pp. 1511–12.

159 A. Mohammadi, M. Kaboli & J.V. López-Bao (2017) Interspecific killing between wolves and golden jackals in Iran, *European Journal of Wildlife Research*, 63(61), 1–5, p. 1.

160 Ibid.

161 Ibid., p. 2.

162 Ibid.

163 A. Stephen, R. Suresh & C. Livingstone (2015) Indian biodiversity: past, present and future. *International Journal of Environment and Natural Sciences*, 7, 13–28, pp. 13–4.

164 J.E. Castelló (2018) *Canids of the World*, Princeton, NJ: Princeton University Press/Princeton Field Guides, pp. 138–43.

165 E.C. Majumder (1950) The Vedic Age, in R.C. Majumder (ed) *The History and Culture of the Indian People, Vol 1*, Bombay: Bharatiya Vidya Bhavan, p. 47.

166 R.S. Sharma (2003) Historical archaeology and problems of Urban History, in H. Siddiqi (ed) *Studies in Archaeology and History*, New Delhi: Rampur Raza Library, 49–58, p. 51.

167 J. Keay (2010) *India a History from the Earliest Civilisations to the Boom of the Twenty-First Century*, London: Harper Collins, Updated edition, pp. xvii–xviii.

168 M. Gadgil & R. Guha (1992) *This Fissured Land, an Ecological History of India*, New Delhi: Oxford University Press, p. 63.

169 Ibid., pp. 64–5.

170 A. Roberts (2018) *Evolution. The Human Story*, London: Dorling Kindersley, p. 209; and A.M. Jukar et al. (2020) Late Quaternary extinctions in the Indian Subcontinent, *Palaeogeography, Palaeoclimatology, Palaeoecology*, https://doi.org/10.1016/j.palaeo.2020.110137, 1–11, p. 7.

171 J.F. Jarridge & R.H. Meadow (1980) The antecedents of civilization in the Indus Valley, *Scientific American*, 243(2), 122–33.

172 Jukar et al., 2020, p. 7.

173 See G.R. Sharma et al. (1980) *The Beginnings of Agriculture*, Allahabad: University of Allahabad Press.

174 K.A. Chowdhury (1977) *Ancient Agriculture and Forestry in Northern India*, Bombay: Asia Publication.

175 Gadgil & Guha, 1992, p, 66.

176 S. Rao (1957) History of our knowledge of the Indian Fauna through the ages, *Bombay Journal of the Natural History Society*, 54, 2, 1957, 251–80, pp. 255–6.

177 S. Singh et al. (2017) Sacred groves: myths, beliefs, and biodiversity conservation—a case study from Western Himalaya, India, *Hindawi International Journal of Ecology*, Article ID 3828609, 1–12, p. 1.

178 Ibid.

179 For more on the use of sacred sits to conserve the environment, species and natural resources and enable sustainable use of them, see D.J. Hughes & S.M.D. Chandran (1997) *Paper Presented in the Workshop on the Role of Sacred Groves in Conservation and Management of Biological Resources*, KFRI, Peechi, India, 1997; and A.K.M.N. Islam, M.A. Islam & A.E. Hoque (1998) Species composition of sacred groves, their diversity and conservation in Bangladesh, in P.S. Ramakrishnan, K.G. Saxena & U.M. Chandrashekara (eds) *Conserving the Sacred for Biodiversity Management*, New Delhi: UNESCO and Oxford-IBH Publishing, pp. 163–5.

180 Singh et al., 2017, pp. 1–2.

181 Y. Gokhale et al. (2011) Sacred landscapes as repositories of biodiversity. A case study from the Hariyali Devi sacred landscape, Uttarakhand, *International Journal of Conservation Science*, 2(1), 37–44, p. 38.

182 Ibid.

183 S. Mitra (2005) *Gir Forest and the Saga of the Asiatic Lion*, New Delhi: Indus, pp. 36 and 43.

184 Ibid.

185 J. Auboyer (1972) Animals in India, Houghton Brodrick, 115–45, p. 127.

186 Keay, 2010, p. 2.

187 Pigott, 1960, pp. 133–5.

188 Stein, 2010, p. 47.

189 Keay, 2010, p. 2.

190 Ibid.

191 D.R. Kinsley (1989) *The Goddesses' Mirror: Visions of the Divine from East and West*. New York: State University of New York Press, pp. 3–4.

192 W.J. Wilkins (1900) *Hindu Mythology*, New Delhi: Rupa Publication (reprint in 2013 of 2nd edition), p. 423.

193 S. Visnu (2006) *The Pancatantra, I London*. Penguin Books Ltd. Kindle Edition.

194 Ibid., loc 262.

195 Ibid., loc 456.

196 Ibid., loc 689–703.

197 Ibid., loc 717 and 760.

198 Ibid., loc 1502.

199 Sarma, 2006, loc 2825–2855.

200 Ibid., loc 2844.

201 Ibid., loc 2855.

202 Ibid., loc 3515.

203 Ibid., loc 3533.

204 Ibid., loc 2094–2115.

205 Rudyard Kipling (no date) *The Jungle Book*, Kindle Edition, loc 70.

206 C.H. Donald (1947) Jackals, *Bombay Journal of the Natural History Society*, LVII, 1947–8, 1–2, 721–9, p. 721.

207 Ibid., p. 725.

208 Mary Frere (no date) Old Deccan Days/The Brahman, the Tiger and the Six Judges, https://en.wikisource.org/wiki/Old_Deccan_Days/The_Brahman,_the_Tiger,_and_the_Six_Judges, accessed 23 February 2021; see also K. Lock & D. Kennett (1994) *The Tiger, the Brahmin and the Jackal*, Flinders Park, South Australia: Martin international.

209 See G. Samuel (2010) *The Origins of Yoga and Tantra*, Cambridge: Cambridge University Press; and Keay, 2010, p. 42.
210 Ibid., p. 72.
211 Kallidaikurichi Aiyah Nilakanta Sastri (1998) Age of the Nanda and Mauryas, Delhi: Motilal Banarsidass, 2nd edition, p. 16.
212 Mitra, 2005, p. 43.
213 Keay, 2010, p. 101.
214 C.B. Asher & C. Talbot (2008) *India before Europe*, Cambridge: Cambridge University press, pp. 50–1; and P. Jackson (2003) *The Delhi Sultanate: A Political and Military History*. Cambridge: Cambridge University Press, p. 28.
215 B. Stein (2010) *A History of India*, Oxford: Blackwell, pp. 158–60.
216 Stein, 2010, p. 159.
217 Gadgil & Guha, 1992, p. 94.
218 Rao, 1957, pp. 268–9.
219 W. Dalrymple (2019) East India Company sent a diplomat to Jahangir & all the Mughal Emperor cared about was beer, *The Print*, 24 August, 2019, https://theprint.in/pageturner/excerpt/east-india-company-sent-a-diplomat-to-jahangir-all-the-mughal-emperor-cared-about-was-beer/281255/ accessed 3 March 2019.
220 W. Dalrymple (2015) The East India Company: The original corporate raiders. *The Guardian*, 4 March 2015 https://www.theguardian.com/world/2015/mar/04/east-india-company-original-corporate-raiders – accessed 3 March 2020.
221 S. Bose & J. Ayesha (2004) *Modern South Asia: History, Culture, Political Economy* (2nd ed.), Abingdon: Routledge, p. 41.
222 R. Lydekker (1896) *A Geographical History of Mammals*, Cambridge: Cambridge University Press, p. 274.
223 R.C. Wroughton (1914) Mammals of India, Burma and Ceylon, *Bombay Journal of the Natural History Society*, XXIII, 1914, 460–81, p. 470.
224 R.I. Meena, S. Kumar & S. Alam (2014) *Action plan for the Conservation of the Asiatic Lion (Pantheraleo persica Meyer, 1826)*, Junagadh: Gujarat Forest Department, p. xi.
225 Ibid., p. 117.
226 J.E. Hughes (2015) Royal tigers and ruling princes: wilderness and wildlife management in the Indian princely states, *Modern Asian Studies*, 49(4), 1210–60, pp. 1214–5.
227 *Proceedings*, 12 July 1831, p. 101.
228 R.C. Wroughton (1915) The Indian jackal, *Bombay Journal of the Natural History Society*, XXIV, 1915–7, p. 649.
229 R.C. Wroughton (1918) Summary of the results from the Indian mammal Survey, Family IC Canidae, *Bombay Journal of the Natural History Society*, XXVI, 1918–21, pp. 338–41.
230 W.T. Blanford (1891) *The Fauna of British India including Ceylon and Burma*, London: Taylor & Francis, Mammalia: Blanford, William Thomas, 1832–1905: Free Download, Borrow, and Streaming: Internet Archive accessed 1 September 2021, pp. 140–2.
231 Col. A.E. Ward (1928) Mammals and birds of Kashmir and the adjacent hill provinces, *Bombay Journal of the Natural History Society*, XXXII, 304, 711–5, p. 714.
232 Anonymous (1929) Game preservation and experiments in India, *Bombay Journal of the Natural History Society*, XXXIII, 1–2, 1929, p. 127.
233 Col K. Guman Singh (1955) Game preservation in Jammu and Kashmir state, *Journal of the Bombay Natural History Society*, 53, 1955–6, 1, 646–50, p. 647.
234 D. Johnson (1822) *Sketches of Field Sports as followed by the4 Natives of India*, London: Longman, https://archive.org/details/sketchesoffields00johnrich/page/78/mode/2up accessed 4 March 2020, pp. 39–40.
235 Johnson, 1822, p. 93.
236 Ibid., pp. 231–3.
237 R.C. Morris (1938) Disappearance of jackals, *Bombay Journal of the Natural History Society*, XL 1938, 1–12, p. 117.

238 J. Juan Spillet (1968) Report on wild life survey in South and West India, *Journal of the Bombay Natural History Society*, 65, 2, 1968, 296–325, p. 318.
239 J. Spillet (1968) Report on Wild Life Survey in South and West India, *Journal of the Bombay Natural History Society*, 65, 3, 1968, 613–63, p. 641.
240 Ibid., p. 661.
241 S. Eardley-Wilmot (1897) The "Kol-Bahlu", and the instinct of fear in wild animals, *Bombay Journal of the Natural History Society*, XI, 1897–8, pp. 548–9.
242 F.J. Hill (1893) The jackal or lion provider, *Bombay Journal of the Natural History Society*, 8, pp. 306–7.
243 V. Menon (2014) *Indian Mammals A Field Guide*, Gurugram, India: Hachette India, p. 278.
244 Y.V. Jhala (2018) Ecology of the golden jackal and wolves in India, in G. Giannatos et al. (eds) *Proceedings of the 2nd International Jackal Symposium, Marathon Bay, Attiki Greece 2018*. Hellenic Zoological Archives, 9, November 2018, 31–2, p. 31.
245 A. Srivathsa et al. (2019) Examining human – carnivore interactions using a socio-ecological framework: sympatric wild canids in India as a case study. *Royal Society Open Science*, 6, 182008, 1–17, p. 1.
246 Ibid., p. 8.
247 G. Singh (2020) Golden jackals make airports their homes, *Mongabay*, 9 December 2020, https://india.mongabay.com accessed 29 March 2022.
248 Naeem Iftikhar Dar (2009) Predicting the patterns, perceptions and causes of human–carnivore conflict in and around Machiara National Park, Pakistan, *Biological Conservation*, doi:10.1016/j.biocon.2009.04.003, 1–7, p. 3.
249 S.S. Bibi (2013) Study of ethno-carnivore relationship of Dhirkot, Azad Jammu and Kashmir (Pakistan), *Journal of Animal & Plant Sciences*, 2393, 854–9, p. 854.
250 F. Akram (2019) Spatiotemporal patterns of wildlife road mortality in the Pothwar Plateau, Pakistan, *Mammalia*, 83(5), 487–95, p. 487.
251 M.M. Jaeger, R.K. Pandit & E. Haque (1996) Seasonal difference in territorial behaviour by golden jackals in Bangladesh: Howling versus confrontation, *Journal of Mammalogy*, 77(3), 768–75, pp. 769–70.
252 Ibid., p. 773.
253 M.A.C. Hinton & T.B. Fry (1923) Mammal Survey of India, Burma and Ceylon, No. 37, Nepal, *Bombay Journal of the Natural History Society*, XXIX, 1–2, 1923, 399–428, p. 413.
254 P. Byrne (2001) *Gone Are the Days: Jungle Hunting for Tiger and Other Game in India and Nepal 1948-1969*, Safari Press. Kindle Edition, loc 3674.
255 Ibid., loc 1709.
256 S.R. Jnawali et al. (2011) *The Status of Nepal's Mammals: The National Red List Series*, Kathmandu, Nepal: Department of National Parks and Wildlife Conservation, https://www.academia.edu/14715540/The_Status_of_Nepal_Mammals_The_National_Red_List_Series?email_work_card=view-paper accessed 1 January 2021, pp. 90–1.
257 J.F. Kamler (2021) Home range, habitat selection, density, and diet of golden jackals in the Eastern Plains Landscape, Cambodia, *Journal of Mammalogy*, 102, 2, April 2021, pp. 636–50, p. 636.
258 Ibid.
259 Ibid.

6

AFRICA FROM COLONISATION TO 1960

As the colonial invasion and occupation of the vast majority of the African continent gained pace in the last half of the 19th century in Africa, exploitation of natural resources increased substantially, including of wildlife for ivory, horn, hides, meat, and trophies. Wildlife was also in the firing line in colonised regions where pastoralism developed, and predators were persecuted, while forests cleared for crop production. Right at the end of the 19th century, the colonial conservation movement emerged, gaining support in the opening decades of the next century, but never entirely divorced from commercial exploitation of wildlife. This involved both regulation of hunting – and often banning of all indigenous hunting, even for crop or livestock protection – and the creation of reserves for the protection of wildlife in general or particular species in some cases. Game conservation was often linked to ensuring future hunting for settlers and visiting hunters, as well as aesthetic appreciation of wildlife by Europeans. In addition to commercial and sports hunting, a colonial introduction that was aimed particularly against jackals and other predators was poison. Across British, French, and Portuguese colonial territories in West, Central, East, and southern Africa, poisons like strychnine were used to kill predators from lions down to jackals and honey badgers, which were suspected of killing livestock and poultry. There were regular campaigns in many countries, which continued well after the end of colonialism.

Pastoral and farming communities in East Africa such as the Maasai, Samburu, Nandi, Barabaig, Kikuyu, Luo, and Luhya, to name but a few, had coexisted with the two jackal species and the golden wolf for centuries. To them, jackals could be a nuisance, killing young livestock and poultry, but were not major threats and might be killed opportunistically. To the incoming settlers, they were vermin to be killed, hunted for sport or viewed with a strange mixture of fascination and contempt. Large numbers of European settlers and hunters

DOI: 10.4324/9781003199793-7

who recorded their experiences got fixated on the vocalisations of the jackals, particularly at night. Kittenberger (a Hungarian hunter and animal collector for zoos and museums) is one such. In his account of hunting and collecting trips, he referred to the "disagreeable" cry of the jackal at night and wrote, "On the plains teeming with game we find them in astonishing numbers, and like the hyena, especially where lions are numerous, because something for the common folk will always fall from the royal table."[1] He does indicate, though, that they had adapted quickly to the presence of hunting expeditions and would circle camps at night looking for the opportunity to feed from the carcasses or hides of animals shot during the day, so Kittenberger would trap them and beat them to death in the traps. He added that the skins of jackals made nice rugs – he doesn't identify which jackal species, but one can guess at them being black-backed jackals.[2]

The British colonial authorities in Kenya labelled jackals – along with other predators – as vermin and encouraged their killing. Unlike with lions, leopards, and cheetahs, jackals were rarely recorded on game shooting returns. This is evident from the annual game returns published by the colonial administration. In 1902, the list included just two jackals killed near Lake Baringo,[3] which, given the thousands of animals killed on licence annually, is entirely unrealistic, especially when you read the accounts by settlers and commercial or sports hunters, in which they routinely and opportunistically kill jackals. In 1905, the official game report noted that jackals (species no given) were damaging to game sought by hunters and to fruit trees grown by settlers. No numbers were given for those killed.

In her books *The Flame Trees of Thika* and *The Mottled Lizard*, about her early life in Kenya, Elspeth Huxley makes scant reference to jackals. In *The Mottled Lizard*, she recalled hunting jackals with a pack of hounds, known as the Makuyu hounds.[4] In her biography of the settler and hunter Lord Delamere, Huxley wrote of him buying up of land to raise cattle, sheep, and poultry, his importing of thousands of merino sheep, and his intention to poison shoot or hunt down with dogs the jackals on his land to prevent predation of stock.[5] As well as being seen as a threat to settlers' livestock, jackals (along with hyenas and leopards) were viewed as a nuisance in towns that grew up under colonial rule, notably Nairobi. The first game warden of Kenya, Blayney Percival, wrote of the constant presence of jackals and hyenas around the town at night foraging for food among refuse and animal remains around butcheries.[6]

The famous, one should perhaps say notorious game warden and predator poisoner, George Adamson, admitted that when he undertook poisoning campaigns against hyenas, he also killed jackals, which he said he rather regretted as they killed Samburu sheep and goats, which Adamson saw as degrading the bush and savanna in Samburu County.[7] The Kenya Game Department records did not list the numbers of jackals killed in this way and made scant reference to them over decades of annual game reports, as they were not hunted for trophies, were not a danger to life, and were a minor nuisance to settlers. There did not seem to be the visceral hatred and desire to exterminate them that was evident in South

Africa. In southern Africa, jackals became the hated vermin to be exterminated wherever possible with whatever means possible – shooting, hunting with dogs, trapping, using gun traps, or poisoning. But they survived and even, in some places, thrived amid the concerted long-term and well-funded persecution. This chapter will concentrate mainly on the black-backed jackals of South and southern Africa and the war conducted against them.

Southern Africa – Malawi to the Cape, Angola to Mozambique

From the early 19th century through to the independence era, in South Africa and, to a lesser extent, Namibia, there was constant war against jackals, primarily the black-backed jackals of the arid and semi-arid savanna and semi-desert regions, which were home to large flocks of sheep and goats. Black-backed jackals were common, despite many decades of persecution by settlers, backed by the South African colonial administration, across the Cape from the Atlantic to the Indian Oceans, in parts of what was then called the Transvaal, the Free State, Natal, and up into Namibia (South West Africa), Botswana, Zimbabwe (Southern Rhodesia), Lesotho, eSwatini, and parts of Mozambique. From the Zambezi northwards, in northern Natal, much of Zimbabwe, northern Botswana, Angola, Mozambique, Malawi, and Zambia, the black-backed was replaced by the Sundevall side-striped jackal.[8]

During expeditions in Angola, the British hunter John Statham reported the presence of jackals in central Angola when he was hunting sable antelope in the 1920s – most probably Sundevall side-striped jackals. They, like hyenas, leopards, and lions, would enter his camp and ruin the carcasses or skins of animals he had shot. On the plateau of central-south Angola, he noted that much of the game had been exterminated but jackals, which could have been side-striped or black-backed, remained as did honey badgers.[9]

To the south in Botswana (called Bechuanaland by the British), the hunter James Chapman wrote that black-backed jackals were very common and that the Tswana hunted them for their skins and made fans from their tails stretched over a wooden handle.[10] Chapman, who hunted there in the middle of the 19th century, also noted that the San he encountered in the Kalahari enthusiastically cooked and ate a dead jackal he gave them.[11] The Tswana also made hats and cloaks from jackals' skins.[12] H.A. Bryden, the hunter and naturalist, said that the Tswana ran an extensive trade in karosses they made from jackal skins, which were sought after because of the luxuriant fur.[13] The Tswana used the karosses themselves and traded them to Europeans. Mackenzie, in his excellent work on hunting and empire, notes that in many parts of southern Africa, including Sotho and Tswana regions, the poor wore antelope skins and the rich and powerful wore jackal and cat furs.[14]

In Namibia, Shortridge found that black-backed jackals were "very plentiful" and widely distributed in the 1920s, despite persecution as sheep and goat killers and hunting for their skins to make karosses.[15] They were particularly numerous in the region between Windhoek and the Botswana border and along the Atlantic coast, only being scarce along the Caprivi Strip, where their place was taken by the side-striped jackal.[16]

South Africa's war against the jackal

Dutch settlers in the Cape in the 17th and 18th centuries increased their farming and livestock husbandry, buying (or stealing) Khoikhoi sheep, goats, and cattle and obtaining Xhosa cattle. They also increasingly imported sheep like merino to increase the production of wool, as the African fat-tailed sheep did not produce high-quality wool. The colonising power changed in 1795, when the British occupied Cape Town during the Napoleonic War, retaining it under the 1814 peace settlement. It was an ideal revictualling station for ships going to British possessions in India. By this period as hunters, farmers, and traders had penetrated eastwards and northwards, lions and other large carnivores, but not black-backed jackals, had been largely exterminated in farming areas of the Cape.[17]

British hunters and traders provided some of the most informative accounts of wildlife in South Africa at this time. Sir John Barrow, private secretary to the first British governor of the Cape, gave extensive accounts of his travels to the Cape.[18] Journeying to Paarl, the Karoo, and then Graaff-Reinet, he described seeing zebra, quagga, gemsbok, kudu, springbok, and ostriches. In the Camdeboo region, near Graaff-Reinet, he noted, "A heap of stones, piled upon the banks of a rivulet, was pointed out to me as the grave of a Hottentot; ... The intention ... was that of preventing the wolves, or jackals, or other ravenous beasts, from tearing up and mangling the dead carcass."[19] Barrow reported that Khoikhoi people, which he in the racist parlance of the time called Hottentots, killed jackals for their skins, which they wore:

> The dress of a Hottentot is very simple. It consists chiefly of a belt made of a thong cut from the skin of some animal. From this belt is suspended in front a kind of case made of the skin of the jackal. The shape is that of a nine-pin cut through the middle longitudinally; the convex and hairy side of which is uppermost.[20]

He also saw jackal skins in the huts of Khoikhoi communities in Namaqualand, suggesting hunting of jackals there to provide skins and fur.[21] But the settlers, from early on in their presence, killed the jackals as vermin, believing them to be a serious threat to their sheep, goats, and calves.

Jackals were killed by shooting, hunting with dogs, trapping, and poisoning. But unlike the Cape lions, cheetahs, spotted hyenas, and wild dogs, jackals survived and, over the next centuries, became the focus of persecution by farmers protecting their stock. In some areas, as the numbers of English-speaking settlers increased, jackals were hunted from horseback using dogs, like British fox hunting. In the 1820s, the Cape governor, Lord Charles Somerset, championed this type of jackal hunting, and some settlers imported foxhounds. In Mafikeng, a hunt was set up by Sir Frederick Carrington that was said to go out after jackals twice or three times a week. It was said by those taking part to "give such excellent sport," according to the keen English foxhunter, H.A. Bryden,[22] though it is doubtful it was hugely effective in reducing jackal numbers. He noted, in particular, the speed and stamina of jackals and recounted one hunt where a black-backed jackal outpaced a pack of hounds and horses for 25 minutes before escaping down a deep burrow.[23]

During travels in southern Africa from Natal to the Kalahari from 1849 to 1863, James Chapman often mentioned seeing large numbers of jackal, lion, leopard, cheetah, and other prepared skins and ivory; some of these had been hunted by the San to trade to the Tswana, who sold them to Europeans.[24] Morton and Hitchcock record that in the mid- to late 19th century, there was "specialised hunting in the Kalahari for high value animals (i.e., animals with high value body parts). They included elephant, rhinoceros, lion, leopard, cheetah, black-backed jackal, ostrich for feathers, gemsbok for horns."[25] Chapman collected live specimens for a natural history museum and zoo on the orders of the governor of Cape Colony, Sir George Grey. When he travelled across South Africa on hunting and animal collecting trips, he noted that the Baharutse community near what is now Zeerust, North West Province, killed jackals to makes karosses, which were "neatly fashioned and sewn, some of which they barter with the traders for cattle."[26] He said that the jackal skins are prepared with care when they were intended to be used for bedding or karosses. He added that some local peoples would boil the skulls of jackals to extract fat from them for use in traditional medicine.[27] When he travelled further north towards present-day Zambia and Barotseland, he said that jackals were very numerous because the Makololo people didn't hunt them as some of the San and Tswana did.[28]

In 1834, many of the descendants of the Dutch settlers, who became known as the trekboers, protested against the abolition of slavery in the British-ruled Cape by undertaking a mass exodus north and east, taking their herds and flocks with them. Meanwhile, British settlers in the Western Cape and around Grahamstown in the Eastern Cape were spreading the settler farming economy, in particular, the rearing of sheep for wool. The latter increased after the introduction of purebred wool-bearing breeds – following the success of their introduction in Australia in 1797 and subsequent profitable export of wool back to Britain.[29] In the 1830s, the necessities of extensive wool production led to the importing of purebred merino sheep on a massive scale. This led to the successful expansion of meat and wool production, helped substantially by the policies of the British colonial administration, which made the Cape economy

> part of an international network of colonial possessions ... There was a wool boom in 1853 ... The need to nurture wool farmers at this time was extremely important because by 1872 the ever-increasing wool exports had

peaked at the huge sum of £3 million [about £187m in modern purchasing power] ... Within a few short decades, woolled sheep were the mainstay of the Cape economy and government protected and supported this industry assiduously.[30]

Walker et al. believe that introduction of merino breed of sheep was a very significant factor in changes in the ecology of the Cape and the accelerated loss of habitat and wildlife.[31] While it took some time before their numbers exceeded those of the fat-tailed indigenous sheep, by 1855, three-quarters of the 6.5 million sheep in the Cape Colony were merino, and by 1891, they numbered 12 million.[32] Beinart detailed the value of valuable wool exports: "Wool exports increased from about 500 000 lb (227,000 kg) in 1838, the first year of reporting, worth £27,000 to 12 million lb in 1855 (£634,000) and 40 million lb in 1875 (£2,855,000)."[33]

The massive expansion of sheep farming was largely responsible for the progressive decline in the huge springbok herds which had survived earlier trekboer and settler hunting, and the ending of the annual migration of hundreds of thousands of them from the Northern Cape and southern Kalahari to the Karoo. They were shot in huge numbers, as were the lions, wild dogs, and other predators preying on them.[34] After the extermination of the springbok and other ungulates, and the wiping out of most large carnivores, jackals survived and thrived on the small mammals, birds, reptiles, and insects remaining, supplemented by killing or scavenging the carcasses of the growing number of sheep being raised.[35] The massive growth of sheep farming across the Western Cape, Karoo, and Eastern Cape initiated the war of attrition against the jackal there that continued almost to the present. It reduced or limited the increase in jackal numbers but failed to exterminate them. The wool industry thrived during the century and, by 1846, accounted for about 60% of Cape exports.[36]

To protect sheep from predation, bounties were paid for the proven killing of predators. In 1814, jackals were rated in the bounty system at 1 rixdaler, compared with 20 for a hyena and 25 for a leopard. In 1815, the year after the British (formally) took over the Cape Colony, the official bounty for a jackal was one twenty-fifth of that for a leopard, but by the end of the century, it had risen to 70% of the bounty for a leopard.[37] Up to 1830, after which they were exterminated, many farmers in the Eastern Cape near Graaf-Reinet thought that African wild dogs were the greatest threat to their sheep rather than jackals.[38] But as the larger predators were killed off or, like leopard, severely reduced in numbers, it was jackals that became Public Enemy No. 1 for sheep farmers.[39] By 1865, a third of the Cape settler population lived in sheep-farming districts, most earning their livings directly or indirectly from sheep. The massive importance of the sheep farming goes some way to explaining the animosity to the jackal, as the chief surviving predator. Sheep and other livestock, long present in the jackal diet, had increased in importance as a food source as settlers exterminated their wild prey.

A variety of methods were used to prosecute the war on jackals – shooting, hunting with dogs, trapping, and poisoning. In the Cape, poisoning was increasingly used in the late 19th century, and in 1884, the first Wild Animal Poisoning Club was formed. The clubs operated not only on farmlands but also in forest reserves, which had been areas protected from hunting and in which jackals, caracals, and other small predators found refuge. The clubs "took it upon themselves to lay poisoned meat in the reserves and to award public bounties for animal destruction, among other initiatives."[40] When the Cape administration moved in 1888 to codify game regulations through the Game Protection Act, jackals were omitted, being counted as vermin. Following on from the limited success in removing jackals up until then, from 1889 through to 1910, a concerted vermin extermination campaign against wild carnivores, but especially jackals and caracals, was waged by smallstock farmers with the backing of the colonial state.[41] This extermination campaign was implemented alongside the drive for the expansion of farm fencing, partly as an anti-predator measure. But a downturn in the global wool market and a fall in prices led to the reduction of resources farmers had available for anti-predator measures and sheep farmers put pressure on the government to provide financial incentives to kill jackals and other stock raiders. In reality, though, livestock losses to predation and theft were low, usually below loss to disease, amounting to 2.3% of sheep and 1.7% of goats in 1893–94, though in their attempts to get government aid in fighting the jackal threat, farmers claimed that losses were about 10% from predation alone and cost the livestock industry £1.6 million a year.[42]

In addition to the poisoning clubs, hunt clubs were also established to pursue and kill jackals and foxes – the Cape Hunt Club used foxhounds imported from Britain, with some farmers keeping their own pack of hounds. Some used crosses between mongrels and the large hunting and livestock protection dogs known as boerboel or boer dogs, bred from mastiffs, with terriers employed to dig jackal pups out of dens – on occasions farmers would blow up deep dens with dynamite. The numbers of dogs kept in the Cape rose from 300,000 in 1891 to 390,000 in 1904, with much of the increase in farming areas, with dogs bred to live alongside sheep and develop a bond with them, in the way that conservationists now encourage farmers afraid of livestock predation to use guarding dog breeds, such as Anatolian shepherd dogs, to live with their flocks of sheep/goats or herds of cattle to protect them from predators.[43] Though the Cape settlers were not interested in the conservation side of using dogs to deter attacks, they wanted their dogs also to kill jackals. But to some farmers, dogs were almost as much of an enemy as jackals and were blamed for stock killing, especially after the abandonment of large numbers of farm dogs during the South African War. Left to their own devices, dogs would willingly kill sheep and goats.

Trapping was used on many farms but seemed to be more successful with leopards and caracals, rather than jackals, which were too cautious, one might say cunning, to be caught in large enough numbers to reduce the population and so the potential threat. Traps were most successful in areas that had vermin-proof fences, with predators often contained within the fences and more susceptible to the allure of baits in traps as they could not forage so widely.[44] But poison, and so the poisoning clubs, remained the weapon of choice for most farmers, as it did not involve time-consuming hunts with dogs or regular checking of traps. The government made strychnine available at cost price through poisoning clubs and from local magistrates, but measures were taken to ensure it did not get into the hands of Khoikhoi or Xhosa herders or any other "untrustworthy person," according to the select committee on the destruction of vermin.[45] The clubs would try to coordinate the use of strychnine in their areas over specific periods to have maximum effect on the local jackal population. Bounties were also distributed through the clubs and became a subsidy for mass poisoning.[46] This made it a cheap, easy answer.

Rich settler farmers had used strychnine since the 1850s, well ahead of the establishment of the poisoning clubs in the 1880s.[47] Strychnine did have drawbacks, as jackals have fine-tuned sensory systems and could detect the bitter taste of the poison. Their developed ability to regurgitate food (similar to wild dogs, coyotes, and wolves) carried back to dens/burrows for their pups enabled them to regurgitate meat laced with strychnine and so, often, avoid death when feeding on a poisoned carcass or bait. Van der Merwe believed that jackals would learn, if they had thrown up a poisoned bait or section of carcass, to avoid poisoned baits, and this learning process produced warier jackals who would be less tempted to eat what could be poisoned meat, with only the less wary, old, or starving jackals eating the meat and so ensuring the survival of the fittest jackals that would be ever harder to poison.[48] This meant farmers had to find ways of delivering the poison so that it would be swallowed without the jackal experiencing the bitterness – such as having the poison in pill form of encasing it in animal fat.[49] Poison also had the drawback of potentially killing raptors, foxes, mongooses, and other small predators that had a beneficial effect by keeping rodents and insect pests in check.

According to Van Sittert, farmers wanted monetary incentives to kill wild animals and weren't interested in compensation schemes that did not involve killing predators.[50] What they didn't realise at the time was that if they exterminated jackals and other small carnivores, then other species such as rats, mice, springhares, baboons, queleas, and locusts would all increase their effect as agricultural pests once they were no longer prey of mesopredators – and these were included on Cape government lists of agricultural pests.[51] One farmer in the Northern Cape told the Cape authorities in 1899 that where jackals had been removed or heavily depleted, the mice they had preyed on had increased and "the whole veld for six miles had been utterly destroyed by the mice" eating the roots of grasses and other plants.[52] Van der Merwe, in his comparison of the coyote and black-backed jackal, said that the farmers on whose testimony he based much of his study had said that when they exterminated jackals on their farms, the numbers of small ungulates, hyraxes, hares, and other rodents increased quickly, indicating that even on sheep farms jackals had a wide diet and were not solely dependent on killing sheep or scavenging their remains.[53] There was little understanding at this time of trophic cascades and so of the effect of removing predators – the removal of large predators and most medium to large ungulates had made the jackals the apex predators, and the removal of jackals would mean a massive increase in smaller mammals, birds, and insects.[54]

The period also saw the growth of itinerant pest controllers who would poison, shoot, or hunt jackals with dogs and live on the proceeds of the bounty given for the production of proof that a jackal had been killed. This included poor members of the Khoikhoi, Xhosa, and Coloured (mixed-race) communities.[55] That bounties could be a source of livelihood is showing by the number of recorded animals killed and tails of which handed in to obtain the payment – with 350,228 jackals killed and bounty claimed between 1889 and 1908. The number each year varied from 1,512 in 1889–90 to a massive 60,863 in 1898–99. The next nearest bounty payments were 103,321 separate payments for baboons between 1889 and 1908 and 23,028 for caracals in the same period;[56] of course, far more jackals and other wildlife referred to as vermin were killed than the official figures show, given that not all were reported as part of a bounty claim and many that were poisoned would not have been recovered by farmers or vermin hunters. But stock farmers then began to complain about how the bounty was making farm labourers too independent by giving them a source of cash income that lessened their desire to work for low wages on farms. Farmers referred to "a great many loafers, who, instead of going to work, go and try to get jackals' tails," accusing them too of stealing from farms when they were killing jackals.[57] Beinart gave the example of one African hunter in Vryburg district in the Northern Cape, who killed 655 jackals, 32 wild dogs, 62 caracals, 17 baboons, and 5 leopards in one year.[58]

While jackals were the main enemy of the smallstock farmer, bounty rates were highest for wild dog (16 points), leopard (16 points), cheetah (16 points), and both hyena and caracal (8 points), with the jackal on 4 points. In 1899, this points system changed to one based on a cash payment, with jackals priced at seven shillings in 1896 but jumping to ten shillings in 1903.[59] The amount paid out in bounty for all predators rose from a bit over £1,000 in the early 1890s to £28,000 in 1898–99 (equivalent to £3.14m); in the latter year, 50,000 separate rewards were paid out for the production of jackal tails.[60] The payment of bounties led to the decline in the poisoning clubs as the desire to get rid of jackals and the lure of cash payments were incentive enough. So much of an incentive, indeed, that fraud was believed to be occurring on a large scale. Van Sittert discovered that there was a strong suspicion at one stage that traders from German South West Africa (Namibia) and Bechuanaland (Botswana), where jackals were plentiful, were bringing jackal tails into the Cape to claim the bounties.[61] He added that "it was clear that vermin were being 'farmed' in the northwest and extensively traded everywhere else, including imports from neighbouring colonies and even abroad to supply the bounty."[62] By "farmed," one must presume that in some areas, jackals were allowed to increase in numbers so that they could be culled regularly to supply tails with which to claim the bounty. To combat fraud, in 1899, the scalp and ears had to be produced along with the tail and, in 1903, the whole skin including ears and tail.

According to the records and testimony of farmers at the time as recorded by Beinart, the various extermination campaigns not only failed to get rid of the threat farmers believed that jackals posed, but the range of jackals also appeared to be spreading, and in areas with mainly cattle farming, little was done to combat the presence of jackals as they had little effect on cattle herds and even with newborn calves were more likely to eat the placenta than attack the calf. On arable farms, jackals were often welcome as they kept rodents and insect pests under control.[63] With the availability of these farming areas with little persecution, eradication in the Cape was never likely to succeed. Sheep-farming areas might have fewer resident jackals at the height of control measure and become sinks for jackals dispersing from "safe areas" – sinks being areas where hunting or other direct or indirect forms of anthropogenic mortality of animals can create the space by removing animals for others to move in when dispersing from natal territories or well-populated regions.[64]

The ability to both evade persecution and disperse into areas where killing had reduced jackal numbers made the expectation of a long-term solution to the jackal problem, as farmers would have termed it, unlikely in the extreme. There was also a problem in many areas of the Cape that the progressive removal of wild ungulate species, not to mention the negative effect of the total disappearance of elephants and rhino, on keeping thick bush down, enabled jackals to have plenty of cover in areas used as grazing for sheep, which enabled them to evade hunters and dogs. Jackals were scarce in the wetter, hillier areas of the Eastern Cape bordering Natal, where Xhosa communities concentrated more on cattle and crop growing.[65]

Fencing and the clear demarcation of farm boundaries increased in the last decade of the 19th century, largely as a result of the need to manage

> the massive increase in domestic livestock populations without any prospect of further significant new land being added to the colony ... The old system of kraals (corrals), herds and pounds that had policed the boundaries of private property in the unenclosed countryside of an expanding colony, was simply no longer adequate to the task

with over 26 million sheep, goats, cattle, horses, and donkeys kept by Cape farmers by 1891, up from 13 million in 1865.[66] The protective enclosure of livestock in kraals overnight and the crowding involved were believed also to be a cause of soil erosion.[67] There were also increased levels of disease in livestock. This led to a shift towards erecting jackal-proof fences around farms, and state subsidies were redirected to this and away from sponsored bounties.[68] The relative failure of attempts from the 1880s to exterminate jackals through bounties and the distribution of subsidised strychnine did lead to enthusiasm for "'fencing the jackal' out around the turn of the century, leading to the erection of 'vermin-proof' fences made of closely strung barbed wire"; the greater use of fencing, it was hoped, would also stop stock theft by "two legged jackals."[69]

These supposedly vermin-proof fences were expensive, and many farmers could not afford the miles of such fencing needed to try to keep out jackals and make extermination within fenced areas easier. Fencing, like the other strategies, was by no means a foolproof deterrent to jackals. Flooding could wash fences away, and snow on higher ground in cold winters could pile up and enable jackals and other predators to cross the fences, and "[w]ild animals were another threat to the integrity of fence lines, antbears [aardvarks] tunnelling under them and baboons lifting them to create ingress for jackals and other stock predators."[70] Many farmers or rural residents hated the fences and would deliberately tear them down. The ripping up of fence posts and cutting of wire got worse during the South African War of 1899–1902, when the British army put up more barbed-wire fences to restrict the movements of Boer guerrilla groups. The mobile warfare that moved back and forth across northern areas of the Cape saw the destruction of new and old fences by Boer guerrilla units but also British troops pursuing them.[71] The three and a half years of combat would have left some farms vacant, led to the destruction of fences and reduction in anti-predator measures, and no doubt provided opportunities for wily operators like jackals to raid livestock and feed from animal carcasses and even human bodies left on the veldt during insurgency and counter-insurgency sweeps. During the war, the limited effects of extermination campaigns against the jackal were reversed in some areas as the fast-breeding and dispersing jackals recolonised now unprotected farms from which they had previously been excluded. After the war, as will be seen, large-scale persecution resumed.

At the end of the 19th and start of the 20th centuries, there was increasing interest in the study of the natural history of the jackals, as an aid to anti-predator measures. But all too often, the accounts of jackal diet, hunting, and general behaviour were drawn from anecdotal sources rather than any form of scientific observation or study of scats, gut contents, or other ways of determining what they ate. Even reputable works on the Cape's natural history followed this general pattern.[72] They generally give accounts received from farmers, who had reason to paint jackals in the worst light.

Fitzsimons, who wrote one of the first major studies of South Africa's wildlife, described the black-backed jackal in South Africa as "very widespread, and in the bush districts it abounds in spite of the most strenuous efforts of farmers to exterminate it,"[73] a conclusion that was correct and prescient about the future of attempts to rid farms of jackals. It was said to be nocturnal in its habits in areas with a settler population but more diurnal in areas far from European habitations. The difference in behaviour a result of the jackal long since learning "to dread the white man with his gun."[74] Fitzsimons called the jackal "a cowardly, treacherous and secretive animal"[75] – on what this is based, other than the prejudices of white farmers, is not clear and is hardly a scientific observation in what is supposed to be a serious examination of South African wildlife. He added that when run down by dogs, the jackal offered little resistance, but then it is described as snapping viciously if attacked by a group of small dogs. Nattrass et al. rightly critique his description of the jackal as chiefly a scavenger that became "a special-ist predator of colonial live-stock," though noting he did accept that the colonial extermination of springbok and other small ungulates played no small part in changes in diet.[76] Fitzsimons came up with the brutal conclusion that "the jackal seeks to make a living by helping himself to the colonists' domestic animals" and that with the arrival of settlers and the end of the jackal's role in clearing up wild game carcasses, "the jackal is no longer needed, and consequently becomes one of those animals which sentence of death must be pronounced upon."[77] He recommended ways of laying poisoned baits that would overcome the caution of the jackal. He also suggested that increased human settlement and the clearing of thorny scrub would help in the eventual demise of the jackal by depriving it of cover. It is emblematic of the views of wildlife at the time that it was not valued in its own right but valued as something to be exploited or persecuted as a threat to human wealth accumulation and that the extermination of a species could be presented in a work on natural history as desirable. Fitzsimons gave little space to the side-striped jackal and admitted he had not seen one in the wild.[78]

In eastern South Africa, British had established an outpost on the Indian Ocean coast, initially to trade with the expanding Zulu kingdom, which was named Port Natal. The Zulu making, Shaka, granted the British an era of land to establish a port and to provide land for settlers. The area that became known as Natal was annexed by Britain to prevent settlement by the trekboers who had left the Cape in the 1830s and moved east and north-east from the Cape Col-ony. The Zulu and other Nguni peoples of the region were engaged in mixed farming – growing crops and keeping cattle. With a wetter climate, less open grassland, deeply cleft valleys in the region were not suitable for sheep farming and also had a substantial large predator population, which would be less easy to wipe out than in the open country of the Cape, though lions were exterminated in British-controlled Natal (though not neighbouring Zululand) by 1865.[79] The defeat of the Zulu kingdom in the 1879 war with the British and the annexation in 1897 of the country by Britain led to increase in settlement and the depletion of the until-then-abundant wildlife, though black-backed jackals remained com-mon and side-striped jackals survived in the north of what had been Zululand. Mackenzie noted that in Natal and eastern Transvaal, most large game had been exterminated in the 1860s. The Boer settlers, he wrote, had "shot out all the an-imals of north-eastern Transvaal for their skins"[80] and to clear the land for cattle and crops. The carcasses had often just been left for predators, like jackals, which would have benefitted in the short term but would have been deprived of a major source of food in the long term, forcing them to adapt to greater predation on smallstock and scavenging cattle carcasses – this gave them a bad name among Boer stock farmers, though persecution was not as organised or as extreme as in the Cape sheep-farming areas.

In north-eastern Transvaal, the Sabi Reserve was formally brought into being along the border with Portuguese Mozambique in 1902, with James Stevenson-Hamilton as the game warden. He served from 1902 to 1946 as warden of the

Sabi and then also the adjacent Shingwedzi Reserve, which in 1926 were joined to become the Kruger National Park. On his first tour of the reserve, Stevenson-Hamilton was disappointed to see very little large game and saw it as his task to turn the area "from a hunter's paradise into an inviolable game sanctuary."[81] He recounted that on this tour in 1902, "it was not until the fourth day ... that we came across a few tracks of zebra, waterbuck and impala. The following morning I saw, in the flesh, a reedbuck ewe, a duiker and two jackals and in the evening was much heartened by the appearance of a herd of nearly thirty impala."[82] Stevenson-Hamilton called for the expansion of Sabi and the creation of the Shingwedzi Reserve to the north. In correspondence with the Society for the Preservation of the Wild Fauna of the Empire, he detailed the wildlife in the reserves and merely referred to jackals (species not specified) as numerous.[83] In the Pongolo Game Reserve in KwaZulu-Natal, Stevenson-Hamilton noted that jackals were common, and there were few other large carnivores, with wild dogs only present in small numbers and no lions or hyenas.[84] But right from the start of his wardenship, partly to placate local farmers who saw the Sabi Reserve as a sanctuary for stock-killing predators such as lions, leopards, wild dogs, hyenas, and jackals, and partly to preserve numbers of large antelopes, which were seen as the main reason to have a protected area, Stevenson-Hamilton oversaw an annual cull of predators, which went on for decades, even after it became clear that tourists visiting the national park from 1927 onwards wanted to see lions and other carnivores. The predators were described as vermin – the list of verminous animals included lions, leopards, cheetahs, wild dogs, crocodiles, jackals, hyenas, and avian raptors.[85] In 1905, a report published in the *Journal* noted that 230 carnivores, crocodiles, and raptors were killed to protect the game – this included 51 jackals.[86] In his report for 1903–04, Stevenson-Hamilton said there had been "a considerable decrease in wild dogs, hyaenas, and jackals" but not in lion numbers, with no reason for the increase suggested. The decades of culling kept carnivore numbers from increasing progressively but did not exterminate them, especially not the resilient jackals. In addition to hunting, they were affected by canine diseases spread by free-running domestic dogs – distemper reducing side-striped jackal numbers in the 1920s and 1930s, with black-backed jackals surviving better and encroaching on areas previously inhabited by the side-striped.[87] A former Kruger game warden and zoologist, Smuts, recorded that between 1903 and 1927, 18,428 mammalian carnivores were culled by the authorities in Kruger, of which 3,133 were jackals and 1,272 lions.[88] But, again, as in the Cape, the black-backed jackal bounced back, and Smuts said that in the 1960s and 1970s, it was very common, though he rarely saw side-striped jackals.[89]

In the Cape, the war against the jackal was renewed in the opening decades of the 20th century. The Cape government pushed hard for livestock farmers to continue to erect fences and to repair those torn down during the war. The pro-fence farmers' lobby – prosperous farmers with extensive landholdings – pushed through an amendment to the existing Fencing Act to convert part of the funding for bounties paid for killing vermin to be put aside to give loans for the building of jackal-proof fences. They complained of what they called "the injustice farmers suffered who were erecting jackal fences, as they had to bear the whole expense, and some men were getting their farms fenced in without cost."[90] But the Cape government in the new Union of South Africa, formed post-war in 1910, reintroduced bounties for jackals and other carnivores, which then stayed in place until 1956;[91] this was despite the fact that poisoning was not seriously combating the jackal presence, between 1889 and 1908, 350,228 jackals had been killed but the problem remained.[92] Magistrates paid out the bounties on provision of proof (in the form of skins or tails) of exterminated vermin. The bounty, poisoning, and fencing programmes, as Van Sittert noted, were driven by "the economics of profit and loss" and bore no relation to any form of environmental concern and were primarily to ensure profits for farmers and income for the state from commerce, chiefly exports of wool.[93] At the start of World War I, to encourage continued wool production, the Cape government started providing subsidies for keeping hunting dogs for vermin control – and hunting clubs around Cradock in the Eastern Cape killed 2,300 black-backed jackals between 1910 and 1924, again without eradicating them from the area.[94] Between 1914 and 1923, a total of 317,787 jackal bounties were claimed in the Cape, though this figure declined gradually over the next few decades, according to Beinart.[95]

The extermination of vermin, with jackals at the top of the list, became compulsory in the Cape in 1917 and remained law until 1957, but without "solving" the jackal problem.[96] Vermin hunting clubs expanded in numbers across the Cape and "were also endowed with the martial law authority of the militia to override the constraints of private property."[97] Farmers who failed to kill vermin on their land could be fined. Between 1917 and 1957, £900,000 (£38m in today's values) was paid from public revenues for more than 6.1 million animal proofs over the four decades.[98] In 1956, the last year of the payment of jackal bounties, bounties were paid for 20,084 jackals, 219,322 hyrax, 15,323 Cape foxes, 8,478 African wild cats, 7,012 baboons, 5,640 crows, 3,408 caracals, 814 mongooses, 359 porcupines, 153 eagles, 121 aardwolves, 99 otters, 90 leopards, and 40 honey badgers.[99] The bounties were ended at the end of that year, as it became evident that they weren't working, that the expansion of predator-proof fences in the interwar years and the growing concern over wildlife conservation rendered the bounties unwanted and, among some, unpopular. The Cape government formed a nature conservation department in 1952, and its first head, Douglas Hey, began to move policy and, to an extent, public opinion away from endless attempts at extermination.[100] Hey appointed a commission to examine the bounties and recommended they be ended, which followed at the end of 1956. Farmers still killed jackals, and some government money subsidised employment of hunters of hunting with hounds – and there were still 110 hunt clubs with hounds in the early 1960s.[101] Farmers were still using poison and obnoxious explosive devices called getters – based on US coyote-getters – baited devices that exploded when bitten and killed or maimed the jackals.[102]

In Southern Rhodesia (Zimbabwe from 1980), wildlife was diverse and widely distributed. As in parts of South Africa, under the colonial rule through the British South Africa Company of Cecil Rhodes, where large cattle and other livestock farms were established, ungulates competing for grazing and water and predators that were potential predators were exterminated or reduced in number. As elsewhere in Africa, jackals (black-backed in the south and west and side-striped in the north and in woodland areas) largely survived the killing of predators. But it was to be campaigns against sleeping sickness and rabies that had the greatest effect on destroying wildlife, with the rabies campaigns killing thousands of jackals. Wildlife policy under company rule and then self-government allowed for the mass slaughter of wildlife for a variety of reasons, including livestock protection, land clearance, rabies prevention, and tsetse fly eradication. Between 1899 and 1912, thousands of predators, jackals, and other wildlife, as well over 100,000 domestic/feral dogs, were killed to prevent the spread of rabies.[103] Over 1 million game animals were killed on and around European farms to create buffer zones to prevent tsetse fly.[104] Jackals were viewed by farmers and the colonial government "as unmitigated nuisances in any stock country," and farmers were encouraged to exterminate them on white-owned farmlands.[105] The only areas unaffected by persecution were reserves or game sanctuaries, with the first established in Matopos in 1926. Over the next seven years, land in Wankie (Hwange), Victoria Falls, and Kazuma Pan was set aside as game reserves, with Matopos becoming a national park.[106] In these areas, hunting was prohibited or severely limited, and the tsetse eradication measures were not applied. The numbers and distribution of protected areas increased in the 1930s, 1940s, and into the 1950s. A national parks board was set up in 1949, and there was an increase in safari tourism involving white settlers, visitors from South Africa, and some from Europe and North America.

Horn of Africa and Sudan

The British explorer and colonial official in Sudan, Wilfred Thesiger, recounted seeing jackals regularly on an expedition in the Danakil Depression in Ethiopia in 1930. He didn't like jackals and hyenas hanging around his camp and shot a jackal with his shotgun. Later in his trip, he reported seeing "a large sandy-coloured jackal, a wolf-like brute and very mangy" among the rocks and found the carcass of a kid, an adult goat, a dead but freshly killed crocodile about 2ft long, and several dead flamingos.[107] Black-backed jackals, side-striped jackals, and North African wolves were common in French, British, and Italian Somaliland during this period but rarely rated a mention in reports on wildlife or hunting.[108] In his survey of the mammals of the region, Drake-Brockman recorded the presence of what he called the grey jackal (*Canis variegatus*). One can speculate that this refers to the golden wolf rather than the side-striped jackal, which is present but not common in Somalia. The grey jackal is described by him as common, especially in the region of Somaliland between Hargeisa and Berbera.

He described it as noisy and aggressive towards hyenas, trying to bite them to get them to go away. But then he said, "[T]hey are very cowardly." He recorded that the black-backed jackal was common across Somaliland and known for attacking goats and young sheep.[109]

Reports by the British colonial authorities in Sudan noted the presence of the golden wolf, which they thought was the "eastern representative of the Senegal jackal" (*Canis lupaster anthus*) but is the separate sub-species, the North Africa golden wolf (*Canis lupaster lupaster*).[110] The Game Preservation Department there viewed jackals as vermin and a threat to agriculture, noting in a report of 1907 that "[a] considerable amount of damage to melon crops seems done by jackals and hyenas in the Halfa Province. Both these animals are largely frugivorous when opportunity offers."[111] In the Darfur region, a visiting hunter reported that the British colonial administration had established a pack of hounds to help horsemen hunt jackals and hyenas for sport and because they were seen as vermin.[112] In the 19th and early 20th centuries, jackals (most likely North African golden wolves or Equatorial side-striped jackals), both species of hyena, lions, and leopards were common across Darfur and parts of northern Sudan.[113] They were progressively exterminated using shooting, trapping, and poisoning, though jackals and hyenas remained when most other large predators were killed.[114]

There is a huge gap in accounts of jackals and their coexistence or conflict with humans in North, Central, and West Africa during this period, and the trail of those populations will be picked up again in the next chapter.

Notes

1 K. Kittenberger (1989) *Big Game Hunting and Collecting in East Africa, 1903-1926*, Reprinted by St Martin's Press, New York, p. 268.

2 Ibid., p. 267. See also A. Chapman (1908) *On Safari: Big Game Hunting in British East Africa*. Longmans, Kindle Edition.– loc 1811.

3 Extract from Sir Charles Eliot's Reports of the British East Africa protectorate for the Years 1902 and 103, *Journal of the Society for the Preservation of the Wild Fauna of the Empire (henceforth, Journal)*, 1904, pp. 49–54, p. 53.

4 E. Huxley (1981) *The Mottled Lizard*, Harmondsworth, Middx: Penguin, pp. 90–1 and 93.

5 E. Huxley (1980) *White Man's Country. Lord Delamere and the Making of Kenya*, London: Chatto and Windus, p. 101.

6 A. Blayney Percival (1924) *A Game Ranger's Note Book*, Read Books Ltd. Kindle Edition, loc 2260.

7 G. Adamson (1968) *Bwana Game*, London: Collins, pp. 74–6.

8 Captain G.C. Shortridge (1934) *The Mammals of South West Africa*, London: William Heinemann, p. 167.

9 J.C.B. Statham (1922) *Through Angola, a Coming Colony*, Edinburgh: William Blackwood, pp. 66 and 160.

10 J. Chapman (1858) *Travels in the Interior of South Africa: Comprising Fifteen Years' Hunting and Trading; with Journeys across the Continent from Natal to Walvis Bay, and ... Lake Ngami and the Victoria Falls Volume 1*, London: Bell & Daldy. Kindle Edition via Hard Press 2017, loc 582.

11 Ibid., p. 971.

12 G. Thompson (1967) *Travels and Adventures in Southern Africa, Vol 1*, Cape Town: Van Rieebeck Society, p. 86.
13 H.A. Bryden (1893) *Gun and Camera in Southern Africa*, London: Edward Stanford, reprinted by Book on Demand Ltd in 2013, p. 62.
14 J.M. Mackenzie (1988) *The Empire of Nature. Hunting Conservation and British Imperialism*, Manchester: University of Manchester Press, p. 67.
15 On Mammals Collected in 1923 by Capt. G.C. Shortridge during the Percy Sladen and Kaffrarian Museum Expedition to South-West Africa, as reported to ZSL by Oldfield Thomas, *Proceedings of the Zoological Society of London (Henceforth Proceedings)*, 1925, 95, 1, p. 221.
16 Shortridge, 1934, p. 167.
17 Guggisberg, 1961, pp. 38–9.
18 S.J. Barrow (1802a) *An Account of Travels into the Interior of Southern Africa in the Years 1799 and 1798*, London: G.F. Hopkins, Kindle edition, loc 154–60.
19 S.J. Barrow (1802b) *Travels into the Interior of Southern Africa: In Which Are Described the Character and the Condition of the Dutch Colonists of the Cape of Good Hope, and ... in the Animal, Mineral and Vegetable*, London: T. Cadell and W. Davies, Kindle Edition, loc 968.
20 Ibid., loc 1538.
21 Ibid., loc 5631.
22 H.A. Bryden (1897) *Nature and Sport in South Africa*, London: Chapman and Hall; reprinted by Forgotten Books, London, p. 111.
23 Bryden, 1897, pp. 118–9.
24 J. Chapman (1971 from 1868 original) *Travels in the Interior of South Africa 1849–1863. Hunting and Trading Journeys from Natal to Walvis Bay, and Visits to Lake Ngami and Victoria Falls, Vol 1*. Cape Town: A.A. Balkema, p. 71.
25 F. Morton & R. Hitchcock (2014) Tswana hunting: continuities and changes in the Transvaal and Kalahari after 1600, *South African Historical Journal*, 66(3), 418–39, p. 433.
26 Chapman, 1858, loc 341.
27 Ibid., loc 1009.
28 Ibid., loc 2625.
29 National Museum of Australia (no date) merino sheep introduced, https://www.nma.gov.au/defining-moments/resources/merino-sheep-introduced#:~:-text=In%201797%20the%20first%20merino%20sheep%20were%20landed%20in%20Australia.&text=The%20first%20Australian%2Dproduced%20fleece,had%20become%20Australia's%20major%20export accessed 5 April 2021.
30 Ibid., pp. 36–7.
31 Walker et al., 2018, p. 164.
32 Ibid.
33 W. Beinart (2018) An overview of themes in the agrarian and environmental history of the Karoo since c.1800. *African Journal of Range & Forage Science*, 35(3–4), 191–202, p. 192.
34 W. Beinart & L. Hughes (2007) *Environment and Empire*, Oxford: Oxford University Press, pp. 71–2.
35 Ibid., p. 72.
36 G.M. Theal (1915) *History of South Africa since 1795, Vol II*, London: Allen & Unwin, pp. 43 and 207.
37 N. Nattrass, M. Drouilly & M. Justin O'Riain (2020) Learning from science and history about black-backed jackals *Canis mesomelas* and their conflict with sheep farmers in South Africa, *Mammal Review*, 50, 101–11, p. 103.
38 N. Nattrass & B. Conradie (2015) Jackal narratives: predator control and contested ecologies in the Karoo, South Africa, *Journal of Southern African Studies*, 41, 4, 753–71, p. 755.
39 Cited by Beinart, 2003, p. 58.
40 D.L. Ogada (2014) The power of poison: pesticide poisoning of Africa's wildlife, *Annals of the New York Academy of Sciences Issue: The Year in Ecology and Conservation Biology*, 1322, 1–20, pp. 1–2.
41 L. van Sittert (1998) "Keeping the enemy at bay": the extermination of wild carnivora in the Cape Colony, 1889- 1910, *Environmental History*, 3, 333–56, p. 335.

42 Ibid., p. 337.
43 L. Marker (2016) *Livestock Guarding Dogs Protect Cheetahs and Wolves, Too!*, Cheetah Conservation Fund, 24 September 2016, https://cheetah.org/ccf-blog/livestock-guarding-dogs/livestock-guarding-dogs-protect-cheetahs-and-wolves-too/ accessed 8 April 2021; see also. Ruaha Carnivore project (no date) *Livestock Guarding Dogs*, https://www.ruahacarnivoreproject.com/protecting-livelihoods/guarding-dogs/ accessed 8 April 2021.
44 Van Sittert, 1998, p. 342.
45 Cape of Good Hope (1899) *Report of the Select Committee on the Destruction of Vermin, 1899*, A10–99, cited by Van Sittert, 1998, p. 342.
46 Van Sittert, 1998, p. 344.
47 Beinart, 2003, p. 209.
48 N.J. van der Merwe (1953) The Jackal, *Flora and Fauna*, 4, pp. 6–7.
49 Ibid., p. 215.
50 Ibid.
51 Ibid., p. 338.
52 Cape of Good Hope (1899) *Report of the Select Committee on the Destruction of Vermin, 1904*, A9–1899, p. 17.
53 N.J. van der Merwe (1953) The coyote and the black-backed jackal: a comparison of certain similar characteristics, *Fauna and Flora*, 3, 45–51.
54 For the role of mesopredators in ecosystems, see – L.R. Prugh et al. (2009) The rise of the mesopredator, *BioScience*, 59, 9, 779–91, p. 780.
55 Cape of Good Hope (1896) *Report of the Select Committee on the Destruction of Vermin, 1896*, A10–96, cited by Van Sittert, 1998, p. 344.
56 Figures collated from official Cape administration reports for the period by Van Sittert, 1998, p. 343.
57 Cape of Good Hope, 1904, A–21904.
58 Beinart, 2003, p. 211.
59 *Agricultural Journal of the Cape of Good Hope*, from 1898 to 1908, cited by Van Sittert, 1998, p. 345.
60 Cape of Good Hope, 1899, A9–1899, p. 8.
61 Van Sittert, 1998, p. 346.
62 Van Sittert, 2005, pp. 282–3.
63 Beinart, 2003, p. 211.
64 See A.J. Loveridge et al. (2016) The landscape of anthropogenic mortality: how African lions respond to spatial variation in risk, *Journal of Applied Ecology*, 54, 815–25, pp. 822–3.
65 Beinart, 2003, p. 213.
66 L. van Sitter (2002) Holding the line: the rural enclosure movement in the Cape Colony 1865-1910, *Journal of African History*, 43, 95–118, p. 98.
67 Ibid.
68 K.G.I. Behrens (2018) Ethical considerations in the management of livestock predation, in G.I.H. Kerley, S.L. Wilson & D. Balfour (eds) *Livestock Predation and its Management in South Africa: A Scientific Assessment*, Port Elizabeth: Centre for African Conservation Ecology, Nelson Mandela University, 82–105, p. 92.
69 Van Sittert, 2002, p. 98.
70 Ibid., p. 100.
71 T. Pakenham (1979) *The Boer War*, London: Weidenfeld and Nicolson, pp. 536–7.
72 F.W. Fitzsimons (1919) *The Natural History of South Africa, Volume II*, Longmans, Green & Co, pp. 92–109.
73 Fitzsimons, 1919, p. 92.
74 Ibid.
75 Ibid., p. 93.
76 N. Nattrass et al. (2017) Understanding the black-backed jackal, *CSSR Working Paper*, No. 399, Institute for Communities and Wildlife in Africa, http://cssr.uct.ac.za/pub/wp/399, accessed 4 June 2021, pp. 5–6.

77 Fitzsimons, 1919, p. 100.
78 Ibid., pp. 107–9.
79 K. Somerville (2020) *Humans and Lions Conflict, Conservation and Coexistence*, Abingdon: Routledge/Earthscan, p. 61.
80 Mackenzie, 1988, p. 110.
81 J. Carruthers (2001) *Wildlife and Warfare: The Life of James Stevenson-Hamilton*, Pietermaritzburg: University of Natal Press, p. 81.
82 J. Stevenson-Hamilton (2012) *South African Eden. From Sabi Game Reserve to Kruger National Park*, London: Penguin (originally printed in 1937), Kindle Edition, loc 446–55.
83 Major T. Stevenson-Hamilton (1905) The game preservation in the Transvaal, *Journal*, II, 1905, p. 30.
84 Ibid., p. 39.
85 Ibid., p. 107.
86 Stevenson-Hamilton, 1905, p. 40.
87 C.A. Spinage (2012) *African Ecology - Benchmarks and Historical Perspectives*, Heidelberg: Springer, pp. 50–1.
88 L. Smuts (1982) *Lion*, Johannesburg: Macmillan South Africa, p. 174.
89 Ibid., pp. 226–7.
90 L. Van Sittert, 2002, p. 105.
91 Van Sittert, 1998, p. 334.
92 Ibid., p. 352.
93 Ibid.
94 Beinart, 2003, pp. 222–3.
95 Ibid., pp. 226–7.
96 L. van Sittert (2016) Routinising genocide: the politics and practice of vermin extermination in the Cape Province c.1889–1994, *Journal of Contemporary African Studies*, 34(1), 111–28, pp. 112–3.
97 Ibid., p. 115.
98 Ibid., p. 116.
99 N. Nattrass & B. Conradie (2015) Jackal narratives: predator control and contested ecologies in the Karoo, South Africa, *Journal of Southern African Studies*, 41, 4, 753–71, pp. 758–9.
100 Van Sittert, 2016, p. 119.
101 Nattrass & Conradie, 2015, p. 759.
102 Van Sittert, 2016, p. 120.
103 R. Mutwira (1989) Southern Rhodesian wildlife policy (1890-1953): a question of condoning game slaughter? *Journal of Southern African Studies*, 15, 2, Special Issue on The Politics of Conservation in Southern Africa, Jan 1989, p. 250.
104 Ibid.
105 H.E. Horny, A note from Southern Rhodesia, *Journal*, 1945, 52, pp. 23–5.
106 Ibid., p. 258.
107 W. Thesiger (1996) *The Danakil Diary Journeys through Abyssinia, 193-34*, London: Flamingo, pp. 15 and 175.
108 Report of H.E.S. Cordeaux, Commissioner of the Somaliland Protectorate, 1901 and 1902, *Journal of the Society for the Preservation of the Wild Fauna of the Empire* (henceforth, *Journal*), 1904, pp. 71–2.
109 R.E. Drake-Brockman (1910) *The Mammals of Somaliland*, London: Hurst and Blackett, Reprinted by Kessinger Legacy Reprints, pp. 45–6.
110 *Proceedings*, Jan-Apr 1903, pp. 295–6.
111 Soudan. Report of Game Preservation Department, 1907, *Journal*, 1909, p. 118.
112 R.D.Q. Henriques (1938) *Death by Moonlight. An Account of a Darfur Journey*, London: Collins Publishers, p. 333.
113 M.B. Nimir (1983) *Wildlife Values and Management in Northern Sudan*, Dissertation submitted to for the Degree of Doctor of Philosophy Colorado State University Fort Collins, Colorado Summer 1983, pp. 40–1.
114 Ibid., p. 45.

7

BLACK-BACKED JACKALS AND RELATED SPECIES IN CONTEMPORARY AFRICA

This chapter looks at the modern distribution, habits, and relationship with humans of the jackal species and African golden wolf. Despite the limited number of studies of black-backed jackals in comparison with lions, leopards, cheetahs, and wild dogs, black-backed jackals "are the most common of the larger carnivores in sub-Saharan Africa ... In southern Africa, black-backed jackals are both abundant and widespread, particularly in the semi-arid regions."[1] The black-backed jackals of southern Africa will be a particular focus of this chapter.

For the African jackals and golden wolves, the long-term threats are habitat loss, human encroachment, direct persecution to protect livestock, or because of traditional hatred, deaths on the roads, in snares set by bushmeat hunters or poachers, from poison used to kill other predators, or transmission of diseases from domestic/feral dogs. As Kaunda warned,

> As human populations continue to encroach on wildlife habitats in sub-Saharan Africa, contact between livestock and jackals will increase. Apart from the implications of jackals' predatory activities on livestock, this trend could have significant implications as jackals are also frequently in contact with other wild carnivores. As such, they could serve as an important link in disease transmission between wild carnivores, domestic livestock, and humans.[2]

In East Africa, livestock predation by carnivores (usually lions, leopards, and spotted hyenas, but with jackals also blamed and often badly affected by trapping or poisoning aimed at the other predators) leads to retaliatory killing, alongside the threat of human encroachment because of expanding pastoralist and farming populations.[3] In these and other areas across Africa, persecution of carnivores

DOI: 10.4324/9781003199793-8

outside protected areas (PAs) can develop into a sink or edge effect, where depletion of predator populations in unprotected areas attracts predators within PAs to move into newly emptied territories, where they are vulnerable to human persecution.[4] These conclusions, derived from research in southern and East Africa, are supported by Sogbohossou et al.'s studies in Pendjari Biosphere Reserve in Benin, West Africa.[5] They highlighted the intensification of human–wildlife conflict, posing a major threat to a range of species, but particularly predators, who are subject to persecution and get killed as a by-product of poaching for meat, hides, or horn.

Poisoning, despite being illegal in many countries, remains an indiscriminate and easily deployed threat to wildlife, especially predators that also scavenge. Strychnine, cattle dip, insecticides, and pesticides are widely used to kill predators – livestock carcasses being laced with the poisons, which then kill any carnivores or raptors that feed from it.[6] As Ogada reported,

> The use of traditional poisons is waning owing to the easy availability of inexpensive, highly toxic agricultural pesticides … All classes of pesticides have been used to poison wildlife, including organochlorines, organophosphates, carbamates, and pyrethrins. Carbofuran is the most widely abused pesticide in Africa … Other commonly abused pesticides include strychnine, aldicarb, diazinon, and monocrotophos.[7]

Jackals and wolves are also killed by shooting, trapping, and hunting with dogs across their whole range.

North and West Africa

The golden wolf is present in much of North Africa, from Mauritania and Western Sahara through Morocco, Algeria, and across to Egypt, and is not believed to be absent from any of the countries in that range. It is likely that *Canis lupaster lupaster* has been present there since at least the middle to late Pleistocene. In recent years, there has been a resurgence of interest in the numbers and distribution of the wolves after the decision, detailed in earlier chapters, to call them golden wolves and not jackals. One naturalist website described the "discovery" of wolves in Algeria and Morocco, notably in the Middle Atlas Mountains, with animals photographed which showed obviously wolflike characteristics – though they had been there for millennia but had been called jackals for the last few hundred years.[8]

Some of the most important research into golden wolves in North Africa has been carried out by Liz Campbell, who has researched the wolves of the Atlas Mountains and refers to "the rediscovery of a forgotten species,"[9] reminding people that Aristotle had described reports of a wolflike animal in North Africa. They, like the golden, black-backed, and side-striped jackals, have proved hugely adaptable and able to live in a variety of habitats, with varying prey/food bases, and the ability to thrive in close proximity to humans. In the Atlas, they live in cedar and oak woodland and wetland. There is a wide range in their colour from rich brown to grey, with a grey-black back.[10] They are able to cope with the cold, snowy winters in the mountains but can also live in hotter grasslands, desert margins, agricultural land, and coastal plains. Unlike in East Africa, the Horn, and parts of West and Central Africa, the North African wolves do not have apex predators with which to compete, following the extermination of the Barbary lion and the North African leopard in the wild, leaving the wolf as the apex predator across North Africa, in some areas sharing that denomination with the striped hyena.[11] Both carnivores are subject to anthropogenic violence across North Africa, especially in Morocco where it the wolf is persecuted as a suspected livestock killer.[12] Farmers in Morocco are not compensated for losses from predation, and periodic livestock predation by wolves results in shepherds hunting them "by shooting, trapping and poisoning as preventive or retaliatory measures, and golden wolves receive no protection."[13] Golden wolves do not only take sheep, goats, and poultry but also hunt small to medium wild ungulates, including the rare Cuvier's gazelle in the Atlas and the mountains of Ait Tamlil and Anghomar.[14]

In Algeria, the golden wolf is in danger of decline, chiefly as a result of habitat loss and persecution. As in Morocco it is the apex predator, again alongside the striped hyena. It, like its jackal cousins, takes full advantage of the opportunities of living near human habitations with the opportunities to forage at waste dumps and the chance to kill sheep, goats, and poultry.[15] In a study of golden wolf diet at Tlemcen in northern Algeria, it was found that in 246 wolf scats, 34 food items were identified, including wild and domestic animals, fruits, leaves, soil, and organic waste. Animal remains represented 84.8% of biomass, with plant material the remainder. Wild boar was the most important prey animal, with considerable amounts of livestock remains (24% of the total biomass – including sheep, cattle, goats, and poultry, though with no clear sign of whether they were predated or scavenged); plant material included juniper berries and seeds, peaches, plums, and melons.[16] Interestingly, during the study by Eddine et al., no farmers reported losses of domestic animals to predation.[17] But there is a strong belief in Algeria that wolves are responsible for killing livestock, and from 2014 to 2016, 70 wolves were killed by trapping, shooting, or poisoning near vicinity of orchards and farms in the study area.[18]

In his detailed work *The Carnivores of West Africa*, Rosevear records that the golden jackal (now called the North African golden wolf, *Canis lupaster lupaster*) and the West African golden wolf (*Canis lupaster anthus*) and the Equatorial side-striped jackal (*Lupulella adustus lateralis*) are present, with the former more common and widely distributed than the latter two.[19] The North African golden wolf is distributed across the whole West African region from Cameroon and Chad west to the Atlantic and then north to Morocco and across the breadth of North Africa. The West African sub-species is limited to Senegal. Rosevear noted a wide variation in colour of wolves across their range. In his description of the side-striped jackal, he uses the name Sundevall (*Lupulella adustus adustus*) for them, though this sub-species is now recorded as being present from Gabon and DR Congo southwards and eastwards, it does not reside in West Africa, where the Equatorial side-striped jackal (*Lupulella adustus lateralis*) is represented.[20] A recent study in the Ferlo Nord Wildlife Reserve in Senegal found that West African wolves (49% of records) and Equatorial side-striped jackals (34%) were the most commonly recorded carnivores. Both species foraged alone, although the wolves lived in family groups averaging five animals, while the side-striped were usually seen in pairs.[21] Apart from striped hyenas, no other large carnivores were seen. The wolves and jackals fed on a wide diversity of mammals, birds, insects, reptiles, and vegetables/fruit materials. In the reserve they predated on the Dorcas gazelles being reintroduced there, as well as scavenging from carcasses of those which died from other causes.[22]

The Horn of Africa, Sudan, and East Africa

The Horn of Africa is home to all three of the species – with the Kaffa side-striped jackal present in Ethiopia, Somalia, parts of northern Kenya, Uganda, and South Sudan; the Equatorial side-striped jackal in Kenya, Sudan, Tanzania, and Uganda; the North African golden wolf in Djibouti, Eritrea, Ethiopia, Somalia, Sudan, and South Sudan; the East African golden wolf in Kenya and Tanzania; and the East African black-backed jackal in Djibouti, Eritrea, Ethiopia, Somalia, Sudan, South Sudan, Kenya, Tanzania, and Uganda.[23]

There is little data on the jackals and wolves in Djibouti, where wildlife surveys have been few and far between. One survey, carried out between October 1999 and January 2000,[24] noted that after a ban on hunting in 1977, pressure on wildlife eased and that the golden wolf was present there and not considered endangered, a status shared by the side-striped and black-backed jackals. As whole, the jackals and wolves were seen as the most abundant predators and occurred widely across the territory, with the golden wolf the most common of the species.[25] In Ethiopia, while the overall outlook for wildlife conservation was not encouraging, due to human expansion and the effects of long-term and widespread military conflict and of poaching, the Kaffa side-striped jackal (the rarest of the jackals/wolves in Ethiopia), the East African black-backed jackal, and the North African golden wolf were all found there and not in imminent danger of serious decline, according to a survey published in 1980.[26] It noted that the side-striped was found on Ethiopia's central plateau and that one reason for its seeming rarity was its physical similarity to the golden wolf, the most commonly occurring species, which was very widespread from Tigre down to the south-central areas (not the Ogaden Desert), and from the border with Sudan and South Sudan across to the Danakil districts. It had been recorded at altitudes of around 3500m in the Bale Mountains.[27] The black-backed jackal is also often confused with the side-striped but is again more common, with a distribution mainly in lowland areas below 1600m in northern, eastern, and southern areas, including arid bush and semi-desert.[28] The black-backed is the most common of the jackals/wolves in the Ogaden in the south.

The major threat to wolves and jackals in Ethiopia is human encroachment on wildlife habitat leading to degradation or fragmentation, also increasing the frequency of human–wildlife conflict. In the dry and relatively infertile highlands of northern Ethiopia, golden wolves, hyenas, and leopards are blamed by livestock keepers for predation on sheep, goats, cattle, donkeys, and domestic dogs – with wolves mainly blamed for killing sheep and goats.[29] Livestock keepers in the region told Yirga that they lost on average $20.2 per household per year to predation – 7% of annual income. A total of 3,122 animals were lost to predators from 2006 to 2011, with golden wolves blamed for 1,067 (34.2%).[30] This is a region with heavily depleted wild prey species, meaning, as Yirga wrote, that, "hyaena, leopard and jackal are presumably highly dependent on anthropogenic food sources. The high human density and depletion of prey base are perhaps the most important causes for human – carnivore conflict in the area."[31] In the Bale Mountains of south-east Ethiopia, livestock predation by carnivores was estimated to have killed 704 domestic animals in 3 years. Jackals (species not identified) were blamed for 16% of fatalities, behind spotted hyenas and leopards, killing mainly goats and sheep.[32] In the Bale Mountains and northern Ethiopia, most jackal/wolf attacks took place in daytime, while animals were grazing away from human settlements. Around the Chebera Churchura National Park (NP), 580km south-west of Addis Ababa, 354 households were surveyed in 2012–14 about wildlife conflict, 30.6% identified black-backed as a serious problem for livestock keepers, 57.4% thought them a minor problem, and 12.0% as no problem; 1,449 predator attacks were reported, but most of them were blamed on lions, leopards, and baboons.[33]

The result of a perception of jackals and golden wolves as inveterate livestock killers has led in many areas to shooting, spearing, snaring, and poisoning of the animals – sometimes, as elsewhere, a by-product of trying to eliminate larger carnivores or snaring for bushmeat. Another reason for the killing of the animals in parts of Ethiopia is use in traditional medicine. In Ethiopia, the use of animal body parts is widespread with the flesh of animals, bones, blood, body fat, and skins used to treat a variety of ailments. Dreje and Chane in their study in southern Ethiopia found that 46 different ailments were treated in this way with parts from 21 different animals (66.64% of them mammals).[34] The skins or body parts of black-backed jackals were among the animals used for medicine, but the tradition healers surveyed did not mention side-striped jackals, golden wolves, or honey badgers.[35]

The almost constant state of warfare in much of Somalia and South Sudan has rendered research and any estimate of mammal numbers and range, which explains this gap in the geographical examination of the presence and nature of coexistence with humans of jackals/wolves in those areas. In Sudan, there have been limited opportunities for research, and Nimir wrote his doctoral thesis on wildlife and its management in northern Sudan, noting that it had been much more diverse and abundant in the past than it was at the time of writing in 1983.[36] Human population expansion, encroachment on wildlife areas, poaching, and inadequate laws, he said, were the cause of the decline in wildlife populations and ranges.[37] Wildlife lacked aesthetic or environmental value to many Sudanese, with economic values through trade in bushmeat, ivory, and other wildlife products predominating. In his survey of wildlife in North and South Kordofan, North and South Darfur, the Blue Nile, the White Nile, the Gezira, Khartoum, Kassala, the Red Sea, the Nile, and the Northern Province, Nimir noted that black-backed jackals were locally common in areas with open woodland and savanna but doesn't refer to side-striped or golden wolves.[38] None of the species were included on the official game licences, which permitted the regulated hunting of named wildlife species. There was mention of jackals (without specifying the species) in reports of predation of wildlife, with them being blamed for some killing of sheep, goats, and poultry. In southern Darfur, 28 out of 188 reports between 1973 and 1980 were of jackals killing livestock or eating crops.[39] I was unable to trace any recent accounts of jackals or golden wolves in Sudan or their role in crop or livestock damage.

The jackals and golden wolves of Kenya and Tanzania are far more heavily researched and documented than those of the Horn and Sudan and are only second in the number of research papers to southern Africa. Fuller et al. studied the black-backed, Kaffa, and Equatorial side-striped jackals and East African golden wolves of the Kenyan Rift Valley. They came to the conclusion that while there was some interaction, black-backed jackals and golden wolves generally occupied different habitat, and competition was low.[40] Black-backed competed more with, and generally dominated, side-striped jackals, but the latter were found mainly in open acacia or euphorbia woodlands, where there were generally fewer black-backed. All of the jackals and golden wolves were more active at dusk and dawn than in the day.[41] Jackals and golden wolves are present in the major PAs of Kenya like the Mara, Tsavo, and Samburu, but the majority are still found outside PAs. They are commonly seen in the NPs, conservancy reserves, and other safari areas, where they are most often (certainly in the case of black-backed jackals) seen around lion, hyena, or other large carnivore kills but also foraging in the vicinity of camps and the park headquarters. Outside the PAs they are found on farming and rangeland areas and around towns.

One of the chief pressures on the jackal and golden wolf populations in Kenya outside PAs (there are no verified population statistics available for them nationally or regionally in Kenya) is persecution as suspected stock-killers. Areas next to PAs can act as sinks for predators with farmers killing animals that moved there from PAs, freeing territories that attract more jackals, wolves, lions, and hyenas dispersing from their natal territories in PAs. On the fringes of the Maasai Mara Reserve and Mara conservancies, large- and medium-sized predators come into conflict with pastoralists, some of whom illegally graze their cattle in PAs. The overlapping of carnivore ranges inside and outside PAs creates conflict that can threaten carnivore populations, because of the prevalence of retaliatory or pre-emptive killing by pastoralists. In 2003, Ogutu, Bhola, and Reid used playback of carnivore calls to assess numbers of black-backed jackals, spotted hyenas, and lions in the Maasai Mara Reserve and to examine effects on livestock.[42] Hyena density was 1.3 times higher on the ranches than in reserves, lion density was eight times lower on ranches than reserves, while jackal densities were almost identical.[43] The researchers set up 35 calling stations in the reserve and at Koyiaki ranch – jackals were seen at 23 (65.7%) of the stations, in answer to playbacks of hyenas mobbing lions on a kill, involved in an inter-clan fight, squabbling on a kill, and the bleats of a dying wildebeest calf.[44] Jackals responded to the calls marginally more often on the ranch than in the reserve. Estimates were given of lion and hyena numbers in the Mara region and on the ranches, but not of jackal numbers.[45]

In cattle and game farming or private safari areas outside PAs in Laikipia, north-central Kenya, there were elements of coexistence and conflict between carnivores and both large and small livestock farmers. A study by Gadd between 1999 and 2002, based on interviews with pastoralists and landowners who combined livestock with wildlife-based tourism, looked at depredations on livestock and crops. The resident predators were lions, leopards, cheetahs, wild dogs, spotted hyenas, and black-backed jackals.[46] On average, 7.5% of farmers interviewed found jackals a problem, compared with 90% for hyena and 50% for leopard. In the Nakuru Wildlife Conservancy, in central Kenya, Ogutu et al. found that local pastoralists found lions and hyenas to be the most likely to kill livestock, with little concern about jackals, perhaps because cattle rather than sheep or goats were the main stock kept.[47]

Tanzania has populations of Equatorial side-striped jackals, East African black-backed jackals, and East African golden wolves. In their book, *Innocent Killers*, Hugo and Jane van Lawick-Goodall related their observations of golden wolves (then called jackals) and black-backed jackals in Ngorongoro Crater and Serengeti NP. The wolves lived in mated pairs with their offspring and generally hunted alone. They also actively defended their territory against intruding wolves. When lions made a kill on their territory, the resident pair would not allow wolves other than family members to feed, even if there was an abundance of food.[48] The golden wolves remained in their territories during the migration, while many black-backed jackals followed the migration, usually jackals that had dispersed from their natal territories.[49] Lamprecht, in 1978, noted that black-backed jackals in the Serengeti took full advantage of the calving seasons of wildebeest and gazelles, not only killing newborn animals but also feeding on the placentas. They also fed very regularly on dung beetles and their larvae and balanites fruit. Golden wolves in the Serengeti also varied their diet to include insects and fruits.[50] Both black-backed jackals and golden wolves would forage around human settlements, such as the research and ranger camps in the Serengeti. A much later study of carnivore biodiversity in Tanzania's PAs, using 430 cameras across the main northern and some coastal NPs or reserves, recorded black-backed jackals regularly – on the camera images, they were seen 13 times across three sites, and side-striped just twice at one site, with no images of golden wolves.[51] Durant's study of the carnivore populations of Serengeti NP, Maswa, Grumeti and Ikorongo Game Reserves, Ngorongoro Conservation Area, and Loliondo Game Controlled Area found abundant evidence of black-backed jackals and some for side-striped jackals and golden wolves, with all three overlapping to a considerable extent, with no major differences apart from a preference of golden wolves for grassland and non-bushy habitats.[52]

As in Kenya, there was considerable human–carnivore conflict arising from predation of livestock – though again (and unlike South Africa) spotted hyenas, lions, and leopards were seeing by pastoralists as the chief culprits, according to Mbise's research in the eastern Serengeti.[53] Maasai pastoralists and Sonjo agro-pastoralists were surveyed about stock loss to predators – of 181 animals lost by the Maasai, 34 (18.8%) were blamed on jackals compared with 28.2% for hyenas and 20.9% for lions; of 64 lost by the Sonjo, only 6 (9.4%) were blamed on jackals with 50% blamed on hyenas and 25% on leopards.[54] Chaka et al. studied predation on livestock by medium and small predators on the Maasai steppe of Tanzania between 2015 and 2017.[55] They sampled pastoralist community views on predation and found that 106 out of 110 Maasai bomas had experienced livestock predation in the study period, with spotted hyenas, lions, leopards, and jackals blamed. Black-backed jackals that came second in the ranking of stock raiders said to have carried 44 (24.2%) of attacks, compared with 51.6% for hyenas, 19.2% for lions, and 4.9% for leopards.[56] Jackals attacked mostly in daylight hours, targeting sheep and goats. Black-backed jackals lived in close proximity to humans and clearly benefitted from the opportunity to prey on stock and proved capable of finding ways to get through thorn fences around bomas.[57]

Livestock losses are a problem for pastoralist communities in Tanzania. Holmern, Nyahongoa, and Røskafta estimated in 2007 that communities around the Serengeti NP lost on average 19.2% of cash income through predation.[58] Livestock were most often killed by spotted hyenas (97.7%), followed by leopards (1.6%), baboons (0.4%), lions (0.1%), and lastly, black-backed jackals (0.1%) – with 11 goats killed by jackals in 2003. Jackals were also blamed for killing dogs around the bomas.[59] In a later study, Holmern and Røskafta found that black-backed jackals were among the predators blamed for extensive killing of poultry kept by agropastoralists living around the north-western boundary of the Serengeti NP. While side-striped jackals and golden wolves inhabited the wider region, black-backed jackals lived closer to human settlements and were blamed for most poultry losses. Farmers in the region kept about 34.8 million fowl, and mortality rates among chicks were very high. Mongooses were blamed for 39.4% of poultry predated, raptors for 33.2%, jackals were believed to take 9%, and honey badgers were blamed for but 0.7% of losses.[60] The Tanzanian wildlife researcher Alaitetei E. Laltaika told me that pastoralists living in the Ngorongoro district were hostile towards jackals, and he heard numerous reports from them of attacks on young goats and sheep. This led to retaliatory attacks or preemptive killing, "resulting in the decline of the population of the black-backed jackals" in localities with large smallstock flocks.[61]

While the above research suggests regular predation on sheep and goats by jackals, there has not been the same level of organised and long-term persecution seen in southern Africa (particularly South Africa). Many pastoralists and some agropastoralists will kill predators, most often by poisoning with agricultural pesticides, but not on the scale seen further south. Bencin, Kioko, and Kiffner found that in the Babati, Monduli, and Karatu districts of northern Tanzania outside PAs near Lake Manyara NP, Tarangire NP, Manyara Ranch Conservancy, Mto Wa Mbu Game Controlled Area, and the Ngorongoro Conservation Area, 30.62% of local people wanted the number of jackals (including black-backed and side-striped jackals and golden wolves – the respondents did not differentiate between them) reduced, 58.72% wanted spotted hyena numbers reduced and 22% wanted lions reduced.[62] Poorer farmers were more likely to want jackals killed, perhaps because they had more sheep and goats than cattle.

Southern Africa

While Sundevall side-striped jackals are found in Angola, Zambia, much of Zimbabwe, Malawi, parts of Mozambique, northern Botswana, north-eastern Namibia, north-eastern South Africa, and eSwatini, southern African/Cape black-backed jackals are common over most of the region except Zambia and north-central Angola. Golden wolves are entirely absent. Black-backed jackals are almost ubiquitous in protected and unprotected rural areas of southern Africa and are often to be found around human settlements as well as in farming or wildlife zones.

In Botswana, black-backed jackals occur across the country in wildlife, farming, and semi-arid regions. While in Chobe, the Okavango, the Central Kalahari Game Reserve, and the Kgalagadi Transfrontier Park (KTP), they are not the apex predators because of the presence of lions, leopards, wild dogs, spotted and brown hyenas, and cheetahs, in some of the extensive semi-arid rangelands, where pastoralism is practised and large predators have been exterminated or severely reduced in numbers, jackals are among the most common and often "the only major mammalian carnivores in most parts of semi-arid Botswana, preying on small livestock and game, in addition to incorporating anthropogenic food items in their diet."[63] This is indicative of their ability to survive close proximity to humans and persecution of livestock-raiding predators. This is because of their plasticity of diet, eclectic choice of habitat, and ability to vary behaviour to opportunistically benefit from the human presence while avoiding some of the severe costs. Kaunda's study of black-backed jackals at Mokolodi Reserve, near Gaborone in south-eastern Botswana, showed that jackals would commute outside the reserve to feed on carcasses of domestic animals, prey on smallstock, and also take advantage of the opportunities to hunt rodents and small, wild ungulates that were common on nearby farms – as a result, some farmers saw Mokolodi as a refuge for stock raiders.[64] When I visited Mokolodi in 2003, the staff there said that attitudes of farmers varied, with those keeping sheep, goats, and poultry generally being hostile to jackals, while grain farmers welcomed them, as they preyed on rodents that damaged the grain crops. Kaunda rightly predicted in 2001 that growing human populations and food needs would increase contact between livestock and jackals and so conflict between humans and jackals.[65] He also noted that persecution had made jackals in the region of Mokolodi very wary – which made it difficult for him to humanely trap them to fit radio collars to track them.[66]

In semi-arid areas of western Botswana, livestock are the mainstay of the local population and predation a concern for them. The region is flat with dry bush savanna, and land use is divided between communal grazing areas, fenced ranches, wildlife management areas, and the KTP. Water for people and their animals comes mainly from boreholes, as there is little permanent water. The grazing areas support 20,000–25,000 head of free-ranging cattle, with some goats and sheep.[67] Although some large predators are found outside the PA of the park, the black-backed jackal is the most common carnivore,[68] and a source of conflict with farmers who keep smallstock or blame jackals for killing young calves. Mark and Delia Owens, in their book on carnivore research in the Central Kalahari Game Reserve, noted the persecution of jackals in Botswana but recorded how quickly the black-backed jackal pair, whose territory included the Owens's camp, became habituated to their presence and regularly foraged in the camp for food, even when they were there and had been seen by the jackals.[69] They reported seeing the jackals searching out burned insects and small mammals after a grass fire and cooperating in killing a large cobra.

In Namibia, black-backed jackals are found across most of the country, including in very arid regions, while Sundevall side-striped jackals are limited to wetter, wooded areas of the north and north-east.[70] Black-backed jackals are viewed by many stock farmers, particularly those raising sheep and goats, as a serious pest. Under South African rule (which ended in 1990), game ordinances required farmers to exterminate jackals on their land. Owen-Smith recounted how one farmer, called Oberholzer, refused to do this and refused to pay fines on the grounds that if he killed jackals, grass-eating rodents would ruin the grazing on his farm, as they would not be controlled by jackal predation. Another farmer, though, was an avid poisoner of jackals and used strychnine so liberally that most carnivores, including harmless ones like bat-eared foxes, were wiped out.[71] But persecution did not seriously reduce jackal numbers, and they persisted from the southern Namib up to the Kunene River and the border with Angola.

Despite its aridity, jackals thrive in the Namib Desert and on the Namib coast. Their dietary adaptability and the presence of abundant food sources on the coast mean they can survive where larger carnivores (brown hyenas excepted) have more difficulty or have been wiped out by human persecution. In inland areas of the Namib, jackals have a diverse diet that gives them a buffer in lean times when some foods are not available; the diet includes small mammals, nestlings and eggs of ground-nesting birds, lizards, tortoises, insects (especially the giant longhorn beetle and locusts),[72] wild fruits, and occasionally grass.[73] On the coast, jackals thrive on hunting seal pups, scavenging the afterbirth following birth of seal pups, sick and dead seals as well as birds like cormorants, and fish or whale offal washed up on the beaches.[74] Hiscocks and Perrin studied the territories and feeding habits of black-backed jackals at Cape Cross Seal Reserve and found that large numbers of jackals used the beach, and to an extent, territoriality broke down because of the "clumped" food sources. The jackals foraged at seal haul-out spots and from the waste produced by a seal-processing factory.[75] The lack of harassment by humans and larger predators – the only other large carnivore regularly seen at Cape Cross is the brown hyena, which because of the abundance of food did not regularly come into conflict with the jackals – meant the need to be crepuscular or nocturnal receded, and jackals regularly foraged or hunted along the beach in daylight.[76] I observed this on a very short visit to Cape Cross in August 2008, when I was able to walk among the seals and get close to jackals on the beach in the middle of the day. Similar behaviour has been monitored and recorded at Van Reenen Bay seal colony, c80km south of Lüderitz, where Kolar observed 33 black-backed jackals foraging on the beach and saw 83 jackal kills of seals and numerous unsuccessful attempts to hunt seal pups.[77] Once they had killed a pup, jackals had great trouble opening up the thick skin and often turned the carcass inside out to get at the meat and organs.

Jackals using the seal reserves or haul-out areas to hunt or forage had territories away from the beaches and would commute up to 20km to the feeding areas. On their home territories, they displayed normal territorial defence with chasing, body-slamming, and other behaviours common to jackals, but territoriality was rare on the beaches except where jackal territories were close to the beach, especially during denning, and little fighting between commuters crossing the territory of other resident jackals to reach the seal colonies was observed by researchers.[78]

In arid and semi-arid areas of south-west Namibia near Aus, land is used for commercial livestock or game ranching, the latter concentrating on springbok and gemsbok. Camera trap surveys by Edwards et al. indicated that jackals, brown hyenas, Cape foxes, aardwolves, honey badgers, and a small number of spotted hyenas and leopards were present on the farms, hunted stock or game animals, or scavenged their carcasses.[79] Farmers in many parts of Namibia blame black-backed jackals for killing sheep, goats, and calves – as a result, deterrence methods and culling are used to protect stock. Jackals are frequently victims of poison put into the carcasses of cattle killed by lions or hyenas, especially in areas north of Windhoek in Damaraland, around Etosha and in the Zambezi region. Fencing rarely worked to keep jackals from farmland or game ranches because warthogs, aardvarks, and other burrowing animals would dig under fences providing entry and exit points for jackals, who were also good at wriggling through gaps in wire.[80]

The use of large guard dogs did have the effect of deterring attacks and reducing livestock losses to jackals and cheetah and increased the number of jackals killed in farms by dogs and farmers combined.[81] It is difficult to assess the costs of jackal predation on stock as most figures given are for overall stock losses to predation, with cattle killed by larger carnivores the biggest drain on farmers' income. Rust and Marker put a figure on US$508,898 for annual losses on communal or resettled land in 2013, noting it was mainly cattle but that on game and smallstock farms, jackals killed goats, sheep, poultry, and smaller game animals.[82] Some farmers tried to both reduce the number of carnivores and garner income by having them shot by trophy hunters. On farms surveyed by Rust and Marker, 13 jackals were removed by hunters bringing in US$100 in trophy fees per jackal, compared with US$3500 for a cheetah and US$14,550 for a lion. They estimated that in the areas they studied, jackal hunting by trophy hunters could bring in US$14,400 a year.[83] There is no evidence that trophy hunting of jackals is on a large enough scale to affect numbers or seriously reduce predation.

South Africa's war against the jackal

South Africa has the most protracted history of conflict between people and black-backed jackals. Minnie et al. estimated that more than 350 years of lethal approaches to managing black-backed jackals had failed to control numbers or end predation of livestock or game animals bred on farms. They are still seen as the chief killer of sheep, goats, and small game ungulates.[84] In a country with about 80% of land devoted to livestock farming, the cost of livestock losses to predation and of preventing predation is a major issue in farming areas, and research into and experimentation with methods of non-lethal and lethal management of black-backed jackals and other predators are very important.[85]

In the early years of the Sabi Reserve, which became Kruger NP, predator culling was a priority for decades. The zoologist and game warden G.L. Smuts recalled that despite the extensive culling of jackals over decades in Kruger NP, their numbers remained high, and they were common throughout the park, including as visitors to the staff village, Park HQ, and tourist accommodation at Skukuza.[86] The culling in NPs has been reduced but not disappeared completely, while outside PAs human–jackal conflict continues with a high level of culling, though with little government role as bounties and loans for anti-predator measures have disappeared, and wildlife laws have devolved responsibility to private landowners or communal farmers.[87]

From the 1960s, there was the start of what James called "a revolution in wildlife management,"[88] with an increase in government-funded PAs and the growth of private-sector game farming to provide meat and game species to be hunted or sold. Through this, wildlife outside PAs that could bring in tourism/hunting income began to take on an economic value, attracting farmers to diversify partially or completely into game.[89] This affected carnivores, mainly smaller ones like jackals in farming areas where larger carnivores had been exterminated, with some game farmers tolerating jackals and others seeing them as a threat to the young of antelope they were breeding. Naturally occurring antelopes such as springbok, impala, kudu, and blue wildebeest were the basis for game meat, hunting, or tourism with rarer species such as blesbok, bontebok, and black wildebeest introduced. Where tourism was part of the mix, persecution of carnivores was limited as they were a potential draw for tourists. Waterholes were provided on game farms, which was of benefit to jackals.[90]

On livestock farms, both in the Western Cape (where conflict with jackals has always been higher) and in Free State, North West, Gauteng, Limpopo, and KwaZulu-Natal, jackals today remain the enemy for many farmers. About 91% of South Africa's land area is arid or semi-arid, and so various forms of pastoralism still dominate in terms of land use, with 69% of land designated rangeland and livestock contributing 47% of South Africa's agricultural gross domestic product, employing 245,000 people.[91] There are an "estimated 13.6 million beef cattle, 1.4 million dairy cattle, 24.6 million sheep, 7 million goats, 3 million farmed game animals, 1.1 million pigs and 1.6 million ostriches in addition to poultry."[92] Sheep, goats, calves, young of game animals, young ostriches, ostrich eggs, and poultry are all potential food for jackals. It is not surprising that they remain abundant on much of South Africa's rangeland and are feared and hated by many farmers. Turpie and Akinyemi estimate that black-backed jackals account for over 65% of predation losses in areas with smallstock and smaller game species. Jackals are also perceived to be the main cattle (chiefly calves) predators in all provinces apart from Limpopo, where leopards are the main threat.[93]

In the Cape, the predominance of sheep and the high jackal population meant that while bounties and poisoning clubs were a thing of the past, hunting clubs (often termed "militias") persisted in the Karoo into the 1990s, with jackal and caracal the main quarry – the clubs were phased out by 1993.[94] Hunting with dogs, shooting, and trapping, once poisoning was made illegal in South Africa, are the main methods of removing jackals. But,

> [w]ith special permission from provincial authorities farmers can use collars "filled with sodium monofluoroacetate(1080) to kill black-backed jackal ... and the collars need to be filled and supplied by licenced suppliers in terms of the Hazardous Substances Act, Act 15 of 1973. The collars should also only be used on small livestock to control black-backed jackal and caracal when other techniques did not work ... Another exception to the rule is Coyote getters – a device used to kill target animals by shooting cyanide into the mouth of the animal when it disturbs the bait. While this is one of the most successful ways to address stray dogs, it is not a selective method and may, therefore, lead to the poisoning of other species, especially smaller animals".[95]

Sodium monofluoroacetate was effective in killing canids, but in an agonising and protracted manner, and started to be used in 1961. The getters were encouraged after 1961 after the Western Cape Department of Nature Conservation's director, Dr David Hey, asked the US Fish and Wildlife Service to develop a getter suitable for use in South Africa. After they started to be used, Hey noted how many jackals managed to "pull" the "getter" without getting killed, adding that "one seldom has a second chance at a smart Jackal!"[96]

Farmers are prohibited from using pesticides, fungicides, and insecticides to poison predators or to use poison other than in the targeted way set out above, to avoid killing wildlife that is not a threat to stock, such as vultures and raptors.[97] There has been encouragement, as conservation objectives have mixed with stock protection, of the use of better fencing (often electric), improved herding techniques, night enclosures, and guardian animals rather than lethal methods.[98] Electric fencing is expensive, with no subsidies now available, but has proved successful in keeping jackals off farms, as has the increased use of guard dogs, and a key deterrent to loss of newborn stock is for farm workers to be present with the animals during lambing or calving.[99] But many farmers cannot afford large stretches of electric fencing or to employ more herders. So, hunting or trapping jackals is still widely practised, often mixed with non-lethal deterrents. Turpie and Akinyemie found that predator control was practised in six provinces on be-tween 37 and 66% of farms, but 60–90% in five provinces used lethal methods – including poisoning, which is illegal other than using toxic collars and getters.[100]

In some areas of the Western Cape, where sheep and goats are still important to the rural economy, "there is extreme intolerance of jackals and they are ag-gressively and continuously hunted to reduce population size and the associated livestock losses … intolerance is so severe that the use of aerial gunning and pro-fessional hunters to reduce jackal densities are becoming commonplace."[101] But jackal numbers remain high, and removal on a farm invariably leads to recovery through dispersal from neighbouring areas – in the Western Cape and the west-ern parts of the Eastern Cape, this can result from jackals dispersing on to farms from the Karoo NP, Mountain Zebra NP, and the two sections of the Addo Elephant NP.[102] But Nattrass et al. cautioned that their studies showed "that distance from the local PA was neither statistically nor substantively significant in predicting livestock predation."[103] They also found that genetic evidence from jackals indicated that there were "no detectable barriers to their dispersal across South Africa."[104]

For sheep farmers, predation by jackals again became a serious problem from the 1990s onwards. It is likely that changing land use enabled jackals to increase in numbers and disperse. As Nattrass and Conradie reported, "Increased pre-dation on South African sheep farms appears to be a consequence of natural 'rewilding' or recolonisation, a process driven by the rise of 'life-style' farmers with little interest in making a living solely from farming and an increase in land allocated to nature reserves."[105] They recorded a steep increase in reports by farmers of stock losses between 2012 and 2014 and a corresponding increase in amount spent on predator control, which more than doubled over the period, from 2,200 to 4,600 rand.[106] Use of poison rose from 21 to 50% of farmers, gin traps from 61 to 68%, cage traps from 50 to 78%, and shooting from 58 to 89%; non-lethal methods all declined as the perception of a serious predation problem increased.[107] But increased culling, according to Nattrass and Conradie, may well be counter-productive in the longer term as culling is associated with higher livestock losses in the following year. Farmers report that most of the lethal and non-lethal managerial options to protect their livestock from predation are becoming less effective over time. This is in line with evidence highlighting the capacity of caracals and especially black-backed jackals to adapt.[108]

Culling of jackals appears to worsen stock losses, given the ability of jackals to breed and disperse very successfully, widely, and rapidly, suggesting that culling should not be the go-to answer and that for all the farmers' doubts about non-lethal methods, they may be the best way forward.[109] These questions are part of the eternal problems facing farmers seeking to reduce stock losses in the face of the ability of jackals to learn quickly to avoid traps and getters and to overcome non-lethal methods – such as light and sound emitters to scare off jackals. They are even said to be attacking sheep from behind if they are wearing protective or toxic collars.[110] Much of this is anecdotal, and it is not clear how fast it occurs or how widespread it is in terms of the avoidance of traps and deterrents. Farmers are also sceptical of suggestions that tolerating a dominant jackal pair on a farm will deter dispersal to there by other jackals and so limit predation of stock.[111]

In his comprehensive analysis of game farm attitudes to jackals in North West Province, James looked at the operations of six game farms which had the species already mentioned as important in game farming plus reedbuck, tsessebe, waterbuck, nyala, gemsbok, Burchell's zebra, eland, giraffe, white and black rhino, nyala, sable, and smaller antelopes like steenbok and dik-dik. They generated income from hunting, meat, and animal sales.[112] The game farms had varying populations of predators including black-backed jackals, brown hyenas, and caracals – two of the sites had vulture feeding stations where animal remains were put out to help encourage the recovery of vulture numbers. These attracted black-backed jackals, which would take advantage of this clumped food source.[113] In his study, some farms were very active in controlling predator numbers, some did so occasionally, and others did not engage in culling. He recorded images of predators on the farms on camera traps over 2.5 years from May 2010. Across all the sites for the whole period, he obtained 4,555 shots of jackals, indicating that they were common across the farms regardless of levels of control.[114]

On the farms with active predator persecution, black-backed jackal populations were not appreciably affected in comparison with farms with low or no persecution.[115] Sites with vulture restaurants had very high abundance of black-backed jackals. James concluded that

> abundance estimates imply no difference in the number of jackal individuals between predator control and predator neutral treatments, and indicate that the current level of removal of *C. mesomelas* implemented in the game farms studied in this investigation are ineffective as a predator management strategy ... evidence suggests that the current rate of removal

of jackal individuals is not only ineffective in reducing conflict, but potentially counter-productive in reducing black-backed jackal numbers in the
private game farm environments studied.[116]

The jackals bounce back through breeding and dispersal, and predator control
may create sinks[117] – where jackals move in, are culled, and more jackals move
in to fill the vacuum. Strong populations in areas with little or no persecution
simply act as source populations for areas where culling creates empty territory
with good food sources.[118] This mirrors the ability of jackals in Cape to weather
centuries of concerted persecution and strongly suggests that methods other than
culling need to be found.

Yarnell, in his comparison of black-backed jackals and brown hyenas in
Pilanesberg NP and on Mankwe Wildlife Reserve in North West Province,
noted the importance of the vulture restaurant established at Mankwe in providing a clumped food source that drew jackals in to feed in groups without serious
conflict, and that the absence of lions, spotted hyenas, and wild dogs at Mankwe
compared with Pilanesberg meant that jackal density was $1.152/km^2$, compared
with $0.37/km^2$ at Pilanesberg. The combination of the absence of large predators and supplementary feeding was a major advantage for the jackals.[119] Thorn
et al. also researched human–carnivore conflict in North West Province, noting that from 2006 to 2008, 3,755 livestock or game on farms were killed by
carnivores, amounting to 2.77% of total holdings.[120] Black-backed jackals and
caracals were most frequently blamed for predation and, according to farmers
interviewed by the researchers, were often killed by them in response. In the
province, 2.4% of the land is protected for wildlife, while 54% is used for grazing
and 30% for crop farming.[121] Game farming is an important business there with
revenue from hunting, sale of live animals, stud breeding, and tourism. Farmers
interviewed during the research owned livestock and game totalling 136,337
animals, and the numbers killed by predators were listed by them as "goat 1412,
sheep 1055, springbok 357, cattle 334, ostrich, 196, impala 155, blesbok 71, horse
52, steenbok 43, common duiker 40, mountain reedbuck 25, poultry 12, roan 6,
sable 4, kudu 3, gemsbok 3, blue wildebeest 2, nyala 2, warthog 1, bushbuck 1
and zebra 1."[122] But predation wasn't the chief cause of losses; poaching (32%),
drought (30%), and disease (8%) accounted for most losses with predation at
19% – farmers attributed 41% of predation to jackals, 20% to caracals, 15% to
leopards, 12% to brown hyenas, 7% to cheetahs, and 3% to spotted hyenas.[123]
Out of 99 farmers questioned, 66 said they used lethal control to limit or stop
predation – 20% used poison, 14% shooting at night, 14% hunted them with
dogs, and 14% trapped and then shot them; 65% used non-lethal methods like
enclosures, guard dogs, and human guards.[124] There was no evidence that killing jackals seriously reduced their numbers in the long term.

On Benfontein Game Farm in Northern Cape Province, managed primarily
for wild ungulates, black-backed jackals were common, and a research study
radio-collared 15 from 8 family groups and tracked their movements. It was

found that during the summer months, 64% of the collared animals moved on to neighbouring livestock farms where they foraged for sheep carcasses and actively hunted sheep and lambs.[125] There was no regular culling of jackals on Benfontein, which had no large carnivores, but jackals were occasionally shot during culling of ungulates. On the neighbouring livestock farms, jackals were heavily persecuted, with territorial jackals being exterminated, only for dispersing animals to take their places or for jackals to temporarily move in during the summer.[126] On Benfontein, the jackals relied heavily, during the springbok birthing seasons in spring and autumn, on killing springbok fawns and eating the afterbirth. The months when springbok fawns were not available were when they hunted sheep.

After the election of the African National Congress government in 1994, subsidies for white commercial sheep farmers were withdrawn, and farmers were limited in the methods they could use to kill jackals. By 2006, the National Wool Growers' Association was describing predation as a major and resurgent threat to smallstock production, calling for more financial, organisational, and technical support from the state, which was not forthcoming. At the same time, there were more conservation groups calling for an end to trapping and hunting of predators with dogs.[127] In late 2008, "to the great consternation of sheep farmers," CapeNature, the government body responsible for managing and maintaining 113 nature reserves and wilderness areas in the Western Cape,[128] announced that, effective from January 2009, various control methods, including night-hunting of jackals, would no longer be allowed, on the grounds that it was "concerned that the unselective hunting of jackal and caracal is the reason for the increase in the population of these species over the last 400 years."[129] While hundreds of years of constant and extensive killing of jackals had not reduced the problem of predation or the jackal presence, it is not clear that it directly caused the increase, though it may have encouraged high breeding rates and dispersal. Nattrass and Conradie referred to the CapeNature approach as part of an new "environmental jackal narrative," noting that it "has strong support from the literature on predator ecology … scientific studies of coyotes (*Canis latrans*) showing that coyote populations adapt to persecution (whether by wolves or humans) by increasing their numbers, adds weight to the suggestion that culling jackals is counter-productive."[130] In 2012, CapeNature published a report by Bothma that concluded that jackal control measures had failed and that the "only effective program to control black-backed jackal numbers would be to exterminate all the jackals nationally" but that this would be "economically impossible and unsound ecologically."[131] This was hardly music to the ears of sheep farmers. Nattrass and Conradie noted that Bothma had conflated eradication and control and that while the latter was possible in some areas, eradication was not feasible or necessarily desirable.[132]

If control results were at the least questionable and eradication impossible or undesirable, one challenge to the environmental jackal narrative opposing culling to reduce numbers was evidence gathered by SANParks (South African

National Parks) that abundance of black-backed jackals in the Karoo and Addo Elephant NPs led to "significant sustained declines in the populations of certain ungulate species, notably springbok"; SANParks opted to limit jackal numbers through culling in the parks.[133] About 340 jackals were killed in 2010 across the two parks and in 2011–13, and another 600 were killed by SANParks. It is far from clear that the policy had any great effect. Since the start of the culling experiment, lion and leopard numbers have increased in both parks (partly through reintroductions), which could play a role in limiting jackal numbers through the restoration of apex predators which had previously been exterminated.[134]

It is clear that on stock and game farms, substantial culling has not worked, fences can be effective but are expensive and can be undermined by burrowing animals, and the use of both human and canine guards has its costs. If over 350 years of persecution and stock protection have not worked, the only option is finding ways of mitigating losses and conflict through a combination of methods and accepting that some losses are inevitable unless a drastic decision is taken to try to totally exterminate jackals outside PAs, and even then a final solution is just not on the cards, because they will simply repopulate from populations elsewhere, particularly from PAs or neighbouring countries with substantial populations not subject to widespread persecution.

Notes

1 S.K.K. Kaunda (2001) Spatial utilization by black-backed jackals in southeastern Botswana, *African Zoology*, 36, 2, 143–52, p. 143.
2 Ibid., p. 150.
3 B.M. Kissui (2008) Livestock predation by lions, leopards, spotted hyenas, and their vulnerability to retaliatory killing in the Maasai steppe, Tanzania, *Animal Conservation*, 11, 422–32, p. 422.
4 Ibid.
5 E.A. Sogbohossou et al. (2011) Human–carnivore conflict around Pendjari Biosphere Reserve, *Northern Benin*, 45, 4, 569–57, p. 569.
6 D.L. Ogada (2014) The power of poison: pesticide poisoning of Africa's wildlife, *Annals of the New York Academy of Sciences Issue: The Year in Ecology and Conservation Biology*, 1322, 1–20, pp. 3–4.
7 Ibid., p. 9.
8 MaghrebOrnitho (2012) African golden wolf discovered in Morocco and Algeria, African golden wolf discovered in Morocco and Algeria - MaghrebOrnitho (magornitho.org), accessed 5 January 2022.
9 L. Campbell (2017) Wolves of the Atlas. Studying Africa's hidden wolf in the Moroccan Atlas Mountains, *Wolf Print*, Autumn/Winter 2017, 14–7, p. 15.
10 Ibid.
11 K. Somerville (2021) *Humans and Hyenas Monster or Misunderstood*, Abingdon, Oxon: Routledge.
12 Campbell, 2017, pp. 16–7.
13 Ibid.
14 A. El Alami (2019) A survey of the vulnerable Cuvier's gazelle (Gazella cuvieri) in the mountains of Ait Tamlil and Anghomar, Central High Atlas of Morocco, *Mammalia*, 83, 1, 74–7, p. 74.
15 M. Amroun, P. Giraudoux & P. Delattre (2006) A comparative study of the diets of two sympatric carnivores – the golden jackal (Canis aureus) and the common genet (Genetta genetta) – in Kabylia, Algeria, *Mammalia*, 70, 247–50, pp. 247–8.

16 A. Eddine et al. (2017) Diet composition of a Newly Recognized Canid Species, the African Golden Wolf (Canis anthus), in Northern Algeria Authors, *Annales Zoologici Fennici*, 54, 5–6, 347–56, p. 347.

17 Ibid.

18 Ibid.

19 D.R. Rosevear (1974) *The Carnivores of West Africa*, London Trustees of the British Museum (Natural History), p. 36.

20 Ibid., pp. 49–50.

21 M.J. Paul (2018) Resource partitioning and social behaviour of sympatric African wolves and side-striped jackals in Senegal (north-west Africa): an approach using camera trapping, in G. Giannatos et al., 21–2, p. 21.

22 T. Abáigar et al. (2016) Social organization and demography of reintroduced Dorcas gazelle (Gazella dorcas neglecta) in North Ferlo Fauna Reserve, Senegal, *Mammalia*, 80, 6, 593–600, p. 596.

23 J.E. Castelló (2018) *Canids of the World*, Princeton, NJ: Princeton University Press/ Princeton Field Guides, pp. 126–64.

24 T. Künzel, H.A. Rayaleh & S. Künzel (2000) *Status Assessment Survey on Wildlife in Djibouti Final Report December 2000*, Munich: Zoological Society for the Conservation of Species and Populations (ZSCSP), p. 7.

25 Ibid., pp. 36–7.

26 D.W. Yalden, M.J. Largen & D. Kock (1980) Catalogue of the mammals of Ethiopia, *Monitore Zoologico Italiano. Supplemento*, 13, 1, 169–272, pp. 181–2.

27 Ibid.

28 Ibid., p. 185.

29 G. Yirga (2012) The ecology of large carnivores in the highlands of northern Ethiopia, *African Journal of Ecology*, 51, 78–86, p. 78.

30 Ibid., pp. 78 and 80.

31 Ibid., p. 78.

32 A. Atickem (2010) Livestock predation in the Bale Mountains, Ethiopia, *African Journal of Ecology*, 48, 1076–82, p. 1076.

33 A. Megaze, M. Balakrishnan & G. Belay (2017) Human–wildlife conflict and attitude of local people towards conservation of wildlife in Chebera Churchura National Park, Ethiopia, *African Zoology*, 52, 1, 1–8, p. 4.

34 W. Yohannes Dereje & M. Chane (2014) Ethnozoological study of traditional medicinal animals used by the kore people in Amaro Woreda, Southern Ethiopia, *International Journal of Molecular Evolution and Biodiversity*, 4, 2, 1–9, p. 1.

35 Ibid., p. 3.

36 M.B. Nimir (1983) *Wildlife values and management in northern Sudan*, Dissertation submitted to for the Degree of Doctor of Philosophy Colorado State University Fort Collins, Colorado, Summer 1983.

37 Ibid., pp. iii–iv.

38 Ibid., p. 21.

39 Ibid., p. 139.

40 T.K. Fuller et al. (1989) The ecology of three sympatric jackal species in the Rift Valley of Kenya. *African Journal of Ecology*, 27, 313–23, p. 313.

41 Fuller et al., 1989, p. 320.

42 J.O. Ogutu, N. Bhola & R. Reid (2005) The effects of pastoralism and protection on the density and distribution of carnivores and their prey in the Mara ecosystem of Kenya, *Journal of Zoology*, 265, 281–93, p. 281.

43 Ibid.

44 Ibid., p. 284.

45 Ibid., p. 290.

46 M.E. Gadd (2005) Conservation outside of parks: attitudes of local people in Laikipia, Kenya, *Environmental Conservation*, 32, 1, 50–63, p. 53.

47 J.O. Ogutu et al. (2017) Wildlife population dynamics in human dominated landscapes under community-based conservation: the example of Nakuru wildlife

conservancy, Kenya. *PLoS ONE*, 121, e0169730. doi:10.1371/journal.pone.0169730, 1–30, p. 19.

48 Hugo & J. van Lawick-Goodall (1970) *Innocent Killers*, London: Collins, p. 127.

49 Ibid., p. 134.

50 J. Lamprecht (1978) On diet, foraging behaviour and interspecific food competition of jackals in the Serengeti National Park, East Africa, *Zeitschrift-Saeugetierkunde*, 43, 210–23, accessed 20 October 2021, p. 213.

51 N. Pettorelli et al. (2010) Carnivore biodiversity in Tanzania: revealing the distribution patterns of secretive mammals using camera traps, *Animal Conservation*, 13, 131–39, p. 132.

52 S. Durant (2010) Does size matter? An investigation of habitat use across a carnivore assemblage in the Serengeti, Tanzania, *Journal of Animal Ecology*, 79, 5, 1012–22, pp. 1015–6.

53 F. Peniel Mbise (2018) Livestock depredation by wild carnivores in the Eastern Serengeti Ecosystem, Tanzania, *International Journal of Biodiversity and Conservation*, 10, 3, 122–30, p. 122.

54 Ibid., p. 126.

55 S.N.M. Chaka et al. (2021) Predicting the fine-scale factors that correlate with multiple carnivore depredation of livestock in their enclosures, *African Journal of Ecology*, 59, 1, 74–87.

56 Ibid., p. 79.

57 Ibid., pp. 80–1.

58 T. Holmern, J. Nyahongoa & E. Røskafta (2007) Livestock loss caused by predators outside the Serengeti National Park, Tanzania, *Biological Conservation*, 135, 518–26, p. 518.

59 Ibid., p. 522.

60 T. Holmern & E. Røskaft (2013) The poultry thief: Subsistence farmers' perceptions of depredation outside the Serengeti National Park, Tanzania, *African Journal of Ecology*, https://onlinelibrary.wiley.com/doi/abs/10.1111/aje.12124 accessed 10 May 2021, 1–9, p. 4.

61 A.E. Laltaika, personal communication, 7 June 2021.

62 H. Bencin, J. Kioko & C. Kiffner (2016) Local people's perceptions of wildlife species in two distinct landscapes of Northern Tanzania, *Journal for Nature Conservation*, 34, 82–92, p. 83.

63 K.K. Kaunda (2001) Spatial utilization by black-backed jackals in southeastern Botswana, *African Zoology*, 36, 2, 143–52, p. 143.

64 Ibid., p. 148.

65 Ibid.

66 S.K.K. Kaunda (2001) Capture and chemical immobilization of black-backed jackals at Mokolodi Nature Reserve, Botswana, *South African Journal of Wildlife Research*, 31, 1, 43–47, p. 45.

67 M. Wallgren (2009) Influence of land use on the abundance of wildlife and livestock in the Kalahari, Botswana, *Journal of Arid Environments*, 73, 314–21, p. 315.

68 Ibid., p. 317.

69 Mark & D. Owens (1986) *The Cry of the Kalahari*, London: Fontana, p. 28.

70 G. Owen-Smith (2010) *An Arid Eden. A Personal Account of Conservation in the Kaokoveld*, Johannesburg: Jonathan ball, p. 3.

71 Ibid., pp. 301–2.

72 M. Goldenburg et al. (2010) Diet composition of black-backed jackals, *Canis mesomelas* in the Namib Desert, *Folia Zoologica*, 59, 2, 93–101, p. 93.

73 C.T. Stuart (1976) Diet of the black backed jackal *Canis Mesomelas* in the central Namib Desert, South West Africa, *Zoologica Africana*, 11, 1, 193–205, p. 194.

74 Ibid., pp. 195–7.

75 M.R. Hiscocks & K. Perrin (1988) Home range and movements of black-backed jackals at Cape Cross Seal Reserve, Namibia. *South African Journal of Wildlife Research*, 18, 97–100, pp. 98–9.

76 Ibid., p. 99.
77 B. Kolar (2005) Black-backed jackals hunt seals on the Diamond Coast, Namibia. *Canid News*, 8, 2 http://www.canids.org/canidnews/8/Black_backed_jackals_ hunt_seals.pdf accessed 14 December 2021, 1–4, p. 2.
78 N. Jenner, J. Groombridge & S.M. Funk (2011) Commuting, territoriality and variation in group and territory size in a black-backed jackal population reliant, p. 235.
79 S. Edwards, A.C. Gange & I. Wiesel (2015) Spatiotemporal resource partitioning of water sources by African carnivores on Namibian commercial farmlands, *Journal of Zoology*, 297, 22–31, p. 22.
80 N. Littlewood et al. (2020) *Terrestrial Mammal Conservation, Global Evidence for the Effects of Interventions for Terrestrial Mammals Excluding Bats and Primates*, Cambridge: Open Book Publishers, Synopses of Conservation Evidence Series, University of Cambridge, pp. 105–6.
81 Ibid.
82 N.A. Rust & L.L. Marker (2013) Cost of carnivore coexistence on communal and resettled land in Namibia, *Environmental Conservation*, 41, 1, 45–53, p. 45.
83 Ibid., p. 47.
84 L. Minnie et al. (2018) Biology and ecology of the black-backed jackal and the caracal, Chapter 7, in G.I.H. Kerley, S.L. Wilson & D. Balfour (eds) *Livestock Predation and Its Management in South Africa: A Scientific Assessment*, Port Elizabeth: Centre for African Conservation Ecology, Nelson Mandela University, 178–204, p. 179.
85 J.J. du Plessis et al. (2018) Past and current management of predation on livestock, in G.I.H. Kerley, S.L. Wilson & D. Balfour (eds) *Livestock Predation and Its Management in South Africa: A Scientific Assessment*, Port Elizabeth: Centre for African Conservation Ecology, Nelson Mandela University, 125–76, p. 126.
86 G.L. Smuts (1982) *Lion*, Johannesburg: Macmillan South Africa, p. 52.
87 Ibid., p. 127.
88 R. James (2014) *The Population Dynamics of the Black-backed Jackal (Canis mesomelas) in Game Farm Ecosystems of South Africa*, Thesis for Doctorate of Philosophy, University of Brighton, November 2014, p. 40.
89 Ibid.
90 Ibid.
91 K. Turpie & B.E. Akinyemi (2018) The socio-economic impacts of livestock predation and its prevention in South Africa, in Kerley et al., 53–81, p. 54.
92 Ibid., pp. 54–5.
93 Ibid., p. 57.
94 L. van Sittert (2016) Routinising genocide: the politics and practice of vermin extermination in the Cape Province c.1889–1994, *Journal of Contemporary African Studies*, 34, 1, 111–28, pp. 122–3.
95 Anonymous (no date) the Use of Poison in predation Management, southafrica. co.za/poison-predation-management.html#:~:text=Since%20none%20has%20 been%20registered,is%20a%20serious%20criminal%20offence accessed 16 March 2022.
96 N. Nattrass & B. Conradie (2015) Jackal narratives: predator control and contested ecologies in the Karoo, South Africa, *Journal of Southern African Studies*, 41, 4, 753–71, p. 759.
97 Ibid.
98 Turpie & Akinyemi, 2018, p. 63.
99 Turpie & Akinyemi, 2018, p. 66.
100 Ibid.
101 Minnie et al., 2018, p. 941.
102 Ibid.
103 N. Nattrass et al. (2019) Culling recolonizing mesopredators increases livestock losses: Evidence from the South African Karoo, *Ambio*, 49, https://doi.org/10.1007/ s13280-019-01260-4 accessed 22 March 2022, no page numbers.

104 Ibid.
105 N. Nattrass & B. Conradie (2018) Predators, livestock losses and poison in the South African Karoo, *Journal of Cleaner Production*, 194, 777–85, p. 777.
106 Ibid., p. 780.
107 Ibid., p. 782.
108 Ibid., p. 784.
109 N. Nattrass et al., 2019.
110 Turpie & Akinyemi, 2018, p. 106.
111 Ibid.
112 Ibid., p. 52.
113 R.W. Yarnell (2013) The influence of large predators on the feeding ecology of two African mesocarnivores: the black-backed jackal and the brown hyaena, *South Africa Journal of Wildlife Research*, 43(2), 155–68, p. 155.
114 James, 2014, p. 66.
115 Ibid., p. 74.
116 Ibid., pp. 75–6.
117 L. Minnie et al. (2018) Spatial variation in anthropogenic mortality induces a source-sink system in a hunted mesopredator, *Oecologia*, 186, 941–51, p. 941.
118 Ibid., p. 76.
119 R.W. Yarnell (2013) The influence of large predators on the feeding ecology of two African mesocarnivores: the black-backed jackal and the brown hyaena, *South Africa Journal of Wildlife Research*, 43(2), 155–68, pp. 156–7.
120 M. Thorn et al. (2012) What drives human–carnivore conflict in the North West Province of South Africa? *Biological Conservation*, 150, 23–32, p. 23.
121 Ibid., p. 24.
122 Ibid., p. 26.
123 Ibid., p. 28.
124 Ibid.
125 J.F. Kamler et al. (2019) Social organization, home ranges, and extraterritorial forays of black-backed jackals, *Journal of Wildlife Management*, 83, 8, 1800–08, p. 1800.
126 Ibid., p. 1802.
127 Nattrass & Conradie, 2015, p. 761.
128 CapeNature (no date) https://www.capenature.co.za/about-us accessed 16 March 2022.
129 Nattrass & Conradie, 2015, p. 761.
130 Ibid., p. 763.
131 Bothma, Literature Review of the Ecology and Control of the Black-Backed Jackal and Caracal in South Africa (2012), available at http://www.capenature.co.za/wp-content/uploads/2014/02/Literature-Review-of-the-Ecology-and-Control-ofblack-backed-jackal-and-caracal-Bothma-2012.pdf accessed 12 May 2021.
132 Nattrass & Conradie, 2015, p. 765.
133 Cited by Nattrass & Conradie, 2015, p. 767.
134 L. Minnie, A. Gaylard & G.I. Kerley (2018) Compensatory life-history responses of black-backed jackals undermine population reduction efforts, *Journal of Applied Ecology*, 53, 2, 379–87, p. 379.

8

HONEY BADGERS

Dramatis personae

Honey badger or ratel

Order: Carnivora
Family: Mustelidae
Sub-family: Mellivorinae
Genus: Mellivora
Species: Mellivora capensis

Sub-species: Twelve sub-species have been suggested, based on geographical factors, but with some reference to size and colouring.[1] Rhodes reported that up to 15, 10 of them in Africa, had been described, but cautioned that the division into sub-species was ambiguous.[2] Begg et al. said that the ten sub-species claimed in the past to be found in Africa are "based primarily on size and pelage (mantle) variation, with only *M. c. capensis* present ... [but] no DNA investigation of sub-species has been completed so far and, therefore, sub-species denoted only by morphometrics, or pelage colour and pattern, are of dubious validity."[3]

The 12 suggested sub-species are:

M. c. Capensis (Cape ratel)
M. c. Indica (Indian ratel)
M. c. Inaurita (Nepalese ratel)
M. c. Leuconota (White-backed ratel – West Africa and southern Morocco)
M. c. Cottoni (Black ratel – Ghana and north-eastern Congo)
M. c. Concisa (Lake Chad ratel)
M. c. Signata (Speckled ratel – Sierra Leone)
M. c. Abyssinica (Ethiopian ratel)

DOI: 10.4324/9781003199793-9

M. c. Wilsoni (Persian ratel)

M. c. Maxwelli (Kenyan ratel)

M. c. Pumilio (Arabian ratel)

M. c. Buechneri (Turkmenian ratel)[4]

Clearly, these sub-species are not universally accepted as valid by zoologists,[5] and geographical/pelage differences of the supposed sub-species do not denote significant differences in diet, foraging, breeding of behaviour, or other clear evidence of major variations.

Size, weight, and physical characteristics

Head and body – 74–96cm; tail –14.3–26cm; height at shoulder – 23–28cm.

Weight – female, 6.2–13.6kg; male, 7.7–14.5kg.[6]

The honey badger is a stocky, powerfully built animal with black on the face, sides, and underparts with a broad white/silver stripe on the top of the head, merging into the silver/grey cape/mantle down the back and tail. They have loose skin, very muscular necks and shoulders, and broad forepaws equipped with massive claws.[7] Honey badgers show quite substantial variations in size, pelage, and colouring across their African and Asian ranges (Figure 8.1).[8]

In eastern Democratic Republic of Congo (DRC) and Gabon, honey badgers have been observed that are completely black, while in parts of West Africa, white-headed badgers occur.[9] Camera trap images from Gabon have shown four completely black honey badgers. The images were recorded by researchers from Panthera, the big cat NGO, which was conducting a wildlife survey near the

FIGURE 8.1 Honey badger in the Central Kalahari Game Reserve, Botswana.

Ivindo National Park (NP).[10] Panthera's Philip Henschel said he had never seen a honey badger in Gabon's dense forest before, but increased use of camera traps to count wildlife numbers had indicated their presence even in very dense forest and that as many as half are completely black. It is suggested that the presence of a significant proportion of black badgers in the overall population suggests that the dark colouration is an "adaptive advantage" in survival terms in the dense, dark forests.[11]

In build, the honey badger is about the size of a small- to medium-size dog. The skin is thick, and around the throat, the honey badger's most vulnerable region, the skin may be as much 6mm thick.[12] Its loose skin protects it during fights with other badgers, with competitors for food (such as jackals) and potential predators, including lions, leopards, and hyenas. In a fight, opponents may seize the badger by its skin and even puncture it, without causing major damage to muscle, bone, major organs, or the nervous system. The thick skin and a strong immune system help honey badgers cope with bee stings, scorpion stings, and even snake bites, though there are reports, recorded by Kingdon, of honey badgers succumbing to multiple bee stings and being found dead inside bee nests or hives.[13] They have entirely internal ears, and they close off their ears when digging, to avoid getting soil and sand in their ears. Kingdon also suggests that, as with the insectivorous pangolin, the lack of external ears could be an advantage in coping with defensive attacks by social insects such as bees, soldier termites, and ants, when honey badgers are raiding nests.[14]

The badgers have a short, pugnacious-looking snout with strong jaws and teeth. They are built for digging, tearing open logs and bee nests, aggression against potential competitors, and intimidatory defence against predators.[15] They have powerful 1.5-inch-long (4cm) claws and teeth, strong enough to crack a tortoise shell. The honey badger has a concealed weapon with which to defend itself. Situated at the base of its tail are two anal glands that are used for scent marking, but which can be inverted to squirt out a foul-smelling liquid that can be detected 40m away, according to the South African National Biodiversity Institute.[16] The animals generally expel the substance to mark their territory, but they will also release a "stink bomb" when threatened.[17]

Honey badgers have large brains for their body size and are credited with being highly intelligent and to have the ability to use objects as tools.[18] Captive honey badgers have been known to work together to unlock gates and use rocks, garden implements, mud, and sticks to escape from their enclosures, as shown in the documentary produced by the BBC, called *Honey Badgers: Masters of Mayhem*[19] – the discerning viewer might be somewhat suspicious that rocks, implements, and logs have been placed in such a way to encourage the honey badger to use them as an escape kit.

The general appearance, pugnaciousness, and digging were what led scientists early on to include them in the badger family. As Long and Killingley noted,

the animal was called a badger because it digs in the ground and has large forepaws and long robust claws ... the honey badger resembled the

American and Old World badgers. After scientists carefully studied and described the badgers, the honey badgers were found to be only superficially allied to the other badgers.[20]

They went on to record the prodigious digging abilities of honey badgers across their range, with evidence of them being able to dig burrows in hard ground and dig out rodents, insects, and reptiles in deep burrows.[21]

Distribution and conservation status

Afghanistan, Algeria, Angola, Benin, Botswana, Burkina Faso, Burundi, Cameroon, Central African Republic, Chad, Congo, DRC, Côte d'Ivoire, Djibouti, Equatorial Guinea, Eritrea, eSwatini (formerly Swaziland), Ethiopia, Gabon, Gambia, Ghana, Guinea, Guinea-Bissau, India, Iran, Iraq, Israel, Jordan, Kazakhstan, Kenya, Kuwait, Lebanon, Liberia, Malawi, Mali, Mauritania, Morocco, Mozambique, Namibia, Nepal, Niger, Nigeria, Oman, Pakistan, Qatar, Russia, Rwanda, Saudi Arabia, Senegal, Sierra Leone, Somalia, South Africa, Sudan, Syria, Tajikistan, Tanzania, Togo, Turkmenistan, Uganda, United Arab Emirates, Uzbekistan, Western Sahara, Yemen, Zambia, Zimbabwe (Figure 8.2).[22]

The honey badger is the only small carnivore species with large proportions of its range in and outside Africa, beings present across the greater part of Africa except for the driest centre of the Sahara and the Mediterranean littoral, the Nile Valley, and parts of South Africa.[23] Although widespread in much of India, its distribution in the southern states is fragmented, and it is rare in the northeast. There are no recent records of it in Afghanistan, but its presence across the Tedzhen, Murghab, and Amu Darya River valleys in Turkmenistan suggests that it may be found in north-western Afghanistan.[24]

Honey badgers can live in a wide variety of habitat, from the dense forests of the Congo Basin to the fringes of the Sahara and in the Namib, and from sea level to alpine steppes, such as in Ethiopia's Bale Mountains and the foothills of the Himalayas. Status across the whole range is unclear, but it appears to have declined in parts of Morocco and Israel,[25] and its ranges are increasingly fragmented outside strongholds in sub-Saharan Africa, because of habitat loss and expanding human populations.

In South Africa, there are questions over the extent and continuity of its range.[26] In some areas there, it may be increasing in numbers, assisted by the growth in the number of farms used for breeding game animals, game cropping, or sport hunting. These can provide ideal habitat for the honey badger, and Begg et al. suggest the game farms are responsible for an increase in the badger population in North West Province.[27] Ecological information, in particular about habitat selection, is scarce, partly due to its relatively wide range, elusive nature, and low densities but also because of the absence of many studies beyond those of the southern African populations.

Terrain and land cover types were found by Sharifi, Malekian, and Shahnaseri to be positively correlated with badger presence, and proximity to road and villages was negatively correlated with their presence.[28] Vegetation cover and food availability were key determinants of suitable habitat, along with avoidance of human populations, although in some areas human-created environments, such as eucalyptus plantations and over agroforestry habitat, provide sufficient

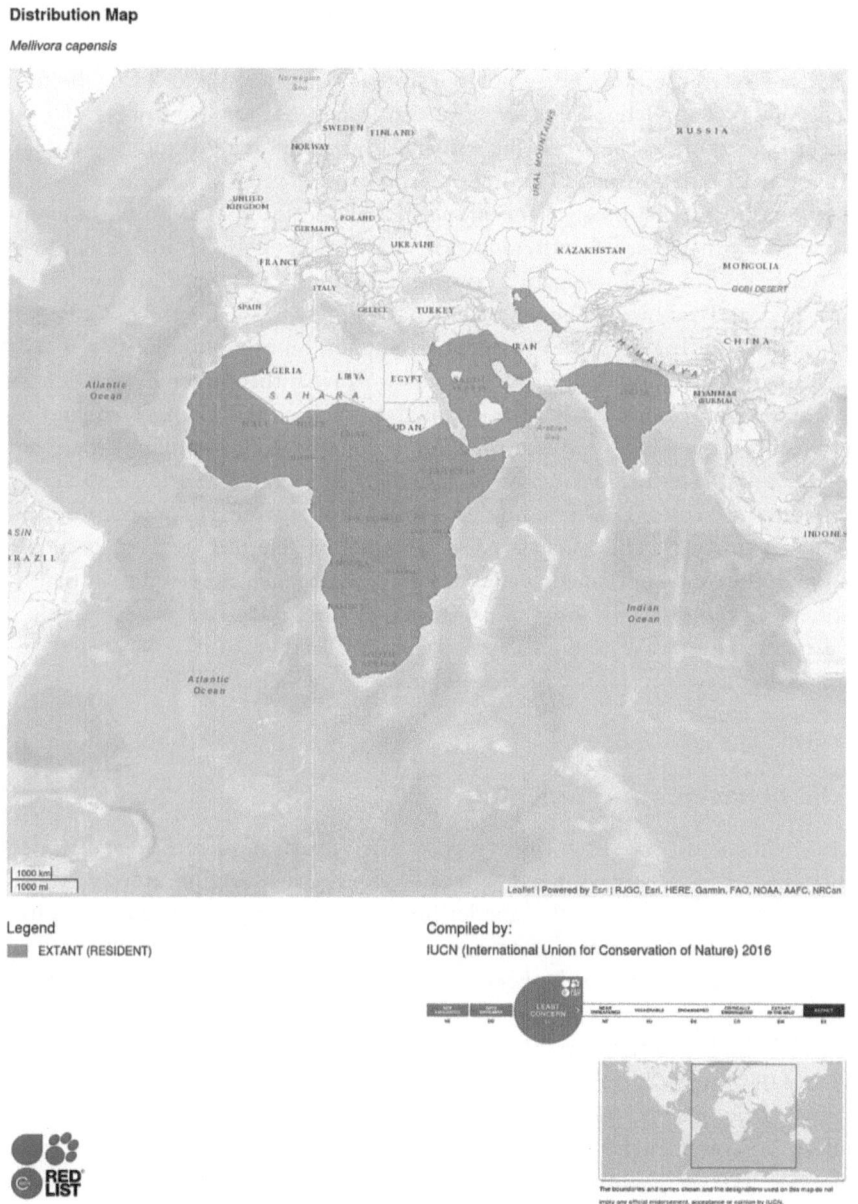

Distribution Map

Mellivora capensis

Legend

▨ EXTANT (RESIDENT)

Compiled by:

IUCN (International Union for Conservation of Nature) 2016

FIGURE 8.2 Map of global distribution of the honey badger. Produced with the kind permission of the IUCN.

cover and foraging opportunities to be attractive.[29] There is scarcity of information on sizes of regional populations and overall global numbers. South Africa is estimated to have a population of 6,000–20,000 (3,960–13,200) mature adults.[30] Reliable regional population estimates are lacking, with no verifiable figures available.[31]

In 2002, the International Union for Conservation of Nature (IUCN) assessed and listed the honey badger as Near Threatened (NT) due to the increased habitat loss and fragmentation, particularly within the interior of South Africa, which suggested the probability of a fall in the global population. But the most recent IUCN appraisal of the status of the honey badger listed it as of Least Concern based on its wide distribution and insufficient data to indicate a major decrease in population size. The honey badgers "are considered rare or to exist at low densities across most of their range," and densities based on night counts have been estimated at 0.1 individual/km² in the Serengeti NP in Tanzania and 0.03 adult/km² in the Kgalagadi Transfrontier Park (KTP), South Africa.[32] The South African Red List of Species included the honey badger in the NT category because of its decline in some areas and fragmentation of populations.[33]

The main threats to honey badger populations are anthropogenic – persecution by people, use of body parts in traditional medicine, and changes in land use leading to habitat loss.[34] Persecution is often done by honey gatherers, beekeepers, and small livestock farmers throughout their range (see the "Interactions with humans" section). Where human-caused deaths are high, populations may decline because "the destruction of suitable habitat coupled to factors such as the small litter size, extended period of cub dependence, increased maternal investment and a relatively short life span makes the honey badger vulnerable to local extinctions."[35] Where local extinctions are known to have occurred, it is possible that translocations to restore populations may be possible in unprotected areas or on game farms.[36]

Social behaviour and breeding

Honey badgers do not show the typical mustelid pattern of intrasexual territoriality; instead, males have massive home ranges that overlap extensively with other males and encompass the smaller home ranges of up to 13 females, with a polygynous mating system.[37] There is conflict between males, which results in the scarring often seen on the backs of older males; generally where territories overlap, scent marking enables avoidance rather there being regular territorial defence.[38] Male home ranges show large variation in size according to terrain, presence of other badgers, and food availability. In the KTP they have been found to range from 229 to 844km²,[39] while in other parts of Africa, they average 500km².[40] Young males in KTP had smaller territories (82–236km²), and females 83–205km².[41] During Begg's study in KTP, she noted that the 13 females within the range of one male were never seen to interact, and their radio-collar locations suggested they never came close than 2.6km, despite having overlapping ranges, the distance between females perhaps being maintained by their regular urination along foraging routes.[42] Females generally did not have fixed den sites within their territories, except when they had young. The radio tracking study showed that the 25 badgers collared (13 female and 12 male) had larger home areas than in less arid parts of the honey badger range.[43] The evidence from tracking badgers in KTP indicated a lack of territoriality, though clear avoidance of each other by females, and the only serious competition (leading to fights which leave wounds on the males' backs – older males have obvious scarring and are known as scarbacks) is over the chance to mate with females.[44] Fighting does occur if dominance posturing does not resolve a conflict. This may involve barging into each other, trying to bite tails, necks, or backs – the latter leading to the scarring mentioned.[45]

Young males have their territories within the larger territories of older males and remain close to their natal ranges for months after they are completely independent. The females disperse away from their mothers on becoming independent and able to forage for themselves – sometimes moving 23–45km from the natal territory.[46] Apart from when females had cubs, neither males nor females concentrated their movements in particular areas of their home ranges. Badgers generally forage alone, though a female might be accompanied by a cub. Observation of groups of more than two or three is unusual. In March 2022, Botswana Predator Conservation posted online camera trap film of seven honey badgers in Moremi Reserve that moved together as a group in single file. It is not clear what the relationship was between these badgers – another video from there appeared to show five male badgers following a female.[47] Apart from when females are denning with cubs, badgers rarely, if ever, sleep in the same place each time they rest but use a different burrow (such as an old warthog or aardvark burrow), or holes they had dug themselves, in rock crevices or even trees, as they are good climbers.[48] Most sleeping is done during the day, as badgers are chiefly nocturnal or crepuscular.

Within their home areas, badgers actively mark to show their presence and delineate territories. They use scent from faeces, urine, and secretions from scent glands (a strong-smelling yellow liquid that is also used to deter attack or while being attacked by predators, snakes, or bees) and may create latrines (as European badgers do); they also roll to leave their scent in chosen locations.[49] As a non-seasonal breeder with territories that overlap, limited aggression, and young males occupying territories within those of mature males, scent marking can serve the purpose of males ascertaining when females are in season and enabling avoidance to prevent conflict. There is a dominance hierarchy between males, and, occasionally, small groups of males (two to five individuals) may at times move around together[50] – as demonstrated by the camera trap evidence noted above. Scent marks may reveal the health and dominance rank of male badgers and serve to maintain female badger avoidance – to accomplish these tasks, the marks will be distributed in a manner that ensures they are found by other badgers.[51] Begg's study in the southern Kalahari revealed a number of methods of scent marking used by the honey badgers there: anal drag, with a badger squatting and dragging the anus on the ground in a straight line or circle; squat mark, very similar to the drag but the badger remains stationary but with up to 52 squats in one place; anal gland secretion at latrines and single-use marking sites; urination, often in holes along regular foraging routes; defecation at latrines and single-use locations on foraging routes; body or neck rubbing on the ground (to release secretions from neck glands but perhaps also to acquire scents already deposited by the badger on the ground through the above methods), often accompanied by vigorous scratching of the ground with the claws.[52] Scent marking will often be accompanied by intensive smelling of the marking site and ground around it. Adult males spent more time scent marking than young males or females.[53] Adult males will mark over the scent marks of other males. Most scent marking is under trees, large bushes, and in holes along foraging routes.[54]

Honey badgers have a number of vocalisations used in social behaviour, threats to other carnivores, or in defence when attacked. The sounds are very guttural, with some like coughs and others as a screaming bark or a haarr-haarr noise. All badgers make a rattling roar in confrontations with leopards, hyenas, and lions. Males use a loud, muttering-type sound to attract females.[55]

Breeding and life expectancy

There are many myths about honey badgers. One is that they must eat honey before mating. This is a complete invention, particularly as badgers only consume honey when eating bee larvae after raiding nests or hives.[56] Breeding is not between mated pairs and has been described as "a relatively promiscuous mating system, with males maximizing the breeding frequency by overlapping their movement ranges and moving faster than female counterparts."[57] There may be competition between males, including adult and young males with the same broad territory, to mate with a receptive female. Honey badgers are thought to reach sexual maturity at 2–3 years old for males and 12–16 months for females.[58] Once sexually mature, males are estimated to spend 15% of their time looking for receptive females, and females may seek out males using scent marks to find a badger that left an attractive scent, presumably in terms of the indications of health, size, and rank.[59] There are clear dominance rankings among males, "loosely based on age, mass and testes size."[60] But research honey badgers, with subordinate males within an adult male's territory being frequently seen to mate with receptive females.[61]

Breeding is not tied to any one season, and cubs are born throughout the year, though in the Kalahari there were slightly more births in the hot-wet season (47%) and hot-dry season (33%) than the cold-dry season.[62] These are dug with a single entrance and an underground chamber about 60cm long and 40cm wide. The burrow contains dry grass or leaves gathered for bedding.[63] Adult males take no part in cub rearing or feeding nursing females with which they have mated. Badgers appear to have only one cub, as two cubs have not been observed once they emerge from the den or are carried by the female to a new den. It is possible that two cubs could be born and one dies before emergence from the natal burrow.[64] Cubs stay with their mothers for 12–16 months as they grow in size and learn to forage by accompanying their mother on increasingly long trips. The long period with the mother and low birth rate mean that populations increase slowly, a problem in areas where numbers are low and they are subject to persecution. They disperse when they are capable of foraging/hunting, perhaps prompted by the arrival of a new cub. Studies have indicated quite a high death rate among cubs – 7 out of 19 monitored by Begg et al. – with starvation, infanticide, and predation the main causes.[65] There is a high rate of mortality among newly independent cubs – 47% – usually a result of starvation or predation.[66] Life expectancy is about 7 years in the wild, though they have lived up to 28 years in captivity.[67]

Diet, foraging, and hunting

Having a wide habitat tolerance, including rainforest, woodland savannas, grasslands, marshes, afro-alpine steppes up to 4,050m, scrub, coastal sandveld, and deserts,[68] their diet is correspondingly eclectic. In the southern Kalahari, diet was found to comprise at least 59 species, dominated by vertebrates (83%), including mammals up to the size of springhare (2kg), reptiles (including pythons and venomous snakes) and birds, birds' eggs, followed by invertebrates (11%, mainly bee larvae), berries, fruit, tsamma melons, honey, and carrion.[69] Begg et al. noted that the hairy-footed gerbil (33%) and barking gecko (19%) were major food items, and large snake species were important in diets of both sexes (being present 42.1% and 48.7% of the time in food intake). Springhares represented 32.7% of the biomass eaten by females, but only 4.2% in males.[70] Badgers will dig out springhares, ground squirrels, rats, and other rodents from their burrows and have been observed climbing up trees to predate tawny eagles and white-backed vulture chicks. They have also been known to predate black-backed jackals, bat-eared foxes, and aardwolf pups and the young of other small carnivores.[71]

It is the breaking into wild bee nests and domestic hives for the larvae that gets the badger its name, rather than eating honey, which is a just a by-product of larvae consumption – the destruction of nests and hives is the cause of most human–badger conflict and persecution of badgers. Begg et al. reported that researchers had seen an adult female with a large cub breaking into 13 hives and making 37 visits to the hives to eat larvae and honeycomb – with about 3,000 larvae of the solitary bee, *Parafidelia friesee*, consumed.[72] Bees are not defenceless, and 61% of the visits ended with the badgers retreating in the face of fierce defence by them. Four male honey badgers were recorded making 81 visits, totalling 53.2 hours, to bee nests, repeatedly digging out the bee larvae.[73] The allure of bee larvae is such that with the decline in many areas of wild bee numbers, honey badgers frequently raid domestic beehives, despite the risks of trapping, poisoning, or shooting that result. They are considered to be the most destructive mammalian predators of honeybees in Africa and can destroy 20 hives in a night; in Tanzania, they were recorded as destroying 2,700 out of 24,000 hives kept by beekeepers in a single year.[74] Keith Begg wrote that the destruction of hives and wild bee nests gave badgers the reputation of surviving mainly on honey and of deriving their "excessive energy … and outbursts of temper" from the large amount of honey consumed – both of which are myths with no scientific backing. Bee larvae only amounted to 14% of food in the stomachs of dead honey badgers examined by researchers.[75] The periodic consumption of young sheep and goats and raiding of poultry are something else which have not endeared badgers to farmers; they have been known to kill adult sheep. They also raid waste sites around villages and safari camps in national parks and reserves, as I observed at Sinamatella Camp in Hwange NP.

One of the tactics said to be used by honey badgers to raid hives without sustaining an unacceptably painful and toxic number of bee stings is the use of their anal secretions. The secretion is said to be to be "unendurable" and acts "like an anaesthetic, causing some bees to flee and others to become moribund," while it has also been alleged that the secretion causes bees to attach themselves to the badger's tail. The badger supposedly takes these bees away from the nest/hive and then returns to take the larvae undisturbed.[76] But 41 observations of badgers raiding hives did not lead to a single verified use of the anal secretions to combat bees.[77] There is also no reliable data on the use of their anal gland secretions to subdue or drive off bees from nests.

Colleen Begg told me:

> [W]e have seen honey badgers break into beehives on numerous occasions in Niassa [northern Mozambique] and Kalahari. We have never seen them fumigate the hive. What generally happens is they break in, get a lot of stings and at some point get very agitated and run away sometimes emitting secretions and sometimes not. They return to the hive on multiple days, each time getting stung less as the hive is weakened. We only saw release of major secretions for lion, leopard and humans. So I think it is simply a defensive mechanism that they release when agitated.[78]

The badgers' loose, tough skin may protect them from bee stings to an extent, and there have been unverified accounts of them caking themselves in mud to avoid being stung. While they may be even partially immune to stings, badgers have been found dead as a result of bee stings in traps set around domestic hives.[79] One badger removed alive from a trap by a beekeeper was found to have 228 stings, while another in the Caledon district had over 1,500.[80] Honey badgers of both sexes forage alone, though a female with an older cub may take it along for it to learn where and what to forage. There have been no occasions when researchers have observed badgers hunting cooperatively. Badgers move around in two main ways – "the slow winding walk with frequent investigation of scent trails and prey burrows used by both sexes when intensively foraging and the faster, directional jog-trot used only by males engaged in social activities."[81] Badgers hunt opportunistically and will eat as much as 60% of their body weight in a day if food is available, one male being observed to eat four adult mole snakes, two adders, and seven mice.[82] Badgers eat tsamma melons in arid areas and other vegetable matter but seem to prefer eating mammals, reptiles, and insects. Drabeck, Dean, and Jansa noted their "near legendary ability to attack venomous snakes" and their ability to recover from highly venomous bites, including from puff adders and even black mamba venom – the resistance to venom may have evolved through repeated exposure and, having developed, enabled badgers to exploit a useful food source.[83] Honey badgers are said to be able to sleep off the effects of snake and scorpion venom and survive relatively unscathed.[84]

Tortoises have also become a regular source of food, notably the tent tortoise in the Karoo. Shells collected there showed that 49% had been cracked open cleanly and the body removed and consumed. Spoor of honey badgers was often found near the broken shells. Lloyd and Stadler examined the shells and presence of predators and concluded that honey badgers were the most likely "suspect," followed by jackals and caracals, though the latter two lacked the powerful jaws and teeth of the badger.[85] The likelihood of honey badgers predating tortoises was reinforced by an eyewitness account from KTP by Professor Angela Fuller of the University of the Witwatersrand, who witnessed a badger catching and spending two hours working out and eventually succeeding in opening the shell to eat the tortoise.[86] Badgers have also gained the reputation, not verified by sightings or research, of attacking and killing large ungulates, including gemsbok and even buffalo. They are ferocious hunters, known to take on powerful animals and are persistent hunters, willing to travel 20 miles (32km) in their pursuit of a meal, a BBC documentary reported.[87] Kingdon reported stories of them hunting the young of large ungulates and to have amazing stamina in pursuing them. He wrote that reports suggested they attacked large prey from beneath the belly and that they tried to tear the scrotum of large antelopes and even buffalo, causing death through massive blood loss. They are even said to attack men in this way – though Kingdon says he has never found anyone who has witnessed or been a victim of this.[88] What is certain and verified by research, and which I have witnessed on two occasions in the KTP, is the immense digging power and stamina of honey badgers – digging down as much as a metre in rock-hard, dry ground to get at rodents, reptiles, or insect larvae.[89] They would dig for a considerable time to capture large snakes, such as mole snakes and cobra, and "[i]n terms of time spent against food biomass, large snakes were the most nutritionally profitable part of the diet."[90]

Honey badgers will spend most of daylight hours inactive, sleeping in burrows or other concealed places. But they will forage, as I witnessed in the KTP in the cold season there, in the day during colder times of the year, when they spend more time foraging than they do in the hot seasons.[91] Males forage over greater distances than females, with the latter often tied to den sites containing a cub. It is estimated that males may forage for up to 27km in a 24-hour period, with the females covering a smaller area of 10km.[92]

Cooperation, commensalism, symbiosis, or kleptoparasitism

One of the honey badger's major relationships with another animal, sustained by popular belief, is the supposedly symbiotic relationship with the honeyguide bird. It has long been said that badgers would follow the honeyguide, which feeds on larvae, bee eggs, waxworms, and even beeswax. The bird located nests and, through its calls and by flying at a height that the badger could see and follow, guides the badgers to nests of the bee, *Apis mellifera*. The badger would then open the nest with its claws and take most of the stings of the bees, enabling the bird to feed on larvae or beeswax that the badger did not consume.[93] The nature of the relationship is open to question – with no absolute proof that the birds do lead badgers to nests or whether badgers respond to the birds' calls.[94] This story has been around for centuries and is replicated in very reputable "standard texts on animal behavior, in popular articles, and in many scientific works on birds and mammals,"[95] such as the most recent authoritative work on the African honeyguide species.[96]

This relationship is also mentioned in the *Princeton Encyclopedia of Mammals*,[97] and in Laurens van der Post's often questionable and over-romanticised accounts of the myths of the San of the Kalahari, which describe the relationship with the honeyguide. Van der Post says that he was told by his San interpreter that the honeyguide sought out honey badgers, flying to find them if they heard the badgers, and that San honey gatherers would imitate the vocalisations of honey badgers to attract honeyguides, which they would then follow, hoping to find bee nests and honey.[98] The first warden of South Africa's Kruger NP, James Stevenson-Hamilton, also reported that he had seen honeyguides and honey badgers together in Kruger and said they had a strange "alliance," but without claiming that he saw the honeyguide clearly guiding the badger or profiting from the badger breaking open nests.

In his book to accompany the television series *The Life of Birds*, the eminent naturalist and documentary presenter David Attenborough gives this colourful but not very scientifically grounded description of the honeyguide–honey badger relationship:

> Honeyguides … eat a variety of insect food but all have a taste, not so much for honey as for bee grubs and bee's wax … African bees … favour more protected sites for their nests – holes in trees or deep crevices between rocks – and to reach those, honeyguides need help. The ratel … has a particular passion for honey and bee grubs … As it [the badger] enters the territory of a honeyguide, the bird will fly down to perch on a bush close to it and call with a distinctive chattering cry. The ratel answers with grunts and ambles towards it. The bird flies off and the ratel follows. Every now and then, the bird stops and calls again … as it waits for the ratel to catch up. Eventually the bird flutters up to a higher perch and gives a different call. That indicates that a bees' nest is close by, perhaps in a tree hole. When

the ratel finds it, it starts excavating. The bees retaliate. The ratel is by no means impervious to their stings … [the badger] rips away the wood to enlarge the entrance hole. Eventually it claws out the combs, dripping with rich deep-brown honey and takes them away to eat. The honeyguide now has its chance and flutters down to feast on what is left of the wrecked nest. The benefits of the partnership to both animals are clear.[99]

Kingdon, in his comprehensive guide to East African mammals, also reported that badgers follow honeyguides to find bee nests and break them open, enabling them to get the larvae and that the birds benefitted from having the nests broken open by eating leftover grubs, any dead bees, and pieces of honeycomb.[100]

Dean et al. examined this narrative and showed great scepticism about the reputed relationship. They said that there was little, if any, scientific evidence to support it. They found the accounts to be based on anecdotal rather than researched or verified information, all of it dating back to before 1955 and contained in the account by Friedman, based on humans and honeyguides,[101] in which Friedman at first seems to deride "native" reports of a relationship between honeyguides and honey badgers but then suggests they were so common that they must have a factual basis:

> The fact that the association of the ratel with the honey-guide is reported consistently by the natives throughout vast portions of southern and eastern Africa is in itself suggestive of the factual basis of Sparrman's account [Swedish naturalist Anders Sparrman, 1748–1820], but by itself cannot be looked upon as constituting definite proof.[102]

Friedman only decided that there was a relationship between the honeyguide and the badger when he collected anecdotal accounts from white hunters, farmers, and colonial officials, some of whom had only seen a single example of the bird and the badger together near a bee nest.[103] Dean et al. said,

> No additional records of Greater Honeyguides guiding honey badgers have been reported since Friedmann's review, despite the great increase in the number of scientists studying the biota in southern and eastern Africa during the past 30 years. If it is a coevolved, mutualistic association, the association would have been incidentally observed.[104]

Colleen Begg, in her doctoral thesis on honey badgers, doesn't go along with the total rejection of the relationship and suggests that the nocturnal foraging behaviour of the honey badger in areas with potential predators and human populations may have had an effect on any symbiotic behaviour with the diurnal honeyguide and that "the association may exist in wilderness areas, but may have disappeared in more populated areas where the honey badger is largely nocturnal."[105] She has also suggested that it may be the case that, as with the relationship between the honey badger and both the black-backed jackal (*Lupulella mesomelas*) and pale chanting goshawk (*Melierax canorus*), that the honeyguide may follow the badger, hoping it will lead it to a bees' nest.[106] This conclusion was also suggested by the observations of Fincham et al. of an orphaned juvenile honey badger and honeyguides in Stone Hills nature sanctuary, near Matobo NP, in south-western Zimbabwe. Honeyguides appeared to follow the badger and to make calls when they were near a bees' nest but with no reaction from the badger. They concluded that "there was innate recognition of the adolescent badger by adult and juvenile honeyguides. It is conceivable that this relates to nutritional benefit for the bird when badgers raid bees' nests."[107]

The other key relationships involving the honey badger and other animals – which I witnessed twice in the KTP on trips two years apart and referred to in the Introduction – is the two-way and three-way interactions between honey badgers, black-backed jackals, and pale chanting goshawks, with the latter two following badgers as they forage, grabbing prey that escaped the badger. This again has been much discussed, and ideas range from symbiosis, through commensalism to kleptoparasitism by the jackal and the hawk. For there to be symbiosis, one would expect some gain for the badger as well as the obvious gain for the jackal and hawk when badger foraging and digging produce rodent, reptile, or insect prey that the other two can eat. Kleptoparasitism would involve theft of prey from the badger or in some way depriving the badger of its just desserts. My observations and those of much more experienced researchers suggest the prey taken by the hawk or jackal had escaped the clutches of the badger. Begg and Begg et al., in my view correctly, identify the relationship as commensal with the jackal and hawk benefitting but with no clear cost or benefit to the badger.[108] This is a relationship that may also occur with ant-eating chats, the greater honeyguide, and crimson-breasted shrikes.[109] Begg concluded that

> [g]oshawks and jackals benefit from the association by increased hunting opportunities and intake rate ... chanting goshawks show increased strike success and expanded prey base when hunting with honey badgers compared to hunting alone. The honey badger does not show any significant differences in capture success, intake rate or predator vigilance when foraging in association compared to foraging alone,

although there was some suggestion of a minor loss to prey for the badger if it was hunting above ground with a following jackal.[110] Begg et al. concluded in 2020 that "honey badgers are aware of associating individuals and react to their behaviour, this seldom results in their own capture of a prey item."[111]

In a later study, Begg et al. provided a fantastically detailed account of badger interactions with other animals and noted that honey badgers preyed or attempted to prey on all mammalian carnivores smaller than themselves, as well as the young of medium-sized carnivores. Lions (*Panthera leo*), leopards (*Panthera pardus*), and probably spotted hyenas (*Crocuta crocuta*) preyed on honey badger adults and cubs, and cubs were killed by black-backed jackals on occasions. They recorded 41 occasions when both jackals and goshawks interacted with foraging badgers, and at times, there were three goshawks and two jackals following a single honey badger.[112] Pale chanting goshawks were present in 36% of daytime sightings of honey badgers and goshawks were seen to benefit by catching 71 prey items when following badgers.[113] There were no indications that goshawks guided badgers to rodent burrows; rather they spotted and followed badgers. In 16% of sightings of badgers, jackals were in attendance – with from one to four jackals present and in all cases the jackals initiated an interaction.[114] Jackals caught 69% of prey animals that escaped a digging badger and on two occasions gave alarm calls denoting predators, but the badger, after a quick look around, went on digging.[115] On 14 occasions a jackal bit the rump of a digging badger or bit a badger cub, at which point the badger aggressively turned on the jackal and it retreated.[116] Both jackals and goshawks would wait near resting badgers and then followed them when they moved on.

Interactions with other carnivores

Honey badgers are formidable foragers and hunters, and have an ability beyond their size to defend themselves and their young against predators such as pythons, leopards, hyenas, and even lions and have even been observed successfully competing with hyenas and lions for access to carcasses. Despite being relatively small, the badger is "extraordinarily muscular,"[117] and the neck and shoulders are strongly muscled, and the front legs are very strong with massive claws, while the thick, loose skin, especially on the neck, gives it very good protection. This does not stop them being predated at times by leopards and killed by determined lions.[118] When faced with larger carnivores, badgers have a formidable display of a rattling roar, rushing at predators and releasing scent from the anal glands that may deter predators in close contact, "and when fighting is inevitable, their coarse, loosely-fitting skin and thick sub-cutaneous fat deposits have an important protection function."[119]

A study in the Serengeti suggested that honey badgers had little fear of larger carnivores, and there was no evidence that they avoided locations with abundant lions, leopards, or hyenas, or that they foraged at times to avoid them.[120] The first head of Nairobi NP in Kenya, Mervyn Cowie, wrote that honey badgers were one of the few animals that did not appear to be intimidated by lions, and he observed three badgers (an unusual event to have three foraging together) approaching a lion kill from which seven lions were feeding. The ratels moved in aggressively, and the lions retreated, enabling the badgers to feed. When the badgers fed and moved off, the lions returned to the kill.[121] Yet it is undoubtedly the case that lions and leopards, and perhaps hyenas too, will kill and consume honey badgers on occasion, with females with young most in danger; jackals have been known to kill unattended badger cubs.[122] Schaller in his study of tigers in India's Kanha NP found that some tiger scats showed that they had killed and eaten honey badgers.[123]

The defensive capabilities, physical strength, and aggression of the badgers may have led to an interesting example of mimicry of badger colouration in the case of cheetah cubs. Eaton pointed out that the young cheetahs "appear to mimic the honey badger or ratel," with the silvery or white band of longer fur running down the back of the cheetah as it does with the honey badger.[124] This appearance is in contrast to most pelage colouration which seeks to be inconspicuous. The band of very conspicuous may suggest that the cheetah cubs have evolved this colour to warn off possible predators on the basis that they are wary of approaching or attacking a honey badger. The mimicry has not been definitively proven to be a successful defence, but it is suggestive of one. It should be added that badgers have been known to prey on very young cheetah cubs.[125]

Interactions with humans

While raiding beekeepers' hives, wild bee nests harvested by honey gatherers, and the accusation of killing poultry, livestock, or farmed game animals are at the centre of the conflictual relationship between humans and honey badgers, there are other, often mythical, elements. Aggression towards humans by badgers are at the top of the list, with stories from Africa and Asia of badgers randomly attacking or doing so in defence or when feeling threatened. Most accounts refer to badgers lunging at the genitalia to castrate the victim, while a few refer to badgers trying to hamstring people.[126] Mervyn Cowie wrote of honey badgers in Kenya easily breaking through chicken wire and killing as many as 30 chickens inside a chicken run and being aggressive towards people, having been known to lunge at men, "leaping at his middle."[127] Reports from India and West Asia speak of badgers successfully fighting off men armed with clubs or supported by hunting dogs. From the same region there are stories, unsubstantiated one must say, of honey badgers digging corpses from recent graves and consuming parts of them.[128]

In southern Africa there is some trophy hunting of honey badgers. The Convention on International Trade in Endangered Species of Wild Fauna and Flora (CITES) data indicates that 16 trophy animals were exported from South Africa annually between 2002 and 2012. The lack of legislation and law enforcement for the protection of the species and the high number of honey badgers reported to have been killed by beekeepers has led to Botswana and Ghana putting them as an Appendix III species on the CITES list, making it mandatory to have a permit for export and import of the species. CITES documents record that there are no established hunting or export quotas for honey badgers.[129] Appendix III "contains species that are protected in at least one country, which has asked other CITES Parties for assistance in controlling the trade. Changes to Appendix III follow a distinct procedure from changes to Appendices I and II, as each Party's is entitled to make unilateral amendments to it."[130]

But it is the badgers' raiding of bee nests and beehives that has brought about the greatest conflict. Persecution by beekeepers has been reported in Jordan, Israel, and Iran, while raids by badgers on hives/apiaries have been reported from Angola, Benin, Botswana, Burundi, Ethiopia, Kenya, Malawi, Mozambique, Nigeria, Senegal, Somalia, South Africa, Tanzania, Togo, Uganda, and Zimbabwe.[131] Kingdon reported a case of a Tanzanian beekeeper near Dodoma using a spear to drive off and then kill three honey badgers raiding his hives. He said that when he was fighting them, they kept trying to go for his "lower parts," as Kingdon puts it.[132] They may also be killed by predator control programmes targeting other species, such as jackals and caracals. There is evidence to suggest they have gone locally extinct in some areas (such as Free State Province in South Africa), which may be through poisoning or trapping, while in others where they are not rare or uncommon, populations may be in decline through persecution, becoming roadkill, or illegal trapping and poisoning.[133]

One survey of carnivore conservation in southern Africa reported that beekeepers in the Western Cape had killed at least 248 badgers and relocated 231 over 15 years.[134] A quarter of farmers in the Western Cape interviewed by Ray, Hunter, and Zigouris admitted trapping badgers and reported killing 30–90 badgers each, one saying that 22 badgers were captured at one hive site alone.[135] Badgers may also die through taking poisoned baits from coyote-getters or eating from poisoned carcasses intended to kill jackals and caracals. At Kruger's Satara Camp, 40 badgers were culled by South African National Parks after raiding fridges, food stores, and bins in the rest camp.[136] The propensity for raiding waste dumps, killing livestock or poultry, and raiding hives may have developed from changes in land use and the decline in natural prey and wild bee nests in some areas.[137]

The meat of the honey badger is not prized as a form of bushmeat, and they are unlikely to be deliberately hunted for food – the meat is said to be unpleasantly sweet. They may be killed and then consumed as a by-catch of snaring for antelopes, warthog, bushpigs, etc.[138] Body parts are highly prized for traditional medicine, and they may be hunted or trapped to obtain skins, claw, and other body parts. Badger paws, skins, fat, bones, and internal organs are used for medical and spiritual purposes, known as *muti*[139] – the badgers' reputation for fearlessness and aggression adds to its allure as a medicine or a form of spiritual protection. CITES recorded in 2002 that fat and other body parts from honey badgers were part of the trade in wildlife products.[140] For further details of the contemporary state of the beekeepers' war against the badger and the trade in body parts, see Chapter 9.

Notes

1 W.C. Wozencraft (2005) Species Mellivora capensis, in D.E. Wilson & D.M. Reeder (eds) *Mammal Species of the World: A Taxonomic and Geographic Reference* (3rd ed.), Baltimore, MD: Johns Hopkins University Press, 532–628, p. 612.

2 J.I. Rhodes (2006) Phylogeographic structure of the honey badger (*Mellivora capensis*), Thesis for Master of Science degree, University of Stellenbosch, December 2006, p. 6.

3 C.M. Begg et al. (2016) A conservation assessment of Mellivora capensis, in M.F. Child et al. (eds) *The Red List of Mammals of South Africa, Swaziland and Lesotho. South African National Biodiversity Institute and Endangered Wildlife Trust, South Africa*, https://www.researchgate.net/publication/319471704_A_conservation_assessment_of_Mellivora_capensis, accessed 21 March 2022; and Gilbert Proulx et al. (2016) World distribution and status of badgers — a review, in G. Proulx & E. Do Linh San (eds) *Badgers: Systematics, Biology, Conservation and Research Techniques*, Sherwood Park, Alberta, Canada: Alpha Wildlife Publications, 31–116, pp. 62–3.

4 G. Baryshnikov (2000) A new subspecies of the honey badger Mellivora capensis from Central Asia. *Acta Theriologica*, 45, 45–55, p. 45.

5 C.M. Begg (2001) Study from July 1996 to Dec 1999 - Feeding ecology and social organisation of honey badgers (*Mellivora capensis*) in the southern Kalahari, University of Pretoria doctoral thesis November 2001, https://repository.up.ac.za/bitstream/handle/2263/29895/complete.pdf accessed 15 September 2020, pp. 1–2.

6 L. Hunter (2011) *Carnivores of the World*, Princeton, NJ: Princeton University Press, p. 166; and Animals Reference – Honey Badger, *National Geographic*, https://www.nationalgeographic.com/animals/mammals/h/honey-badger/ accessed 1 October 2020.

7 L. Hunter (2011) *Carnivores of the World*, Princeton, NJ: Princeton University Press, p. 166.

8 Rhodes, 2016, p. 6.

9 J. Kingdon (1977) *East African Mammals an Atlas of Evolution in Africa, Vol. IIIA*, London: Academic Press; see also N. Oosthuizen & T. Felmore (2019) Honey badger – Africa's most fearless and tenacious carnivore, *Africa Geographic*, 29 March 2019, p. 102.

10 Team Africa Geographic (2020) Black honey badgers spotted in Gabon, *Africa Geographic*, 12 February 2020, https://africageographic.com/stories/black-honey-badgers-spotted-in-gabon/?mc_cid=d44e8a0350&mc_eid=a920cc2294&fbclid=IwAR0OX-qqbz-QZ4BCW0SC2J6KezR_I99Y7S4y3sT5xM7Huc08oGLndyIUPmAE accessed 30 June 2021.

11 Ibid.

12 Kingdon, 1977, p. 87.

13 Ibid., p. 90.

14 Ibid., p. 87.

15 B. Panesar (2020) Honey badgers: adorable but fierce little mammals, *Live Science*, 22 August 2020, https://www.livescience.com/honey-badger.html accessed 3 November 2020.

16 SANBI (no date) The honey badger, South African National Biodiversity Institute, https://www.sanbi.org/animal-of-the-week/honey-badger/ accessed 4 November 2020.

17 J. Kingdon (1979) East African Mammals: v. 3B: An Atlas of Evolution in Africa, San Diego, CA: Academic Press.

18 Panesar, 2020.

19 BBC Natural World (2014) *Honey Badgers: Masters of Mayhem. Natural World*, https://www.bbc.co.uk/programmes/b0418x7x accessed 4 November 2020.

20 C.A. Long & C.A. Killingley (1983) *The Badgers of the World*, Springfield, IL: Charles C. Thomas Publisher, p. 30.

21 Ibid., pp. 33–4.

22 S.R. Jnawali et al. (2011) *The Status of Nepal's Mammals: The National Red List Series*, Kathmandu, Nepal: Department of National Parks and Wildlife Conservation, https://www.academia.edu/14715540/The_Status_of_Nepal_Mammals_The_National_Red_List_Series?email_work_card=view-paper accessed 1 January 2021, p. 69.

23 Proulx et al., 2016, p. 63.

24 Ibid.
25 Rhodes, 2006, p. 12.
26 Proulx et al., 2016, p. 63.
27 Begg et al., 2016, p. 5.
28 H. Sharifi, M. Malekian & G. Shahnaseri (2020) Habitat selection of honey badgers: are they at the risk of an ecological trap? *Hystrix, the Italian Journal of Mammalogy*, 7, 31, 131–6, p. 131.
29 Ibid.
30 Begg et al., 2016, p. 5.
31 Proulx et al., 2016, pp. 64–5.
32 D.L. San et al. (2016) *Mellivora Capensis. The IUCN Red List of Threatened Species* 2016: e.T41629A45210107. https://dx.doi.org/10.2305/IUCN.UK.2016-1.RLTS. T41629A45210107.en accessed 1 October 2020; and C.M. Begg et al. (2016) Mellivora capensis – Honey Badger, *The Red List of Mammals of South Africa, Lesotho and Swaziland*, South African National Biodiversity Institute and Endangered Wildlife Trust, p. 3.
33 C.M. Begg et al. (2016).
34 Rhodes, 2006, p. 12.
35 Ibid.
36 Colleen Begg personal communication, cited by Rhodes, 2006, p. 12.
37 Begg, 2001, p. 162.
38 Hunter, 2011, p. 166.
39 Begg, 2001, p. 155.
40 National Geographic (no date) Honey badger, Animals, Reference, https://www. nationalgeographic.com/animals/mammals/h/honey-badger/ accessed 4 November 2020.
41 Begg, 2001, p. 156.
42 Ibid., p. 162.
43 C.M. Begg (2005) Spatial organization of the honey badger Mellivora capensis in the southern Kalahari: home-range size and movement patterns, *Journal of Zoology*, 265, 23–35, p. 23.
44 Begg, 2005, p. 28.
45 Y. Yaniv & I. Golani (1987) Superiority and Inferiority: a Morphological Analysis of Free and Stimulus Bound Behaviour in Honey Badger (*Mellivora capensis*) Interactions, *Ethology*, 74, 2, 89–116, p. 96.
46 Ibid.
47 Facebook, 23 March 2022.
48 Panesar, 2020 and Begg, 2005, p. 28.
49 C.M. Begg (2003) Scent-marking behaviour of the honey badger, Mellivora capensis (Mustelidae), in the southern Kalahari, *Animal Behaviour*, 66, 917–29, p. 917.
50 Ibid., p. 918.
51 Ibid., pp. 918–9.
52 Ibid., p. 921.
53 Ibid., p. 922.
54 Begg, 2001, p. 168.
55 SANBI (no date) The honey badger, South African National Biodiversity Institute, https://www.sanbi.org/animal-of-the-week/honey-badger/ accessed 4 November 2020.
56 Kingdon, 1977, p. 102.
57 S. Carter et al. (2017) The honey badger in South Africa: biology and conservation, *International Journal Avian & Wildlife Biology*, 2, 2, https://medcraveonline. com/IJAWB/the-honey-badger-in-south-africa-biology-and-conservation.html accessed 14 October 2021, no page numbers.
58 Ibid.

59 Begg, 2001, p. 98.
60 Begg, 2001, p. iv.
61 R. Verwey et al. (2004) A microsatellite perspective on the reproductive success of subordinate male honey badgers, Mellivora capensis, *African Zoology*, 39, 2, 1–4, p. 1.
62 Begg et al., 2005, p. 19.
63 Long & Killingley, 1983, p. 34.
64 Ibid.
65 Begg et al., 2005, pp. 19–20.
66 Carter et al., 2017.
67 Begg et al., 2005, pp. 19–20.
68 J.C. Ray, L. Hunter & J. Zigouris (2005) *Setting Conservation and Research Priorities for Larger African Carnivores*. New York: WCS Working Paper No. 24. Wildlife Conservation Society, pp. 125–8.
69 C.M. Begg et al. (2003) Sexual and seasonal variation in the diet and foraging behaviour of a sexually dimorphic carnivore, the honey badger (*Mellivora capensis*), *Journal of Zoology*, 260, 301–16, pp. 304–5.
70 Ibid., pp. 305–6.
71 Ibid.
72 Ibid.
73 Ibid., p. 305.
74 K. Begg (2001) *Report on the Conflict between Beekeepers and Honey Badgers Mellivora Capensis, with Reference to their Conservation Status and Distribution in South Africa*, Johannesburg: Endangered Wildlife Trust, p. 9.
75 Ibid.
76 Ibid.
77 Keith & C. Begg (2002) The conflict between beekeepers and honey badgers in South Africa: a Western cape perspective, *Open Country*, 4, 25–36, pp. 28–9.
78 Personal communication with Colleen Begg, 8 October 2020.
79 Keith & Begg, 2002, pp. 28–9.
80 Begg, 2001, p. 32.
81 Begg et al., 2003, p. 310.
82 Ibid.
83 D.H. Drabeck, A.M. Dean & S.A. Jansa (2015) Why the honey badger don't care: Convergent evolution of venom targeted nicotinic acetylcholine receptors in mammals that survive venomous snake bites, *Toxicon*, 99, 68–72, p. 68.
84 N. Oosthuizen & T. Felmore (2019) Honey badger – Africa's Most fearless and Tenacious Carnivore, *Africa Geographic*, https://africageographic.com/stories/the-honey-badger/ accessed 4 November 2020.
85 P. Lloyd & D.A. Stadler (1998) Predation on the tent tortoise Psammobatestentorius: a whodunit with the honey badger Mellivoracapensis as prime suspect, *South African Journal of Zoology*, 33, 4, 200–2, pp. 201–2.
86 Personal communication and Andrea Fuller on Twitter: https://t.co/2JskmxPPvB accessed 10 January 2022.
87 Panesar, 2020.
88 Kingdon, 1977, p. 98; and Long & Killingley, 1983, p. 352.
89 Begg, 2001, p. 34.
90 Ibid., p. 35.
91 Ibid., pp. 7–8.
92 SANBI (no date) The honey badger, South African National Biodiversity Institute, https://www.sanbi.org/animal-of-the-week/honey-badger/ accessed 4 November 2020.
93 F.W. Marlowe (2010) *The Hadza Hunter-Gatherers of Tanzania*, Berkeley: University of California Press, p. 117.
94 Ibid.
95 W.R.J. Dean, W. Roy Siegfried & I.A.W. MacDonald (1990) The fallacy, fact, and fate of guiding behavior in the greater honeyguide, *Conservation Biology*, 4, 1, 99–101, p. 99.

96 See for example, C.H.S. Fry, S. Keith & E.K. Urban (1988) *The birds of Africa. Volume 3.* London: Academic Press.

97 D. Macdonald (ed) (2009) *Princeton Encyclopedia of Mammals,* Princeton, NJ: Princeton University Press, p. 500.

98 L. van der Post (1965) *The Heart of the Hunter,* Harmondsworth, Middx.: Penguin, p. 69.

99 D. Attenborough (1998) *The Life of Birds,* London: BBC Books, pp. 161–3.

100 Kingdon, 1977, pp. 91–3.

101 W.R.J. Dean, W. Roy Siegfried & I.A.W. MacDonald (1990) The fallacy, fact, and fate of guiding behavior in the greater honeyguide, *Conservation Biology,* 4, 1, 99–105, p. 99.

102 H. Friedmann (1955) The honey-guides, *Bulletin of the United States National Museum,* 1–292, p. 42.

103 Ibid., pp. 43–6.

104 Dean et al., 1990, p. 100.

105 Begg, 2001, p. 102.

106 K.S. Begg & C.M. Begg (2017) The honey badger: associations, http://www.honeybadger.com/associations.htmlhttp://www.honeybadger.com/associations.html accessed 8 October 2020.

107 J.E. Fincham, R. Peek & M.B. Markus (2017) The greater honeyguide: reciprocal signalling and innate recognition of a honey badger, *Biodiversity Observations,* 8, 12, 1–6, p. 5.

108 Begg, 2001, pp. iv–v.

109 Begg et al., 2016, p. 4.

110 Begg, 2001, p. 103.

111 C. Begg et al. (2022) Interactions between honey badgers and other predators in the Southern Kalahari: Intraguild Predation and Facilitation, (no volume or issue numbers available yet) 323–46, p. 326.

112 Ibid., p. 327.

113 Ibid., p. 329.

114 Ibid., p. 333.

115 Ibid., p. 334.

116 Ibid.

117 Kingdon, 1977, p. 87.

118 Begg et al., 2016, p. 305.

119 Begg et al., 2016, p. 4.

120 M.L. Allen, B. Peterson & M. Krofel (2018) No respect for apex carnivores: Distribution and activity patterns of honey badgers in the Serengeti, *Mammalian Biology,* 89, 89–94, p. 89.

121 M. Cowie (1966) *The African Lion,* London: Arthur Barker, p. 56.

122 Begg, 2001, p. 99.

123 G.B. Schaller (1967) *The Deer and the Tiger A Study of Wildlife in India,* Chicago, IL: University of Chicago Press, p. 284.

124 R.L. Eaton (1976) A Possible Case of Mimicry in Larger Mammals, *Evolution,* 30, 4, pp. 853–6, p. 853.

125 Eaton, 1976, p. 854.

126 Begg, 2001, p. 6.

127 Cowie, 1966, p. 56.

128 Long & Killingley, 1983, p. 354.

129 CITES (no date) *Mellivora Capensis,* https://speciesplus.net/species#/taxon_concepts/9613/legal accessed 21 March 2022.

130 CITES (no date) *How CITES Works,* https://cites.org/eng/disc/how.php#:~:-text=Appendix%20III&text=A%20specimen%20of%20a%20CITES,port%20of%20entry%20or%20exit accessed 21 March 2022.

131 Proulx et al., 2016, pp. 64–5.

132 Kingdon, 1977, p. 98.

133 Proulx et al., 2016, pp. 64–5.

134 J.C. Ray, L. Hunter & J. Zigouris (2005) *Setting Conservation and Research Priorities for Larger African Carnivores*. New York: WCS Working Paper No. 24, p. 126. See also Begg, 2001.
135 Ibid.
136 Ibid., p. 127.
137 Ibid., p. 128.
138 Ibid.
139 Begg et al., 2016, p. 4.
140 CITES (2002) *Eighteenth meeting of the Animals Committee San José (Costa Rica), 8-12 April 2002 Implementation of Decision 11.165 on trade in traditional medicines: List of species traded for medicinal purposes*, p. 8.

9

ORIGINS, EVOLUTION, AND HISTORY OF THE HONEY BADGER

This chapter will deal with what is known of the evolution of the honey badger, accounts of its historical range, and its relationship with humans over millennia. Much of the narrative relates to Africa, southern Africa in particular, the area where most research has been done and from where there are the most hunters' and naturalists' accounts, but every effort has been made to include West and South Asia.

Evolution of the honey badger

From the start it should be noted that the honey badger and its genera (*Mellivorinae/Mellivora*), of which it is the only surviving member, does not just slot easily into the evolutionary path of the badger sub-family, *Melinae*, and its genera *Meles*; the honey badger being included in the latter at first before being assigned its own genera.[1] There are similarities in size, colouration, physical capabilities, and behaviour between honey badgers and true badgers, but they are not that closely related in evolutionary terms, and there is little overlap in their ranges. The first categorisation of the honey badger or ratel was by Schreiber in 1776, with it listed as *Viverra capensis*, thereby linking it to civets, genets, and binturong in the *Viverra* genus. It was changed by Storr in 1780 to *Mellivora*, reflecting its differences from civets/genets and its relationship with honey.[2] It was placed within the family *Mustelidae*, because of its similarities to the badgers of Europe, North America, and Asia.

My narrative starts with the early mammalian prototype carnivores, which have been described in Chapter 2. The order *Carnivora* evolved over millions of years from the *Cimolestes* order of early mammals, which were small animals preying on insects and small vertebrates.[3] They diversified into a wide range of species in size, habitat, and diet, including the *Mustelidae* or *Mustelids*, the family

DOI: 10.4324/9781003199793-10

to which honey badgers belong. The first true carnivores, belonging to the *Viverravidae*, a family now entirely extinct, appeared around 65–60ma.[4] But another group of early carnivores, labelled *Miacids*, evolved into a variety of forms,[5] including the *Vulpavine* group, from which came the dogs (*Canidae*), bears (*Ursidae*), and *Musteloidea*, which encompasses raccoons, coatis, kinkajou, ringtails and cacomistles (*Procyonidae*), red pandas (*Ailuridae*), skunks and stink badgers (*Mephitidae*), and the weasels and badgers (*Mustelidae*).[6] These appeared during the Oligocene and evolved through that epoch into the Miocene.

Studies of the evolution of the *Mustelid* species during the mid- to late Miocene (14–11.4ma) reveal shifts towards small, elongated bodies "that may have facilitated diversification by allowing *mustelids* to chase prey in burrows and small crevices."[7] The evolution may have been prompted by the cooling of the climate in North America and Eurasia, with shrinking of denser forest, development of open areas in woodland, and expansion of savannas/grasslands. These environments supported large numbers of small herbivores which used burrows for shelter and to escape from predators.[8] The development of slimmer bodies or strong front limbs/claws able to rapidly dig into burrows enabled *mustelids* to prey on burrow-dwelling small animals, mostly rodents but also insects and reptiles; this evolution did not create uniformity of body shapes – as you can see in extant *mustelids, ranging in size* from weasels and stoats to badgers and wolverines.[9]

Some *mustelids* retained a large body size. Around 22ma, *Aelurocyon brevifacies*, the largest recorded species of *mustelid* evolved. It was the size of a puma and had broad feet with curving claws. The omnivorous diet of *mustelids* meant that most badgers of Eurasia lost their meat-shearing carnassial teeth in favour of broad, grinding back teeth capable of grinding vegetable foods.[10] The two most powerful surviving members of the *mustelids* are the wolverine (*Gulo gulo*) and the honey badger. There have been suggestions that the honey badger is closely related to the wolverine (*Gulo gulo*) – while they are both *mustelids* and have similarities of behaviour, diet, and aggressiveness, they are not in the same sub-families and not that closely linked, and the honey badger is the sole member of its genus.[11]

The African fossil record of *mustelids*, related but not necessarily direct ancestors of the honey badger, is diverse with a long evolutionary history. According to Valenciano and Govender,[12] it includes the middle–late Miocene *Eomellivora tugenensis* from the Ngorora Formation, Kenya (c12ma), and *Mellivora* fossils found in the same formation and dating from 10ma; the late Miocene *Howellictis valentini* from Toros Menalla 192 in Chad (c7ma); three distinct forms in Lothagam, Kenya, including the very large *Ekorus ekakeran*, identified by Werdelin in 2003 at the Lower Nawata Formation (c7ma); the smaller *Erokomellivora lothagamensis* and the remain of a *Mellivorinae* that is between those two in size. Finally, the fossil record of *Mellivora* includes the Mio–Pliocene *Mellivora benfieldi* from Langebaanweg, South Africa (early Pliocene, c5.2ma), and the late Pliocene and modern *Mellivora capensis*. They also noted that the sub-family occurs in Brisighella, Italy (c5.4ma), with the remains of *M. benfieldi*, and in

Eurasia and North America with the remains of the giant mustelid *Eomellivora Zdansky*. Fragmentary remains have been described in India and Pakistan from the Siwalik Hills, with three forms: *Sivamellivora necrophila* (c14–11.2ma), *Promellivora punjabensis* from the late Miocene sediments of the Dhok Pathan Formation (c10.2–5.3ma), and *Mellivora sivalensis* from the upper part of the Siwaliks in the Pleistocene (c2.58–1.7ma), "which is morphologically similar to *M. capensis*."[13]

Recent research has shown that "[f]ive million years ago, dangerous carnivores – such as giant wolverines and otters, bears, sabertooth cats, and large hyaenids – prowled the West Coast of South Africa. Today we can confirm that, among them, fearlessly roamed a smaller relative of the honey badger."[14] The extinct honey badger from Langebaanweg, *Mellivora benfieldi*, was found and described by Hendey over 40 years ago,[15] based on a few broken mandibles. Valenciano and Govender re-analysed the *Mellivora benfieldi* material and new material housed at the Iziko South African Museum. The honey badger fossils they analysed

> triple the number of known fossils and gives us a unique glimpse into its lifestyle and relationship to other similar *mustelids*. These new fossils demonstrate that this South African species is distinct from the late Miocene forms from Central Africa (*Howellictis*) and East Africa (*Erokomellivora*), as well as from the extant honey badger,

according to Valenciano.[16] It was slightly smaller, but the fossils suggest it would have foraged much like modern honey badgers. Remains from the Langebaanweg area indicated that in addition to the honey badger-like animal, the area had a remarkably rich and diverse mammal assemblage from the Pliocene, "including saber-toothed cats, bears, hyaenas, jackals, mongoose, as well as relatives of the living giraffes, elephants, rhinoceroses, wild pigs, and a variety of birds, fishes and marine mammals."[17] This all fits with evidence that *Mellivorinae* have been traced back to the end of the Tertiary and beginning of the Quaternary periods, which coincide with the end of the Micoene (which lasted from 23 to 5.33ma), Pliocene (5.33–2.58ma), and the beginning of the Pleistocene epoch (2.58ma).[18]

The research at Langebaanweg is important for piecing together the evolution of this group of *mustelids* in Africa during the last 7 million years and reinforces the evidence of the existence among ancient *mustelids* of "a unique group named *Eomellivorini*" and of a diversity of honey badger-type species, which have become extinct apart from the one surviving species. Valenciano and Govender support the existence of two distinct groups of honey badger-type *mustelids* – the *Mellivorini* (comprising the living honey badger, the one from Langebaanweg and several other honey badger/ratel-like relatives) as well as the *Eomellivorini* which are characterised by gigantic proportions.[19] *Eomellivora tugenensis* from the Ngorora Formation represents the oldest African *mellivorine*, from around 12ma. It could, Valenciano and Govender believe, constitute an ancestral form of *Eomellivora*, adding that *Mellivora benfieldi* shares some primitive traits with it but is smaller.[20] They concluded that a sister–group relationship between *M. benfieldi* and *M. capensis* existed and believed there was a "tribe" within *Mellivorinae* that included *Eomellivorini (Eomellivora spp. + Ekorus)* and that *Mellivorini* includes *M. benfieldi, M. capensis, H. valentini, Er. lothagamensis,* and the Indian taxon *Promellivora punjabiensis.*[21]

Heptner and Sludski suggested that the modern honey badger first appeared in the mid-Pliocene (perhaps 3.6ma) in Asia,[22] from where it dispersed into the Middle East and Africa. It evolved to live in a diversity of habitats, inhabiting deserts, semi-deserts, dry savannas, steppes, scrub, wetlands, open woodlands, and dense forests. Honey badger remains are recorded from the late Pleistocene (126–11.7ka – 1,000 years ago) in South Africa.[23] Remains of the now extinct *Mellivora* species, *Mellivora punjabiensis*, have been found in 45 areas in Kenya, Morocco, South Africa, Sudan, Swaziland, Tanzania, and Zambia, dating from the late Miocene and the Pliocene, with one set of remains found in Italy dating from the Miocene.[24] *Punjabiensis* remains were also found at Langebaanweg in 'E' Quarry.[25] This site is "a palaeontological wonderland dating back 5 million years" and included an extinct African relative of the wolverine, which was the size of a leopard.[26] At Swartkrans in South Africa, *Mellivora* remains from 1.8 to 1ma were believed to be those of a honey badger similar to remains found at Pinjor in the Siwalik foothills of the Himalayas, in northern India and neighbouring areas of Pakistan, which were labelled *Mellivora sivalensis.*[27] Rhodes has noted that there appears to have been no clear gene flow between honey badger populations in Africa and Asia since the late Pleistocene, perhaps 12,000 years ago.[28] The Asian branch of the honey badger species expanded its range in the Levant and West Asia in the Pleistocene prior to the closure of the Sinai land bridge between North Africa and Asia.[29]

From the estimated first appearance of honey badger species around 3.6ma, dispersal from Asia into southern Europe, the Middle East, and Africa led to the early and the modern species establishing itself across most of the range it occupies today (though it has long been absent from southern Europe). Its likely arrival in Africa between 2.5 and 1.8ma was at the time when the hominin species *Homo habilis* appeared in Africa, followed by *Homo erectus* around 2.3ma.[30] Van Valkenburgh believes the honey badger was present as part of a large carnivore guild in the Serengeti and surrounding regions from over 2mya – the carnivore diversity declined over time from 22 down to 13 members as many of the larger carnivores and ancient felids and hyaenidae became extinct.[31]

As humans evolved and began to use tools and then establish settlements, fossil evidence from Africa indicates they interacted in diverse ways with the fauna whose descendants are found there today. Excavations from the Middle (96–51ka) and Later Stone Age (2–1.5ka) sites at Die Kelder and Pinnacle Point in south-western Cape province, South Africa, revealed honey badger and black-backed jackal remains at cave sites occupied during these periods by early humans; they were also discovered from Middle Stone Age sites at Klasies River in the southern Cape.[32] The fossils were among large numbers of small animal remains which were suggestive of hunting and consumption of small mammals, including carnivores, by other carnivores, raptors, and by humans, during this period. Small numbers of badger and jackal bones were found among large amounts of mole-rat, hyrax, hare and wildcat, mongoose, and polecat fossils.[33] It is not certain whether badgers had been hunted and consumed or whether they scavenged other animal carcasses at human-inhabited caves – though some small carnivore remains did have cut or burn marks indicative of human consumption.[34]

In northern Algeria, honey badger remains were found at the archaeological site where early human remains, named *Homo mauritanicus* (probably a North African example of *Homo erectus*), were discovered and dated to 1.8–774ma. The cranial features of the honey badger remains found are consistent with those of the modern *Mellivora capensis*.[35] Remains from about 510,000–391,000 years ago from near Casablanca and Temara in Morocco also confirm the presence of honey badgers there in bone assemblages that are in close proximity to evidence of resident toolmaking humans.[36] The fossil record for the dispersal of *Mellivora* is not particularly extensive, and interpretation of them is quite limited compared with the *Eomellivora*. In the late Miocene, *Eomellivora ursogulo*, referred to as closely related to honey badgers, and *Eomellivora wimani* were found in Africa, Eurasia, and North America.[37] The first records of species of *Eomellivora* are from Europe and Asia dating around the start of the late Miocene (12–13ma), and fossils from Ukraine have been dated at 11ma; they first appeared in Africa about 7–8ma, as evidenced by finds at Ngorora, Kenya, and in the late Miocene. From Eurasia, this genus expanded into North America.[38]

Honey badgers would have inhabited similar terrain and climatic regions as early humans, excepting the desert regions in which the badgers could and still do survive, but which would not have supported early humans. There is little early evidence of contact, conflict, or anything other than a distant coexistence in similar habitat. It is possible that the dispersal of the honey badger from Central and South Asia into the Middle East, Eurasia, and Africa resulted from the post-glacial desertification of millions of square miles of land across northern and western Asia – from the Thar Desert of India, through Baluchistan in Pakistan, into Iran, up into Central Asia and Mongolia and west towards the Caspian Sea.[39] While honey badgers are capable of surviving in true deserts, the spread of large areas of very arid desert, often with few freshwater sources, may have encouraged dispersal into more promising areas.

Bee larvae, badgers, and beekeepers

One of the main modern areas of conflict between humans and honey badgers is competition for honeybee nests/hives, with destruction of nests and hives enraging honey gatherers and beekeepers. From early in the hunter-gatherer phase of the human development, destruction of nests would have created a potential area of conflict, with early human foragers seeking honey (and perhaps consuming the larvae) and honey badgers frequently beating them to the nests and destroying them before they could be harvested by people. Marlowe et al., in their study of the contemporary Hadza hunter-gatherers of central Tanzania, suggest that "[o]ur closest living relatives, the great apes, take honey when they can. We suggest that honey has been part of the diet of our ancestors dating back to at least the earliest hominins."[40] Numerous papers have been written on great ape consumption of honey and raiding of nests, some examining chimpanzee, gorilla, bonobo, and orangutang methods of accessing the honey, including the use and fashioning of twigs or sticks by chimpanzees.[41] Early hominins would have had only sticks as tools to break open nests, as they did not use fire – the use of fire may have occurred between 1 and 1.5ma, judging by archaeological evidence from South Africa.[42] Once hunter-gatherers learned how to make and control fire, they then

> learned that they could smoke bees and increase the amount of honey they acquire … Hominins likely already had tools like hand-axes that could be used to gain access to a hive … As long as fire has been controlled, our ancestors may have been able to raid the hives of *A. mellifera* [the aggressive stinging African bee] successfully by smoking the bees.[43]

The Hadza value honey above all other foods and have access to honey from seven different bee species. Honey gathering is divided between men and women, with Hadza women usually taking honey in nests close to the ground, while men climb tall baobab trees to raid the largest beehives with stinging bees. Once gathered, it is very highly prized and jealously guarded,[44] which makes it probable that they would wish to keep other animals that eat honey away from nests once they have been located; badgers would be on the list, though there is no recorded evidence of conflict with them. Marlowe et al. believe honey accounts for a substantial proportion of the kilocalories in the Hadza diet and that "virtually all warm-climate foragers consume honey," providing a map showing 36 different recognised hunter-gatherer communities across the globe, with 29 of them engaging in honey gathering on a regular basis.[45] From early on, for hunter-gatherer communities, judging by studies of extant hunter-gatherer peoples, honey was a much-valued food. As with Marlowe's study of the Hadza, Ichikawa found that among the Mbuti hunter-gatherers of the Ituri forest in eastern Democratic Republic of Congo (DRC). "Honey is the most favourite food of the Mbuti. In the honey season they lead a life mainly relying on honey."[46] Honey remains an important dietary item, and it is still gathered from the wild, as well as being commercially produced, across the world. Dunne et al. recorded that "[w]ild honey is also known to be widely collected by foragers globally, except in environments such as the Arctic and Subarctic where bees do not survive."[47]

It is hard, if not impossible, to date the first human harvesting of honey. It is likely to have started millennia before the first representations of honey gathering in early human rock paintings. Rock paintings from the central Sahara, Zimbabwe, and South Africa suggest humans have collected honey in Africa for at least 20,000 years, and today, honey still contributes significantly to the diets of many African people who supplement other sources of food with foraging.[48] These paintings suggest a relationship between human foragers and animals that feed on bee larvae and waxworms, such as the honeyguide bird.

From the Pleistocene to the colonial period

It is hard to trace the history of the honey badger over the period from about 14,000 years ago to the present. As shown above, we can place it in various locations using fossil records, but it is not, unlike lions, tigers, elephants, wolves and other large, dangerous of charismatic species, regularly mentioned in the historical accounts of the wildlife with which humans coexisted or came into conflict. Monsarrat and Kerley have analysed historical references to mammals in southern Africa. They concluded that dangerous or charismatic large mammals – lions, buffalo, elephants, and leopards – were over-reported in comparison with what was known of their historical abundance, while

> [s]even species were not recorded in the historical dataset despite being historically present in the study area (aardvark *Orycteropus afer*, aardwolf *Proteles cristata*, African wild cat *Felis silvestris lybica*, Cape fox *Vulpes chama*, honey badger *Mellivora capensis*, mountain reedbuck *Redunca fulvorufula* and small spotted cat *Felis nigripes*).[49]

Honey badgers do not appear in records despite an abundance score of 2186, compared with lions which had an abundance level of 1759 but 750 references.[50]

Following Reid's model of the soft and developing hard boundaries between humans and wildlife of the hunter-gatherer period of early human existence outlined in Chapter 3,[51] honey badgers, though peripherally placed in terms of human–wildlife coexistence and conflict, would have experienced growing interaction and growing conflict as humans expanded their settlements with the increase in population and moved from gathering and hunting to cultivation, domestication of animals (including poultry, which were particularly attractive to honey badgers), and, of course, moving from collecting wild honey to putting logs and other possible nesting places in trees to attract bees (and so also honey badgers) keeping hives, another target for foraging honey badgers. Honey is a major cause of competition and conflict between humans and honey badgers. The possibility of humans snaring, spearing, shooting with arrows, or, much later, poisoning of badgers became a violent factor in human–badger interactions, which had been very limited prior to this. Badgers have been hunted for meat and perhaps skins on an opportunistic but minor basis.

With the advent of domestication of livestock and poultry and the growth in beekeeping, honey badgers would have been seen as potential hive and poultry raiders, with humans conceivably killing or trying to prevent badgers raiding nests and domestic poultry. The clearing of woodland or bush, fencing of cultivated land, and accompanying attempts to exclude wildlife from fields or enclosures for livestock, protection of resources (including hives), and the beginnings of irrigation made the human–wildlife border harder. The boundary between farmed/settled areas and wildlife habitat became more exclusionary.[52] Honey badgers were able because of their size and foraging habitats to avoid contact with people most of the time, but their raiding of hives, natural bee nests, and periodic poultry killing made them more vulnerable to human retaliation.

North, West, Central, and East Africa

Evidence of early settled communities suggests honey gathering and then establishment of hives to *domesticate* bees lost none of its importance, as new, reliable sources of food were developed. The Nok people of Nigeria, living around 3,500 years ago, known for their terracotta utensils and figurines and for being an early iron-working culture, left few clues about their diet, but examination of lipid deposits in shards of vessels they used indicated that a third of the vessels had contained beeswax.[53] Dunne believes this did not indicate large-scale use of wax but that vessels were used to melt honeycombs to obtain honey and that this and the storing of honey resulted in traces of beeswax. It is also possible that pots were used as hives. She adds that substantial use of honey was an important part of calorific intake for the Nok as it has also been for the Mbuti and Efe of Ituri, the Hadza, and the Okiek hunter-gatherers of Kenya. She noted, too, that the Berber scholar and explorer Ibn Battuta recorded in 1352 that the people of Mauritania used honey extensively and made drinks of honey and sour milk.[54]

There is no verifiable evidence at this early stage of human settlement of conflict with honey badgers or hunting of them. Robertshaw and Siiriäinen carried out excavations in Lakes Province, South Sudan, of settlements and cattle camps used by the ancestors of and then by the Dinka people – with evidence of human settlement going back at least to the start of the first millennium BCE.[55] Cattle would be moved across the grasslands and floodplains of the region, being gathered in camps situated near permanent rivers during the dry season to take advantage of remaining grazing and access to water.[56] Excavations of the camps indicated that those on the floodplains were used annually and had middens containing evidence of the disposal of waste, including livestock, fish, and wild ungulate bones. Remains of honey badgers were found in what appeared to be waste middens at two of the sites – suggesting they had been killed and perhaps consumed, as it seems unlikely they would have just died in a midden.[57]

In Kenya's Mau Forest and around Mt Elgon, the Okiek and their ancestors have for millennia lived as hunter-gatherers though gradually engaging in cultivation of crops, too. Presence of the Okiek in Kenya predates the Bantu migrations from West and Central Africa. Honey gathering and establishing hives were and still are an important part of their livelihood – being a source not just of food but also of income from the sale or bartering of honey and beeswax. They both gather honey from wild bee nests and hollow out large logs, which they suspend from trees to attract bees.[58] They still hunt extensively using spears, poison-tipped arrows, and dogs. But honey is important, nutritionally and symbolically, permeating all aspects of community life. It makes up just 15% of food intake and is also used to make alcoholic drinks as for ceremonial purposes. Blackburn believes that "[t]o a great extent, Okiek adaptation to a foraging way of life derives from their desire to acquire and use honey."[59] Strips of the land over which they forage will be traditionally reserved for particular families and their lineages – stealing honey from someone's foraging area would lead to conflict. Lineages try to exclude others and protect their sources of honey.[60] In such circumstances, although Blackburn and others do not specifically mention honey badgers, it is hard to believe that the Okiek would not have found ways to deter badgers from raiding hives or nests and would probably, given their hunting skills, have killed them in response to hive destruction or possibly pre-emptively.

Southern Africa

During the last millennium BCE and the first 1600 years CE, southern and south-eastern Africa contained a range of habitats – forest, woodland savanna, dry savanna, montane forest, arid steppe, and desert,[61] with ungulate species present in large numbers, including all the current species plus the extinct bloubok and quagga, which were found in large numbers in South Africa's Cape. The large herds supported substantial populations of spotted hyenas, lions, leopards, cheetahs, wild dogs, and jackals. The honey badger, given its current distribution, was likely to have been present in suitable habitat across the whole region.

The San were the first modern human inhabitants of the region, with evidence of their presence going back to 20,000BCE.[62] Around 12,000–10,000BCE, Khoikhoi peoples moved south into the region from near Lake Malawi through eastern Namibia and the Okavango, settling in Namibia and western Botswana or from the northern end of Lake Malawi via the Zambezi and Limpopo into eastern Botswana, Zimbabwe, and South Africa.[63] The San and Khoikhoi both had cultures of cooperative foraging and hunting, including scavenging from carnivore kills. Later contact with early Bantu migrants influenced Khoikhoi adoption of pastoralism, becoming their main form of economic activity.[64] The San remained dependent on hunting and foraging for wild foods. Bantu migrants – Tswana speakers in the case of Botswana – settled in the better-watered north-eastern and eastern areas of the country, and it was not until the 1870s, according to Lee, that they began to penetrate the San-inhabited regions to the west of what is now the Central Kalahari Game Reserve.[65] This area, despite its aridity, had and still has a wide variety of mammals, with at least 13 ungulate species that would have been hunted by the !Kung San people; there were also wild bee nests which the !Kung harvested for honey, wax, and larvae, as did the resident honey badgers, which were common.[66]

In the first millennia and a half of the CE period, the migration into the region of Bantu-speaking communities brought iron-working, crop cultivation, and pastoralism, with larger settlements forming by the end of the 2nd century CE.[67] By 700CE, Bantu-speaking communities had established settled communities in south-west Zimbabwe/north-eastern Botswana/northern South Africa, combining pastoralism and cultivation. By 1075, the Leopard's Kopje communities had grown into a kingdom based on Mapungubwe on the border of Botswana, Zimbabwe, and South Africa. The Leopard's Kopje people hunted as a supplement to livestock and arable farming, but there is no convincing evidence that they seriously depleted wildlife or permanently excluded it from their territory. The effect on foragers, predators, and scavengers like honey badgers would have been negligible, but there may have been competition between honey gatherers/beekeepers and badgers.

As agriculture became more developed, settlements expanded, and land was increasingly cleared of forest for cultivation or grazing – the forging of iron tools enabled more efficient forest clearance and the establishment of larger settlements in south-east Africa by the 3rd century CE.[68] The migrant communities grew in size and developed into large states like the Mapungubwe kingdom of northern South Africa and eastern Botswana, the Monomotapa of Zimbabwe and eventually the Zulu Kingdom, its Ndebele offshoot, the Xhosa chieftaincies of the Eastern Cape, and the Tswana chieftaincies of South Africa and Botswana. They hunted to supplement food production and, increasingly, for sport. But considerable numbers of wildlife and a diversity of species survived across southern Africa, in substantial numbers in some areas and with no evidence of extinctions. It was only the arrival of European settlers and hunters from the mid-17th century onwards that led to the extermination of game in much of the Cape, Natal, Free State, and Transvaal (as the regions became known under Dutch and then British/Afrikaner rule) and increased conflict between humans and carnivores, including the honey badgers, with firearms and poison some of the chief weapons used to combat honey thieves or poultry killers.

Little is recorded of conflict or any form of interactions of badgers with European settlers, hunters, or explorers in the period between the establishment of Cape Town by the Dutch in 1652 and the era of European hunting, exploration, and amateur naturalists in the mid-19th century. In 1864, for example, the naturalist, physician, and colleague of Livingstone, Dr John Kirk, gave a talk to the Zoological Society of London (ZSL), in which he reported on the aggressiveness of the honey badger when threatened. He said that "when wounded it makes for the tendo Achillis, which it cuts; it is considered in that way a dangerous animal."[69] The hunter Frederick Vaughan Kirby, in his 1894 account of hunting expeditions in Malawi and Mozambique, says in his zoological notes that he only saw one badger during his travels and wasn't able to shoot it, though he was given the damaged skin of another by a Mozambican.[70] In 1899, Sclater presented to a meeting of the ZSL skins and skeletons of mammals collected in British Central Africa. They included a honey badger, though sadly no further details were given of where it was collected and of their regional distribution and abundance.[71]

In South Africa, the extensive settlement, exploration, and interest in archaeology of early colonial officials and amateur archaeologists led to discoveries of San and Khoikhoi rock paintings of bee nests and hives, indicating a tradition of honey collecting stretching back thousands of years.[72] Honey was a food, an ingredient for brewing, and something traded with incoming Bantu communities and then with European settlers and traders. There is no clear reference in works on the Khoisan communities of conflict with badgers, but it is known that they were killed for their skins and body parts which were used by traditional healers and diviners.[73] The early European accounts of the fauna of the Cape and other regions of South Africa, though, make little, if any, reference to honey badgers. Monsarrat and Kerley's examination of 780 historical occurrence records for mammals, assembled from letters, journals, diaries, or books written by literate pioneers in southern Africa, makes not a single reference to honey badgers.[74] But in the early 1800s, the increasing settlement of the Cape and introduction of sheep farming led to persecution of badgers as actual or potential stock-killers, as I charted in earlier chapters in relation to jackals.[75] Keith Begg noted this persecution and said that there were authenticated records of badgers killing or maiming livestock but that the overall effect on farmers was negligible compared with jackals and caracals, despite their known poultry raiding and attacks on young ostriches on the growing number of ostrich farms.[76] The hunt clubs, formed in the Cape to exterminate jackals on farmlands, dug out badgers or hunted them down as they were believed to be a threat to the clubs' dog packs, being capable of mauling hunting dogs.[77] Persecution and the loss of habitat as livestock and arable farms spread explain why honey badgers "have experienced considerable population declines across the interior of South Africa, and possibly have vanished altogether from some portions of their former range," with fragmentation of habitat a serious problem for a mammal with a low breeding rate.[78] Historical sources cited by Beggs indicate that the Western Cape had a healthy honey badger population and that as late as 1876, they were reported to be found in "great profusion at the Cape of Good Hope."

Honey badgers were also caught or killed to be collected as specimens (alive and dead) to be exhibited in zoos and natural history museums. The ZSL reported on 12 November 1861 that its animal collector in the Cape Colony had presented to the zoo a male honey badger caught in the Cape.[79] The Society recorded that another two honey badgers had been given to the zoo at the end of 1861, though the sex and country of origin were not given.[80] A female honey badger was presented to the zoo in August 1866; another originating in India was presented in March 1870. The following year, the naturalist P.L. Sclater told at a meeting of the ZSL that he had been given a "new species" of honey badger by an animal dealer in Liverpool. He called it *Mellivora leuconota* and said it was from West Africa. Sclater explained that he thought it was different from both the Indian and southern African species, being smaller and having a white rather than grey/silver back.[81] This is now considered to be the white-backed honey badger sub-species (*Mellivora capensis leuconota*) of West Africa and southern Morocco.

By 1871, the zoo's honey badger had grown, and its pelage had darkened some-what. Sclater now told the Society that he thought it was the same species as the Indian and other African honey badgers, being a species that spanned much of Africa and Asia, including India, with regional variations rather than distinct species.[82] In his account of the mammals of South Africa published in 1900, Sclater described the honey badger as occurring throughout southern Africa, across suitable habitats in South Africa, and then north to the French Congo in the west and Sudan in the east. It is "plentiful" in South Africa but not often seen due to its retiring nature.[83]

West, Central, and South Asia

The honey badger was found across most of the Middle East, West, southern Central Asia (including in the past southern Russian areas bordering Kazakh-stan), and South Asia (including Pakistan, Nepal, and India).[84] As human settle-ments expanded, agriculture became a primary land use in many areas, and the badger range gradually contracted to what it is today, with disappearance from southern Russia, the foothills of the Caucasus, and most of the Mediterranean – in the latter, the International Union for Conservation of Nature Red List notes:

> Mediterranean regional assessment: Near Threatened (approaching C2). There is little data on population size and trends in this secretive species. However it seems reasonable to suppose that the total Mediterranean pop-ulation is below 10,000 individuals. The species faces threats including poisoning and persecution, which are presumed to be causing continuing decline. It is not known whether or not there would be a significant rescue effect from populations outside the region.[85]

In Europe, there was no evidence of the honey badger having been recent res-ident there or in Turkey, although in 1877, a report to the ZSL suggested that a pale coloured badger spotted in Turkey's Taurus Mountains could be a honey badger and that it was believed honey badgers were resident there and effectively part of the wider population that was found in Mesopotamia. No skeletal or other evidence was presented, and some doubt was expressed by ZSL members that honey badgers were still present in Turkey.[86]

Early European attempts to locate the Indian honey badger (*Mellivora capensis indica*) within the wider *Mellivora* species included examination of a specimen of what was called an Indian badger, described by the exhibitor Mr Burton as being of the *Ratelus* species, at a meeting of the ZSL in London in August 1835. Burton believed it to have many similarities to the South African honey badger but to be distinct in some ways, notably by having fewer teeth.[87] Burton described it as being superbly adapted for digging and said this is where it might have received in India the nickname of the Gravedigger, even though it was not a regular consumer of buried human bodies as far as he knew. The specimen was from

Bengal, where it was said to be rare. Burton said that the previous labelling of *Ursus indicus* did not seem appropriate for what he said was a ratel, even though he did not believe it was the same as the African *Mellivora*.[88] The Indian honey badger is now, of course, treated as a sub-species of *Mellivora capensis* – *Mellivora capensis indica*.[89]

There is no need to repeat the history of wildlife in India set out in Chapter 5, but it is worth recapping that India is estimated to have 100,000–150,000 sacred groves, which were important protected areas of natural habitat. Gokhale said, "The 'sacred groves' are … the 'reserve forests' of the local tribes/communities who maintain/conserve these patches of woodlands in a religious manner. They act as natural gene pool reservations and serve as an example of habitat preservation through community participation."[90] The areas studied by Gokhale et al. are in Uttarakhand's Rudraprayag region, which is still rich in wildlife and whose species include the honey badger. In ancient times and right up to the present, people from local communities would have gathered plants and other forest products for food and medicinal uses, but they are believed to have controlled foraging and hunting of animals in sacred forests, thereby protecting biodiversity.[91] It is worth noting that communities in India gather wild honey using smoke to drive away bees or reduce the defensive responses of guard bees to harvesting of honey from wild bee nests. But in sacred groves or forests, they do not use fire or smoke as it may kill or harm the bees and so is taboo. Madhav Gadgil in his examination of the role of sacred groves noted that the Jenu Kuruba honey gatherers of the Nilgiri Hills were an example of a community that avoided using potentially harmful measures on wild bee nests in sacred areas.[92]

There is almost nothing about honey badgers in the 18th and 19th centuries by British colonial officials, hunters, and early conservationists, beyond what has been noted above from the ZSL reports on the stories about badgers being grave robbers. Begg recorded this in her PhD thesis but said an obvious explanation was not that badgers were digging up and consuming bodies but were eating beetle larvae around the graves.[93]

From 1900 to 1960 – honey badgers in colonial Africa and Asia

References to honey badgers in this period are very scarce, despite extensive European settlement, a thriving safari hunting sector, and the propensity of hunters to record their activities in detail. On their long and very destructive safari in 1909–10 – 512 mammals killed – Theodore and Kermit Roosevelt killed one honey badger and paid it no great attention in the account of the safari, merely recording its colour.[94] Loveridge gave a little more space in his report to the ZSL in 1923 – he described the honey badger in Africa as widely distributed, plentiful, but rarely seen. He noted that "a native" had complained to him that a badger had dug through a hardened mud wall to get into a chicken enclosure at night, where it killed several birds. The man set up a gun trap by the hole, but the badger then dug another hole in the wall and killed three more chickens.[95]

The badger was eventually caught in a trap and killed, its gut containing chicken bones, feet, and other indigestible parts. Shortridge in his compendious *Mammals of South West Africa* reported that badgers were found from the Cape north to Senegal and across the continent to Sudan, Ethiopia, and Somalia, and had been observed in parts of Morocco, Algeria, and Egypt.[96] Shortridge obtained a specimen in Namibia, where he believed them to spread across the whole territory but were nocturnal and rarely seen. He said they were dangerous to disturb, and it would "require a good dog to tackle one singly."[97]

In the Cape, as noted above, they were considered vermin. District records of honey badgers destroyed as vermin over the period 1892–1955 give a total of 744 honey badgers killed as vermin – 61 in 1892, 132 from 1931 to 1939, 332 from 1940 to 1949, and 219 from 1950 to 1955. Far more were believed killed than were reported, according to Begg.[98]

West and South Asia

In 1920, a report to the Bombay Natural History Society from Iraq recorded the presence of honey badgers there, noting that the correspondent, R.E. Cheesman, thought it to be a different species to those found in India, because of variations in size and colouring.[99] Heptner and Naumov reported its presence in the Central Asian states of the former Soviet Union from the Caspian Sea east to the Amu Darya River in Tajikistan, being found in Turkmenistan, Uzbekistan, Tajikistan, and southern Kazakhstan.[100] They also occurred in Iran and the Arabian peninsula, according to reports to the ZSL, as well as parts of the Levant and Palestine.[101]

In India, they were widespread and locally common, if not hugely abundant. Blanford reported them as present in the Siwalik region of what is now Pakistan, and from the foothills of the Himalayas in Uttarakhand down to southern India, excepting the Malabar Coast and lower Bengal.[102] He commented that throughout India, they had a reputation as grave robbers. The *Journal* of the Bombay Natural history Society had a lengthy correspondence between L.E. Clifford Hurst, F.W. Champion, and others about whether the honey badger dug up and ate corpses, but no verifiable evidence was presented, only anecdotal reports.[103] One of those contributing to the correspondence, Dunbar Brander, was convinced that the stories of grave robbing and corpse consumption were true, basing in on the "accepted knowledge" that they will scavenge from buffalo and other animal carcasses, but he presents no conclusive proof.[104] There is no reason why a badger might not opportunistically feed from a human corpse but the grave-robbing accusation currently has no conclusive evidence to support it.

Daniel Johnson wrote about hunting in India in 1882, extolling the virtues of sports hunting there.[105] He makes only one reference to a badger and that, by the description, appears to be the honey badger:

> Badgers are scarce but are occasionally to be met with in the hills. In their nature they very much resemble the bear, and what is singular they are

called by the natives of Ramghur Badger-Ball, Ball being the Hindoosta-
nee word for Bear. Captain Williamson calls bears, balloos, which I believe
is a corruption. Badgers in India are marked exactly like those in England,
but they are larger and taller, are exceedingly fierce and will attack a number of dogs; I have seen dogs that would attack an hyena or wolf, afraid to
encounter them.[106]

Johnson's observations on the honey badger being present but rarely seen were
repeated regularly in contributions to the leading natural history journal in India
during British occupation, the *Bombay Journal of the Natural History Society*.[107]
Another contributor to the journal noted that in 43 years of hunting in India, he
had seen only a honey badger on two occasions.[108]

Notes

1 C.A. Long & C.A. Killingley (1983) *The Badgers of the World*, Springfield, IL: Charles
    C. Thomas Publisher, pp. 76–7.
2 J.I. Rhodes (2006) Phylogeographic structure of the honey badger (*Mellivora capensis*),
    Thesis for Master of Science degree, University of Stellenbosch, December 2006, p. 3.
3 D. Macdonald (1992) *The Velvet Claw. A Natural History of the Carnivores*, London:
    BBC, pp. 22–3.
4 X. Wang & R.H. Tedford (2008) *Dogs Their Fossil Relatives and Evolutionary History*,
    New York: Columbia University Press, p. 8.
5 Evolution of ancestral carnivores leading to the modern wild cat family, http://
    www.catsurvivaltrust.org/evolution.aspx accessed 28 September 2020.
6 Macdonald, 1992, p. 34; and G.D. Welsey-Hunt & J.J. Flynn (2005) Phylogeny of
    the Carnivora: basal relationships among the Carnivoramorphans, and assessment of
    the position of 'Miacoidea' relative to Carnivora. *Journal of Systematic Palaeontology*,
    3, 1, 1–28, pp. 9–12.
7 C.J. Law (2019) Evolutionary shifts in extant mustelid (*Mustelidae: Carnivora*)
    cranial shape, body size and body shape coincide with the Mid-Miocene Climate
    Transition, *Biolog. Letters*, 15, no page numbers.
8 Macdonald, 1992, p. 182.
9 Ibid.
10 Ibid.
11 V.G. Heptner & A.A. Sludskii (2002) *Mammals of the Soviet Union. Vol. II, part 1b,
    Carnivores (Mustelidae)*. Washington, DC: Smithsonian Institution Libraries and
    National Science Foundation.
12 A.V. Vaquero & R. Govender (2020) New Fossils of *Mellivora Benfieldi* (Mamma-
    lia, Carnivora, Mustelidae) from Langebaanweg, 'E' Quarry (South Africa, Early
    Pliocene): Re-Evaluation of the African Neogene Mellivorines, *Journal of Vertebrate
    Paleontology*, published online 2 November 2020, no page numbers.
13 Ibid.
14 Iziko Museums of South Africa (2020) New research reports on discovery of
    5-million-year-old honey badger-like animal from West Coast of South Africa,
    https://www.iziko.org.za/news/new-research-reports-discovery-5-million-year-
    old-honey-badger-animal-west-coast-south-africa accessed 6 November 2020.
15 See Q.B. Hendey (1976) The Pliocene fossil occurrences in 'E' Quarry, Lange-
    baanweg, South Africa. *Annals of the South African Museum*, 69, 215–47; and, Q.B.
    Hendey (1978) Late Tertiary Mustelidae (Mammalia, Carnivora) from Langebaan-
    weg, South Africa. *Annals of the South African Museum*, 76, 329–57.

16 Iziko Museums of South Africa, 2020.

17 Ibid.

18 Rhodes, 2006, pp. 3–4.

19 Iziko Museums of South Africa, 2020.

20 Valenciano and Romala Govender, 2020.

21 Ibid.

22 Heptner & Sludski, 2002, p. 1210.

23 R.J.G. Savage (1978) Carnivora, in V.J. Maglio & H.B.S. Cooke (eds) *Evolution of African Mammals*, Cambridge, MA: Harvard University Press, 249–67, pp. 255–6.

24 Mellivora, http://fossilworks.org/bridge.pl?a=taxonInfo&taxon_no=41132 accessed 29 September 2020.

25 C.K. Brain (1981) *The Hunters or the Hunted? An Introduction to African Cave Taphonomy*, Chicago, IL: University of Chicago Press, p. 166.

26 A.V. Vaquero & R. Govender (2020) Gigantic wolverines, otters the size of wolves: Fossils offer fresh insights into the past, Phys.org 22 June 2020, https://phys.org/news/2020-06-gigantic-wolverines-otters-size-wolves.html accessed 1 October 2020.

27 A.C. Nanda (2008) Comments on the Pinjor Mammalian Fauna of the Siwalik Group in relation to the post-Siwalik faunas of Peninsular India and Indo-Gangetic Plain. *Quaternary International*, 192, 6–13, p. 7.

28 Rhodes, 2006, p. 15.

29 M. Belmake (2010) Early Pleistocene faunal connections between Africa and Eurasia: an ecological perspective, J.G. Fleagle et al. (eds) *Out of Africa I: The First Hominin Colonization of Eurasia, Vertebrate Paleobiology and Paleoanthropology*, Heidelberg: Springer Dordrecht, 183–205, pp. 197–8.

30 E. O'Brien & C.R. Peters (1999) Landforms, climate, Ecographic mosaics, and the potential for hominid diversity in Pliocene Africa, in T. Bromage & F. Schrenk (ed) *African Biography, Climate Change, and Human Evolution*, New York: Oxford University Press, 115–37, pp. 134–5.

31 B. Van Valkenburgh (1988) Trophic diversity in past and present guilds of large predatory mammals, *Paleobiology*, 14, 2, 155–73, p. 166.

32 R.G. Klein (1976) The Mammalian Fauna of the Klasies River Mouth Sites, Southern Cape Province, South Africa, *South African Archaeological Bulletin*, 31, 123/124, 75–98, p. 75.

33 A. Armstrong (2016) Small mammal utilization by Middle Stone Age humans at Die Kelders Cave 1 and Pinnacle Point Site 5–6, Western Cape Province, South Africa, *Journal of Human Evolution*, 101, 17–44, p. 20.

34 Ibid., p. 28.

35 D. Geraads (2016) Pleistocene Carnivora (Mammalia) from Tighennif (Ternifine), Algeria, *Geobios*, 49, 445–58, p. 454.

36 J.-P. Raynal (2010) Hominid Cave at Thomas Quarry I (Casablanca, Morocco): Recent findings and their context, *Quaternary International*, 223–24, 369–82, p. 375; and E. Campmas (2016) Initial insights into Aterian hunter-gatherer settlements on coastal landscapes: The example of Unit 8 of El Mnasra Cave (Temara, Morocco), *Quaternary International*, 413, 5–20, pp. 8–10.

37 R. Uchytel (no date), *Prehistoric fauna*, https://prehistoric-fauna.com/Eomellivora-ursogulo accessed 1 October 2020; and, A.V. Lavrova & D.O. Gimranov (2018) First finding of a representative of giant mustelids of the genus *Eomellivora* (Carnivora, Mustelidae) in Russia (Tuva, Upper Miocene), *Doklady Biological Sciences*, 480, pp. 82–4, p. 82.

38 Lavrova & Gimranov, 2018, p. 82.

39 D.N. Wadia (1960) *The Post-glacial Desertification of Central Asia and the Evolution of the arid Zone of Asia*, New Delhi: National Institute of Sciences of India, p. 1.

40 F.W. Marlowe et al. (2014) Honey, Hadza, hunter-gatherers, and human evolution, *Journal of Human Evolution*, 71, 119–28, p. 119.

41 See, for example, S.M. Brewer & W.C. McGrew (1990) Chimpanzee use of a tool-set to get honey, *Folia Primatol*, 54, 100–4, p. 100; and M.R. McLennan (2011) Tool-use to obtain honey by chimpanzees at Bulindi: new record from Uganda, *Primates*, 52, 315–22, p. 315.

42 F. Berna et al. (2012) Microstratigraphic evidence of in situ fire in the Acheulean strata of Wonderwerk Cave, Northern Cape province, South Africa, *Proceedings of the National Academy of Science*, 109, 2, E1215–E1220.

43 Marlowe et al., 2014, p. 128.

44 F.W. Marlowe (2010) *The Hadza Hunter-Gatherers of Tanzania*, Berkeley: University of California Press, p. 253.

45 Ibid., pp. 119–20.

46 M. Ichikawa (1981) Ecological and sociological importance of honey to the Mbuti Net Hunters, Easter Zaire, *African Study Monographs*, 1, 55–68, p. 55–6.

47 J. Dunne et al. (2021) Honey-collecting in prehistoric West Africa from 3500 years ago, *Nature Communications*, 12, 2227, no page numbers.

48 H.A. Isack & H.U. Reyer (1989) Honeyguides and honey gatherers: interspecific communication in a symbiotic relationship, *Nature*, 10 March 1989, 243, p. 1343.

49 S. Monsarrat & G.I.H. Kerley (2018) Charismatic species of the past: biases in reporting of large mammals in historical written sources, *Biological Conservation*, 68–75, p. 69.

50 Ibid., p. 72.

51 R.S. Reid (2012) *Savannas of Our Birth. People, Wildlife and Change in East Africa*, Berkeley: University of California Press, 2012, p. 95.

52 Ibid., p. 63.

53 J. Dunne (2021) Beeswax in Nok pots provides evidence of early West African honey use, *The Conversation*, 27 May 2021, http://theconversation.com/beeswax-in-nok-pots-provides-evidence-of-early-west-african-honey-use-161197 accessed 1 June 2021.

54 Ibid.

55 P. Robertshaw & A. Siiriäinen (1985) Excavations in Lakes Province, Southern Sudan, *Azania: Archaeological Research in Africa*, 20, 1, 89–161, p. 89.

56 Ibid., p. 91.

57 Ibid., p. 92.

58 R.H. Blackburn (1982) In the land of milk and honey: Okiek adaptations to their forests and neighbors, in by E. Leacock & R. Lee (eds) *Politics and History in Band Societies*, Cambridge: Cambridge University Press, 283–305, p. 285.

59 Ibid., p. 288.

60 Ibid.

61 C. Ehret (2016) *The Civilisations of Africa. A History to 1800*, Charlottesville: University of Virginia Press, pp. 212–3.

62 R.B. Lee (1979) *The Dobe !Kung*, Cambridge: Cambridge University Press, p. 17.

63 A. Barnard (1992) *Hunters and Herders of Southern Africa. A Comparative Ethnography of the Khoisan peoples*, Cambridge: Cambridge University Press, 1992, p. 34.

64 Ibid., pp. 30–2.

65 Lee, 1979, p. 17.

66 Ibid., p. 24.

67 J. Ki-Zerbo & D.T. Niane, *Africa from the Twelfth to the Sixteenth Century. General History of Africa IV*, London: James Currey/UNESCO, 1997, p. 209.

68 Ibid.

69 *Proceedings of the Zoological Society of London* (henceforth referred to as *Proceedings*), 13 December 1864, p. 651.

70 F. Vaughan Kirby (1899) *Sport in East Central Africa. Being an Account of Hunting Trips in Portuguese and Other Districts of East and Central Africa*, London: Rowland Ward, p. 324.

71 *Proceedings*, 2 May 1899, p. 553.

72 J.C. Hollmann (2015) Bees, honey and brood: southern African hunter-gatherer rock paintings of bees and bees' nests, uKhahlamba-Drakensberg, KwaZulu-Natal, South Africa, *Azania: Archaeological Research in Africa*, 2015 50, 3, 343–71, p. 344.

73 A. le Roux & S. Badenhorst (2016) Iron Age fauna from Sibudu Cave in KwaZulu-Natal, South Africa, *Azania: Archaeological Research in Africa*, 51(3), 307–26, pp. 307–8.

74 Monsarrat & Kerley, 2018, p. 68.

75 C.M. Begg et al. (2016) Mellivora capensis – Honey Badger, *The Red List of Mammals of South Africa, Lesotho and Swaziland*, South African National Biodiversity Institute and Endangered Wildlife Trust, p. 5.

76 K. Begg (2001) *Report on the Conflict between Beekeepers and Honey Badgers Mellivora Capensis, with Reference to their Conservation Status and Distribution in South Africa*, Johannesburg: Endangered Wildlife Trust, p. 13.

77 Ibid.

78 Keith & C. Begg (2002) The conflict between beekeepers and honey badgers in South Africa: a Western cape perspective, *Open Country*, 4, 25–36, p. 25.

79 *Proceedings*, 12 November 1861, p. 307.

80 Ibid., 28 January 1862, p. 22.

81 *Proceedings*, 24 January 1867, p. 98.

82 Ibid., 7 March 1871, p. 232.

83 W.L. Sclater (1900) *The Mammals of South Africa*, London: H. Porter, reprinted by Sagwan Press (no date), pp. 111–2.

84 The Honey Badger: Fact File, http://www.honeybadger.com/facts.html accessed 20 October 2020.

85 T. Jdeidi et al. (2010) *Mellivora capensis. The IUCN Red List of Threatened Species*, e.T41629A10522349, https://www.iucnredlist.org/species/41629/10522349#population accessed 20 October 2020.

86 *Proceedings*, 20 March 1877, p. 274.

87 Ibid., 11 August 1835, pp. 113–6.

88 Ibid., p. 116.

89 W.C. Wozencraft (2005) Species Mellivora capensis, in D.E. Wilson & D.M. Reeder (eds) *Mammal Species of the World: A Taxonomic and Geographic Reference* (3rd ed.), Baltimore, MD: Johns Hopkins University Press, 532–628, p. 612.

90 Y. Gokhale et al. (2011) Sacred landscapes as repositories of biodiversity. A case study from the Hariyali Devi sacred landscape, Uttarakhand, *International Journal of Conservation Science*, 2, 1, 37–44, p. 38.

91 Ibid., p. 42.

92 M. Gadgil (2018) Sacred groves: an ancient tradition of nature conservation. Indian villagers are reviving an ancient tradition to enjoy the ecological benefits it confers, *Scientific American*, 1 December 2018, https://www.scientificamerican.com/article/sacred-groves-an-ancient-tradition-of-nature-conservation/ accessed 23 February 2021.

93 Begg, 2001, p. 5.

94 T. Roosevelt (1910) *African Game Trails*, New York Charles Scribner's Sons. Kindle Edition, loc 2523.

95 *Proceedings*, 1923, 3–4, p. 712.

96 Captain G.C. Shortridge (1934) *The Mammals of South West Africa*, London: William Heinemann, p. 194.

97 On Mammals Collected in 1923 by Capt. G.C. Shortridge during the Percy Sladen and Kaffrarian Museum Expedition to South-West Africa, as reported to ZSL by Oldfield Thomas, *Proceedings*, 1925, 95, 1, p. 231.

98 Begg, 2001, p. 13.

99 R.E. Cheesman (1920) Reports on the Mammals of Mesopotamia, *Bombay Journal of the Natural History Society*, XXVII, 1920–22, 323–46, p. 325.

100 V.G. Heptner & N.P. Naumov (1967) Mammals of the Soviet Union, Volume II, Part 1b, Moscow: Vysshaya Shkola Publishers, Mammals of the Soviet Union (archive.org) accessed 28 June 2021, pp. 1218–21.

101 *Proceedings of the Zoological Society of London*, 6 February 1900, p. 95; and *Proceedings*, 1905, May to December, p. 523.

102 W.T. Blanford (1891) *The Fauna of British India including Ceylon and Burma*, London: Taylor & Francis, Internet Archive accessed 1 September 2021, pp. 175–77.

103 L.E. Clifford Hurst (1935) On Ratels and corpses, *Bombay Journal of the Natural History Society*, XXXVII, pp. 390–1.

104 A.A. Dunbar Brander (1934) Ratel and Corpses, *Bombay Journal of the Natural History Society*, XXXVII, 719–22, pp. 720–1.

105 D. Johnson (1822) *Sketches of Field Sports as Followed by the Natives of India*, London: Longman, pp. 39–40.

106 Ibid., pp. 55–6.

107 See for example, Lt-Col. A.H.E. Mosse (1930) The panther as I have known him, Part III, *Bombay Journal of the Natural History Society*, XXXIV, 3–4, 673–93, p. 682.

108 Anonymous review in *Bombay Journal of the Natural History Society*, XXXVI, 1934, p. 209.

10

HONEY BADGERS IN THE CONTEMPORARY WORLD

Honey badgers remain widely distributed across most of sub-Saharan Africa, in parts of north-west Africa, and in West, Central, and South Asia. There appears to be no major shrinking of the range, but there is concern that they are becoming scarce or even disappearing in some localities. On the International Union for Conservation of Nature (IUCN) Red List they are of Least Concern but decreasing in numbers, and it points out, they "are considered rare or to exist at low densities across most of their range." Their lack of "ecological specialization precludes any obvious, potential range-wide declines," but there are "localized declines recognized, though currently insufficient to warrant listing in a higher category of threat."[1] The decrease is due partly to persecution but also, according to the IUCN, "continuing decline in areas, extent and/or quality of habitat."[2] The IUCN listing for the Mediterranean region of the Middle East is Near Threatened.[3]

In 2012, honey badgers were listed as Near Threatened in the South African Red Data List, assessed by the South African Endangered Wildlife Trust (EWT).[4] In Western and Eastern Cape Provinces of South Africa, badgers are a "Schedule 2 protected wild animal and a permit is required to kill or relocate them ... but they are essentially unprotected outside of game reserves and national parks in other provinces of South Africa."[5] The 2016 version of the Red List published by EWT downgraded the listing to Least Concern[6] (see the "Southern Africa" section for details).

On the listing of mammals by the Convention on International Trade in Endangered Species of Wild Fauna and Flora (CITES), honey badgers only appear as listed in Appendix III by Botswana and Ghana; they are unlisted for the remaining 176 CITES member states, including all other honey badger range states. It was the lack of legislation and law enforcement for protection of the honey badgers and the high number believed to have been killed by beekeepers

DOI: 10.4324/9781003199793-11

that led to Botswana and Ghana putting them as an Appendix III, making it mandatory to have a permit for export and import of the species. Appendix III "contains species that are protected in at least one country, which has asked other CITES Parties for assistance in controlling the trade."[7] CITES documents show that there are no established hunting or export quotas for honey badgers,[8] but the organisation recorded in 2002 that fat and other body parts from honey badgers were part of the trade in wildlife products.[9]

There is no reliable estimate of how many honey badgers there are globally or within individual range states or regions. San Diego Zoo, in its fact sheets on global fauna, says it is "almost impossible to say" what the global population is because of the lack of verifiable figures.[10] Its fact sheet on the honey badger notes that honey badgers occur at low densities across their whole range, but with densities varying from the Serengeti National Park (NP), Tanzania, at 10/100km^2 to Sariska Tiger reserve, India, at 5.48–6.43/100km^2 to Kgalagadi Transfrontier Park, South Africa, at 3/100km^2. Colleen and Keith Begg, the leading specialists on honey badgers, emphasised that

> [b]adgers are particularly difficult to locate during conventional mammal surveys and, as a result, only coarse estimates are available of their distribution and abundance. Their relatively small size, frequently solitary and nocturnal behaviour make them difficult to record even in areas where they are well represented. In the southern Kalahari and Mana Pools National Park conventional survey techniques (spotlight counts and day transects) were poor indicators of honey badger density … In addition, badgers are not often dazzled by vehicle headlights or killed on roads … Recent improvements in camera trapping techniques offer the best chance of surveying the density of honey badgers. This has not yet been done.[11]

The chapter will examine the presence, status and coexistence/conflict between honey badgers and humans across their global range.

Sub-Saharan Africa – from the Horn to the Cape and the Red Sea to the Atlantic

North Africa

The honey badger is absent from most of North Africa, only occurring in parts of Algeria, Morocco, and northern Mauritania. It has not been recently recorded in Tunisia, Libya, or Egypt. There is no contiguous population across North Africa's Mediterranean coast or between populations in Chad, Sudan, and the Horn of Africa and those in the Middle East. In Algeria, the badger is recorded as rare. It was observed along the southern border with Morocco in 1980, and in 2018, one was filmed and another found dead on the roadside in south-west Algeria near Tindouf.[12] Most historical records are from south-west Algeria.

A stuffed badger can be found at the HQ of the Algerian Federation of Hunting Weapons in Algiers.[13] The badger is considered rare and threatened in Morocco and can only be found in the central and eastern High Atlas Mountains, along the Dra'a River in Tafilat and in the Atlantic Sahara coastal region.[14] It is found in both high mountain areas, including elevations of 3,000m, and on sea-level plains.[15]

Cherkaoui and Bouajaj reported that in Morocco, there was evidence of human–badger conflict, involving badger raids on beehives, leading to persecution of the badgers. They carried out interviews with farmers, beekeepers, shepherds, and hunters in southern Morocco in 2015, in an area with honey badgers, African golden wolves, wild cats, gazelles, and wild boar. Fig tree plantations provided habitat and sources of food and prey. Interviewees said that honey badgers were quite common in the Akerkat region of southern Morocco, near the border with Western Sahara, with two to five attacks a year on hives reported by beekeepers.[16] Traditionally, hives are sited on the ground, rendering them very vulnerable to badger raids. In 2015, beekeepers trapped and killed a badger raider, but only five badgers were killed between 2005 and 2015.[17] There is insufficient data on the size and exact distribution of the honey badger population in Morocco to assess whether the killing could affect its survival there.[18] To the south, in Mauritania, badgers are listed as present,[19] with four sightings recorded by survey teams between 2002 and 2009 in the southwestern grasslands, and no verified sightings in the arid north and east.[20]

Central Africa

There is relatively little published research on honey badgers, their ecology, and interactions with humans in Central Africa. The owner of the Sangha Lodge in the Dzanga-Sangha Protected Area of the Central African Republic, Rod Cassidy, said that camera traps had recorded the presence of honey badgers, including almost wholly black ones, at Dja Faunal Reserve in southern Cameroon, while pale-backed badgers were often seen near his lodge.[21] In the Ituri forest region of eastern Democratic Republic of Congo, honey badgers are present, despite the dense forest and areas of swamp, and are among the wide variety of mammals hunted by the Mbuti hunter-gatherers. A survey of the Mbuti and their attitudes towards the badgers showed they had respect for its ability to defend itself against leopards and golden cats. The Mbuti hunted it with bows and arrows because of its defensive capabilities and both ate its meat and used its skin for making drums.[22] In Gabon, honey badgers are known to be present in many habitats, including dense rainforests. Several have been observed that are completely black – camera traps sited in Ivindo captured images of four completely black honey badgers.[23] Panthera's Dr Philip Henschel said that he had never seen a honey badger in Gabon's dense forest before, but increased use of camera traps to count wildlife numbers had indicated their presence even in thick rainforests.[24]

West Africa

There is comparably little verified survey material or research on honey badgers in West Africa. Rosevear in his work on the carnivores of the region said honey badgers occurred in there, including all-black badgers, as in parts of Central Africa.[25] He theorised that the black ones were a different sub-species, *Mellivora capensis cottoni*, to the *Mellivora capensis leuconota* sub-species,[26] but as described at the start of Chapter 8, the division into sub-species is a contentious and far from settled subject. He reported that in West Africa, honey badgers were believed to be poultry killers and were said to be able to rip through thick planks and even burrow under stone foundations to get into henhouses.[27] A survey in northern Ghana's Mole NP in 2006–9, using camera traps, direct sightings, and interviews with local communities, came up with little information. The respondents among local people could not identify smaller carnivores and referred to them as genet or mongoose, so the presence of honey badgers could not be verified.[28] In Nigeria, a study of traditional medicine recorded honey badgers as rare but used in various traditional cures.[29] In 1953, Rosevear in his atlas of Nigerian mammals had said that badgers were present there and were found particularly in the north of Nigeria and around Lake Chad – favouring Sahel savanna but were found areas further south in the Guinea savanna zone.[30] In Togo, they are present and are killed for medicinal purposes rather than for meat or in response to damage to hives or poultry. D'cruze et al.'s 2020 survey of the main traditional medicine market in Lomé, Marché des Fétiches, found that 300 honey badgers had been recently traded in the market, second only to chameleon and viper – both over 500.[31] I failed to find records from the rest of West Africa, but one paper said they are found in the less heavily populated areas of Senegal and are present in protected areas in the east, notably the North Ferlo Fauna Reserve.[32]

Horn of Africa and Sudan

They are believed to be present, though in fragmented and low-density populations, across the Horn of Africa and Sudan. Künzel et al. noted that they were scarce in Djibouti, with only two records – one in 1984 and the other for 2000, involving a single badger taken to a wildlife rescue organisation there.[33] In Ethiopia, they are present in many parts of the country – with sightings in diverse regions, suggesting they are widespread,[34] but with no major studies, estimates of population, or precise distribution. Not surprisingly, given the decades of conflict there, records for Somalia are absent. Sudan and South Sudan are very similar, though we do have Nimir's doctoral thesis on wildlife in northern Sudan, which recorded the honey badger as present in areas of open woodland and savanna with reasonable rainfall.[35] It is present in Sudan's Dinder NP, on the border with Ethiopia's Alatash NP, where the habitat and wildlife are threatened by human encroachment by cattle herders, people gathering wood for fuel and building, and honey gatherers harvesting wild bee nests.[36] There are no direct reports of persecution of honey badgers, though human encroachment is likely to be damaging, especially as RAMSAR (Ramsar Convention on Wetlands of International Importance Especially as Waterfowl Habitat) reports that in Dinder,

> [h]oney collection starts in the dry season, usually in the months of January–March. During this period many uncontrolled fires are usually caused by the poachers and honey-gatherers. To dislodge the bees they burn portions of the tree resulting in accidental fires. In the process not only the tree, but the hive is destroyed also.[37]

East Africa

Honey badgers are present across large areas of Kenya, Tanzania, and Uganda. They haven't been extensively studied there and are likely to be widely distributed but with low population densities. Kingdon, in his 1977 survey of East African mammals, noted their presence in northern Kenya and Uganda, where they fed seasonally on bee larvae and dug into dung beetle and ant and termite nests and had been observed digging into dung piles and waste dumps searching for insect larvae.[38] He also recorded the belief among some Kenyan pastoral communities of the badger's use of secretions from its anal glands to stupefy or deter bees while raiding their nests – a herder, he quotes, says that some of the bees were killed by the secretions. He also reported Maasai having told him they had found badgers dead from hundreds of stings, inside bee nests, and badgers could also be killed by the toxins sprayed by soldier ants.[39]

Honey gatherers and beekeepers in Kenya are known to come into conflict with badgers, though little is recorded about methods used to kill or deter raiding badgers. The New Zealand-based non-governmental organisation Save the Wild has been working with communities in southern Kenya to encourage conservation of black rhino. Part of the project is a beekeeping and honey harvesting scheme at the Kimana Conservancy, between Amboseli NP and the Tanzanian border. The objective is to help the Maasai community there to develop livelihoods without damage to the habitat supporting the rhinos.[40] The project has helped establish hives housing 100 bee colonies. Honey badgers are a problem for the project, as they burrow under fences to get to the hives. One badger succeeded in crossing the fence and breaking open a hive, to eat the bee larvae. Over time, the badger (or perhaps others) returned and broke into 14 hives. All the honey from these hives was lost to the project.[41] As the conservancy rangers and beekeepers worked to construct a badger-proof fence with concrete footings, another badger raid took place – some hives were damaged and the bees left, but the beekeepers were able to recover the colonies and establish them in undamaged hives. Eventually, all the hives were protected with concrete-based fences, but despite their best efforts, the honey badgers keep breaking through. Extra cost for the project, but, as Save the Wild admitted, "One can't help but admire the tenacity of honey badgers,"[42] and, one might add, the forbearance of the beekeepers and rangers in not trapping, shooting, or poisoning the badgers, despite the losses.

In Tanzania, there has been little research into the range or habits of honey badgers. Nearly 50 years ago, Kingdon said that a survey of 56 beekeepers with 24,000 hives revealed the destruction of 2,700 hives by honey badgers.[43] Honey is gathered and bees kept in hives across Tanzania and is an important part of the diet of many people, including the Maasai, Barabaig, Hadza, and Sukuma. It is used as food, to sweeten tea and brew alcoholic drinks.[44] The conservation researcher Alaitetei E. Laltaika, based in the Ngorongoro Conservation Area (NCA), told me honey badgers were quite common but not often seen. He said the Maasai there were convinced that honeyguide birds led badgers to wild bee nests. Alaitetei was sceptical, as he believed that badgers, lacking external ears, did not have good hearing and would not hear the honeyguide until very close to it. He said that there was obvious conflict between beekeepers and honey gatherers and badgers, recalling that when he worked as Park Ranger in the Ministry of Nature and Tourism on anti-poaching operations within NCA, he once had to free a honey badger caught in a snare set by beekeepers. Usually, trapped badgers were killed by local people, he said.[45]

One study of carnivore diversity in Tanzania reported that camera traps at 430 locations in Arusha NP, Lake Manyara NP, Serengeti NP, Kilimanjaro NP and Forest Reserve, Mahale Mountains NP, Tarangire NP, Minziro FR, NCA, Biharamulo-BurigiKimisi Game Reserve, and Zoraninge Forest Reserve and Tanga Coastal Forests recorded 16 images of honey badgers across six of the locations – though locations were not identified.[46] A survey of small carnivore species in the Mt Rungwe Nature Reserve and Kitulo NP in south-west Tanzania sought to establish the species present and the local communities' attitudes towards them. Their results suggested that illegal hunting was affecting honey badgers and civets.[47] Large carnivores, buffalo, zebras, elands, and elephants are all extinct in the reserves (or occasional transient visitors), with the exception of leopards. Honey badgers were present but were rare and shy, mainly inhabiting forests, seemingly avoiding crop fields and villages.[48] But villagers said that badgers attacked their beehives and may have taken poultry. About 64% of villagers said that honey badgers were hunted with dogs or traps. The badgers caught were eaten, and body parts may have been used by traditional healers and diviners. Badger skins were said to be used to cure pain and treat people who had suffered trauma.[49] In the Katavi region of western Tanzania, beekeepers in the Mlele Beekeeping Zone, north of Katavi NP, and communities near the Northern Rukwa Game Reserve came into conflict with badgers over raids on hives. A survey in the regions revealed a healthy population of honey badgers, as evidenced by camera trap sightings at five points across the two areas.[50]

In addition to conflict over badger consumption of bee larvae and related hive destruction, some communities in Tanzania believe honey badgers are regular poultry thieves, breaking into enclosures and henhouses. In an area adjacent to the Serengeti NP, Grumeti Game Reserve, Mugeta Open Area to the west, the

Ikona Wildlife Management Area, and the Ikorongo Game Reserve, Holmern and Røskaft investigated raids on domestic poultry kept by agropastoral communities.[51] In 2003–4, the region's farmers owned 5,161 birds (chiefly chickens but some ducks and guinea fowl), with the average number of birds per household 12.9. Badgers were blamed for some predation with three birds taken by them in the study period – mongooses, hawks and eagles, and black-backed jackals were the main thieves, taking 352 birds between them from a total of 428 birds lost.[52] The badgers account for just 0.7% of losses, hardly evidence of them being hardened chicken killers.

Southern Africa

Honey badgers are widely distributed across southern Africa, being present in most types of terrain, though absent in the Free State. Badgers occur at low densities, and many are in fragmented populations, which means that accurate estimates of numbers are impossible.[53] Attempts at surveys in the southern Kalahari and Mana Pools NP (Zimbabwe) had poor results, while a night survey in South Africa's Hluhluwe-iMfolozi Game Reserve provided only two sightings from 3,381km of travel.[54] The lack of sightings from which to make population estimates may be because badgers can travel up to 40km a night when foraging and don't have strongly delineated territories.

Angola, particularly the southern areas bordering Namibia and the south-east bordering Zambia – the Longa-Mavinga NP and Luengue-Luiana NP and Iona NP protected areas and the surrounding Kalahari-type arid grassland and bush – is likely to have a healthy population of badgers. Almost constant war between 1961 and 2002 and the continued perils of conservation resulting from the presence of landmines left over from the war mean that surveys of smaller carnivore species have been impossible. The snaring of wildlife for food by people left poor and lacking livelihoods because of the ravages of 41 years of war and the disruption to agriculture caused by the landmines mean that carnivores, including badgers, often become victims of bushmeat hunters alongside the ungulates they are targeting. Honey badgers also figure in the list of animals killed for body parts for traditional medicine. Braga-Pereira carried out a study in Quiçama NP in central Angola, south of Luanda.[55] The park was badly affected by the war, and impoverishment of rural communities has encouraged extensive poaching rife. There is a lack of funding for conservation and anti-poaching, with human encroachment through grazing of cattle, collecting wood, and the establishment of small villages within the park area a major problem for conservation and efforts to reduce bushmeat poaching.[56] Surveys carried out in 2014 and 2015 found a diversity of carnivores, including honey badgers but excluding lions, wild dogs, and side-striped jackals.[57] The survey carried out for Groom et al.'s 2018 UNDP/ZSL report found honey badgers to be present and also obtained camera trap evidence of side-striped jackals.[58] The honey badger images in camera traps showed they were mainly found in the north of the park; 90.24% of the photos were at night, showing a high level of nocturnal activity, possibly a result of human hunting and encroachment. Poaching is well organised in the NP, with poachers from different communities appearing to respect each other's hunting areas.[59] Braga-Pereira's study revealed that while most bushmeat snaring or hunting with guns was for meat, some was to provide body parts for medicine or spiritual uses. If bodies of honey badgers were obtained, the bones were burned, ground down to powder, and used as a cure for what local healers referred to as "general weakness" – because of the honey badger's reputation for strength and the density of its bones.[60]

In Botswana, honey badgers are widespread from the arid southern and central Kalahari to the lusher Okavango Delta and Chobe River. They seem less nocturnal, and I have had repeated sightings in daylight, often just before dusk, in the Central Kalahari Game Reserve (where I twice observed the badger–jackal–goshawk interactions I detailed in the Introduction), Linyanti, Savuti, and the Tuli region. In the semi-arid south-west of the country, a wildlife survey found a regular honey badger presence, even in areas with human activity (mainly pastoralism). Six honey badgers were identified in the Ma\tsheng region during daylight hours driving along transects.[61] A survey in the eastern Moremi Game Reserve and adjacent wildlife management areas in the Okavango Delta region found resident honey badgers, though the researchers noted that they appeared on camera traps far more often in the dry than in the wet season, suggesting less need to forage over greater distances when rain meant more prey availability.[62] During a survey of bushmeat hunting in the Okavango, Rogan et al. reported that poachers believed honey badgers were in decline in the Delta and were seen or trapped less often.[63] In Chobe, honey badgers are clearly still present and exhibit great curiosity when confronted with something unusual. The Trans-Kalahari Predator Programme has camera traps in Chobe and caught an image of two badgers closely examining one of the cameras.

The Botswana Predator Conservation Trust (BPCT) also uses camera traps to survey carnivore presence at Santiwani on the north-eastern fringes of the Okavango. On 17 May 2017, one of their traps, which often records honey badgers, videoed ten honey badgers in a group moving along a track – there was no obvious indication of why so many were together, though males seeking out a female in season is a possible explanation.[64] The most honey badgers seen at one time is 12 – foraging around a cattle boma in Kenya searching for dung beetle larvae.[65] In March 2021, BPCT cameras recorded seven badgers travelling together in single file along a track.[66] The BPCT's Dr Peter Apps told me that the Santiwani area appears to be "the honey badger capital of the world."[67]

In Mozambique, conservation and wildlife protection, as in Angola, were rendered all but impossible by decades of liberation and then civil war. But the regeneration of protected areas like Gorongosa NP has enabled surveys to take place for the first time in decades. In Gorongosa, camera traps revealed the presence in several parts of the park of honey badgers, particularly in forested areas.[68] Colleen Begg, the leading honey badger specialist, is the managing director of the Niassa Carnivore Project in northern Mozambique. She posted photos on Facebook page of the project assisting the local community's beekeepers to construct honey badger–proof hives – they need to be more than 1.2m above ground, in a cage, and with metal sheets to prevent badgers climbing on to the hive. The project encourages beekeeping not only to bring in income but also to establish bee barriers to discourage elephant incursions into farmlands or villages.[69]

Namibia has a substantial, widespread, and low-density honey badger population, including in parts of the Namib Desert and the arid regions of Damaraland and the western Kalahari. There appears to have been little change in their distribution since Shortridge's surveys in 1923 and the early 1930s.[70] He believed them to spread across the whole territory, but because of their mainly nocturnal habits, they were rarely seen and numbers hard to assess.[71] In their study of Namibian carnivores, Edwards et al. looked at the carnivores on two cattle and game farms near the Tsau//Khaeb (Sperrgebiet) NP in the Aus region in southwest Namibia, a mix of Namib Desert and Nama Karoo biomes, with mountains and grassy plains, few trees, and only low shrubs.[72] Springbok and gemsbok were the main game animals on one farm, while there were cattle, some springbok, and gemsbok on the other. Fencing on the farms was intended to stop ungulate movement but was not predator-proof and allowed movement of carnivores like honey badgers. Surveys were carried out using camera traps. A total of 11 carnivore species were detected on 4,507 camera traps at night. Black-backed jackal were the most numerous, with 4,482 images captured, while there were 46 honey badger images.[73] The badgers were seen mainly between early evening and midnight and around sunrise. In the area occupied by the Ju/'hoansi San in the Omaheke region of east-central Namibia, much natural prey had gone due to shooting by white farmers over decades and because of the erection of fences, but both honey badgers and black-backed jackals had proved adaptable in moving to anthropogenic sources of food, foraging for live poultry, animal remains, and other human waste around human settlements, and were regularly detected.[74]

South Africa – beekeepers battle badgers

Across South Africa and in eSwatini lowveld regions, honey badgers are widely distributed but are, as already noted, absent from the Free State and also from Lesotho. Precise distribution across South Africa is not known, and in some areas, the honey badger may be locally extinct because of habitat loss, hunting, or because habitat is unsuitable, as in the Free State, being open, steppe-like terrain with little cover and poor foraging opportunities.[75] There are no definitive figures for the overall population in South Africa. Begg et al. estimate that it could be in the range of 6,000–20,000 individuals, including 3,600–13,200 mature animals.[76] Low density of populations means that they are vulnerable in some areas, especially where they are directly persecuted or where they may be killed in traps or by poison intended for black-backed jackals and caracals. In North West Province, there appears to have been range expansion by badgers on to highveld grasslands.[77] Farmers in dry areas of the North West Province have reported a significant increase in honey badger presence, increasing from occurrence in 8% of the area in the 1970s to 40% in 2012, perhaps a result of bush encroachment in grassland areas, which has provided cover and foraging opportunities.[78] Begg et al. also reported signs of increased presence in Gauteng, while in the Eastern Cape, numbers have remained low.[79] The largest population groups are in Mpumalanga, Limpopo, the Kalahari areas of Northern Cape, and the Western Cape's coastal lowlands. In the latter, badger presence is confirmed by the large number of beekeepers who admit to killing badgers, even though they have a protected status in the province.[80] Badgers were certified as present in five of Western Cape's 16 protected reserves and but were rare.[81] This is not entirely surprising as historical records from the Cape indicate that they had been treated as vermin and a bounty paid when they were killed. Over a 23-year period, dates not given, 744 badgers were killed in 27 districts of the Cape.[82]

The Northern Cape seems to have the highest concentrations of honey badgers. Surveys in the 1970s and early 1980s cited by Colleen Begg, in which 44,000 South African landowners were questioned, revealed that badgers were common and widely distributed in all 91 divisional districts of the then Cape province (now divided into Northern Cape, Western Cape, and Eastern Cape), with the greatest densities of population in Namaqualand, Calvinia, and Williston.[83] They were also found in large numbers along the southern coast near Knysna and Uniondale. Lower densities were found on the west coast north of Cape Town – in the Piketberg, Vredenburg, Hopefield, and Malmesbury districts, where badgers were viewed as problem animals by commercial beekeepers.[84] A survey in the Eastern Cape in the mid-1980s found them to be present, but their status was "potentially critical," and they were killed using traps or by hunting with dogs.[85] Between 1977 and 1986, the Department of Nature Conservation's Problem Animal Control in South Africa between 1977 and 1986 used dogs, coyote-getters, and traps to kill 47 problem badgers.[86] The attempt by farmers to deal with the badger problem is not entirely surprising, though extermination rather than protecting hives is not an ideal approach, given that in the Western Cape and Mpumalanga, estimated badger damage to the beekeepers is evaluated at R500,000pa.[87] The South African beekeeping industry contributed an estimated R3.2 billion to South Africa's agricultural economy through pollination alone before 2016, with an additional R100m through honey and bee products, creating direct employment in the sector for about 3,000 people and indirectly for 300,000–500,000 people.[88] It is said that badgers can destroy more than 20 hives in a night.[89] Where persecution has reduced numbers, their survival is threatened as "they have a slow recolonisation rate and currently only a small percentage of South African nature reserves are large enough to sustain viable subpopulations of these animals, leaving the larger part of South Africa's Honey Badger population unprotected."[90]

Persecution by beekeepers is by far the greatest threat to badgers, especially where populations are fragmented and there is no recruitment into declining populations by dispersal from other populations and given the low reproductive rate. Colleen Begg's research at the turn of the last century showed a serious problem, particularly in the Western Cape, where she found that a large majority of 82 beekeepers surveyed, 50 of whom had over 24,000 hives, had problems with badgers. Of them, 82% said they had problems of hive raiding and 50% of the beekeepers admitted killing badgers, even though they knew they were a protected species. About 78% had taken protective measures to stop badgers raiding hives. An estimated 231 badgers were killed in the area studied by Begg in 15 years before 2001[91] – while this is a fraction of the jackals killed by farmers in that period, it is likely to have had a greater population effect because of the low density and slow breeding rate of badgers. Nearly a quarter of the beekeepers kept killing badgers, despite the high cost of doing so, even when cost-effective deterrents were available.[92] The availability of traps and the anger of beekeepers at losses have ensured that 100–200 badgers were being killed annually in the Western Cape, demonstrating, as Begg wrote, that "beekeepers are a significant threat to the conservation of honey badgers (particularly in the Western Cape) and badgers are being needlessly persecuted in a most inhumane way."[93] In the Red List study, Begg and her co-authors said that hive damage could be reduced from 24 to 1% with the help of hive-protection methods; for example, by securing beehives 1m or more above the ground on a stand, by minimising conflicts. It was economically more viable for beekeepers to be "Badger Friendly" rather than expending resources and time trying to kill them.[94] But many beekeepers prefer to take the problem animal/vermin approach rather than using non-lethal methods.[95] They believed that there were too many badgers in the Western Cape and that killing was better than these areas and questioned how they could constitute such a problem.[96]

In 2002, the Badger–Beekeeper Extension Programme was established to educate beekeepers about hive-protection measures and badger conservation, and to increase public awareness of the issue.[97] This included having a Badger Friendly label (BFL) on honey produced through the programme – encouraging the production of honey using non-lethal methods to deter badger raids. Those taking part in the programme had fewer instances of damage – one assessment recorded 46 damaged hives in 2009 compared with 179 in a 2001 report.[98] Begg et al. reported optimistically in 2016 that "the problem of Honey Badgers raiding beehives is decreasing in intensity or, at worst, staying constant, with no beekeepers mentioning that the conflict was increasing in intensity or frequency of hive damage."[99] On the downside, the consumer awareness of the BFL programme was low with only 2.8% of consumers saying in 2009 that they would prioritise buying honey with a BFL and the number of stickers sold to retailers declining between 2005 and 2008.[100] Carter et al. concluded in 2017 that success had been achieved through the Badger Friendly campaign, but more work needed to be done evaluating the long-term effects on honey badger conservation.[101]

Another anthropogenic cause of honey badger mortality, though one which is hardly to evaluate, is the use of body parts, such as skin, claws, body fat, and internal organs, in traditional medicine. Williams and Whiting demonstrated that animal body parts are "important ingredients in the preparation of curative, protective and preventative medicines for such purposes as immunity from disease, protection against bad luck and witches, aphrodisiacs and potency."[102] Badger body parts are sought because of their reputation for strength, courage, and survivability. In KwaZulu-Natal, badger body parts are used as charms to supposedly increase the strength of hunting dogs.[103] Surprisingly, surveys of traditional medicine markets in the Western Cape and Gauteng came up with the use of body parts of black-backed jackals but not so much honey badgers.[104] Nieman et al. found that among Xhosa inhabitants of the Western Cape, 47.1% of people interviewed said that they knew of or had used honey badger body parts for medical or protective reasons.[105] Whiting, Williams, and Hibbitts said at the Faraday market in Johannesburg, only one of 32 traditional medicine dealers had badger parts for sale.[106]

Middle East and West Asia

Precise numbers and ranges of honey badgers in the Middle East/West Asia are hard to calculate. Masseti, for example, noted that while honey badgers were present in neighbouring countries, their presence had not been confirmed in Syria. He wrote that

> [t]his is not to say, however, that the species is to be considered as completely absent from the country, otherwise it would be impossible to explain its fairly well-documented occurrence in neighbouring Jordan and Israel. In this case too, it cannot be ruled out that the absence of specific sightings/reports for Syria could plausibly be attributed to the absence of targeted studies.

He also pointed out that "in the entire Near East this species is frequently persecuted, being poisoned and hunted by bee farmers because it of the damage it causes to beehives."[107]

In Arabia, Yemen, and the Gulf states, honey badgers are found along the coasts and on the fringes of the deserts. But wildlife is constantly under threat, and biodiversity in the whole of the Arabian peninsula is "and continuously affected by human pressures such as hunting and habitat destruction."[108] Khorozyan et al. found that badgers were present in eastern Yemen in the Hawf District of Al Mahrah Governorate which borders Oman, along with the rare Arabian leopard, caracals, wild cats, striped hyenas, and wolves. Honey badgers were caught on camera traps 57 times – 76.4% at night and across the 2010–12 survey period were widely distributed in Yemen and relatively common.[109] Their presence in Saudi Arabia was demonstrated by their predation of young houbara bustards being raised in fenced enclosures at the Mahazat as-Sayd Protected Area, Makkah (Mecca) Province, on the central Red Sea coast. The bustards, now rare in Arabia because of over-hunting, were being prepared for release into the protected area, but a badger or badgers got into the enclosure and killed 29 bustards. The fence was 2m high topped with barbed wire, and traps to catch predators were set up around the enclosure. Foxes (Rüppell's and red), sand cats, wild cats, feral cats, and badgers were often caught in the traps.[110] The deadly badger raid occurred on 8 December 2009, when a badger got into the enclosure and killed 6 of the 75 bustards there, with another 23 dying in the panic that ensued. The badger appeared to have climbed the wire fence and torn holes in nets enclosing the birds. Attempts to trap the badger or badgers failed, and twice a badger broke its way out of a trap. One badger was finally trapped in February 2010 and relocated.[111] Evidence of badgers attacking and killing foxes trapped around the enclosure was also found, with clear signs that the badgers had eaten part of the foxes attacked. Zafar-ul Islam et al., in their report on these events, said that the honey badgers were widespread in Saudi Arabia, but sightings were relatively rare, and that they were protected there under the 1977 National Hunting Law.[112]

Russia, Central Asia, and Iran

Most of the references of honey badgers in these regions date back to the Soviet period, during which they were found in areas of southern Russia bordering the Soviet Central Asian Republics on stony hill or mountain sides, in river valleys, and in sandy desert or steppe areas.[113] In their comprehensive survey of mammals of the Soviet Union, Heptner and Naumov said that the badgers were found in areas of southern Russia near the Caspian Sea and bordering the now-independent republics of Central Asia but that the distribution in Russia was not large and only constituted a small part of the badger range in the wider region.[114] Across the borders to the south, the honey badger is widely distributed across Kazakhstan, Turkmenia, and the Karakalpakia in Uzbekistan. It is found along the Amu Darya River which forms the northern border of Afghanistan with Turkmenistan, Uzbekistan, and Tajikistan. The IUCN Red List does not record it in Afghanistan, though its absence seems strange as it is found in the Central Asia republics to the north, Iran to the west, and Pakistan to the south and south-east, and the distribution of this species within the borders of our country (the former Soviet Union and its Central Asian republics) is more or less continuous.[115] It is found even in the Karakum Desert of Turkmenistan. Baryshnikov confirmed the badger presence in this Central Asian range in 2000.[116]

In Iran, the honey badger is distributed in southwestern, southern, central, and north-eastern parts of Iran within the provinces of Khuzestan, Fars, Kerman, Yazd, and Golestan.[117] A survey looking at written records from 1959 to 2012 and new records from the Iranian Department of the Environment from 2012 to 2014 noted the regional presence but also that poaching of the species had occurred in protected and unprotected areas, with its fat used in traditional medicine. No estimates of numbers were given.[118] Sharifi, Malekin, and Shahnaseri said in 2020 that honey badgers were among the rarest mammals in Iran, and their main strongholds were in the south, particularly the Zagros Mountains, the Khuzestan Plain, and the coast of the Persian Gulf.[119] They reported that in 2016, a badger carcass was found in Baluchistan, which was the first record there.

South Asia

In India, the honey badger range spreads across the majority of the country and into neighbouring countries to the west and the north. Able to adapt to most terrains apart from the highest mountains, it is mainly found in areas that are more arid, with hilly, stony areas, and in moist and dry forests – Menon says it is uncommon in most areas but is not under threat.[120] In some areas, like the protected Gir Forest, the honey badger is common and not under threat from human activity, because of the protection accorded to the region because of the Asian lion population. Nationally, the distribution is wide, but densities are low, and the honey badger is listed in Schedule I, Part I of the Indian Wildlife (Protection) Act, 1972.[121] But, as Pati, Vijayan, and Mehra noted, overall in India, sightings and knowledge of honey badgers are very limited, partly because there, the badger "is a secretive animal and scanty information is available on this species."[122] Even in the Gir Forest area Gujarat, home to the last surviving wild Asian lions, which is well protected and with few problems of human–wildlife conflict, the honey badger is rarely sighted and is chiefly nocturnal. Pati, Vijayan, and Mehra studied the badgers in the Gir region and found they liked riverine habitats for foraging but dry, deciduous forests for dens and temporary burrows.[123] It not only hunted small mammals, reptiles, and insects but also foraged for fruit and tubers. During a year's monitoring of the Gir area, the researchers saw only 13 badgers, 7 of them at night. Presence was also detected by searching out latrine sites and claw marks. In the riverine areas of Gir, "they preyed frequently on freshwater creatures such as crabs, frogs, molluscs, fish."[124] There was no evidence to support the stories that honey badgers dug up human graves and consumed corpses.

In Nepal, the honey badger is listed as endangered, and the Nepalese population may be as low as 100.[125] Where found, it inhabits the dense forests, grasslands, undulating terrain, and scrub forests of the Himalayan foothills, where wildlife is threatened by habitat loss due to logging, poaching, and persecution.

Endnote

That final detail from Nepal rather sums up the global situation of the honey badger. A tough, adaptable, and resilient mammal, it is nonetheless found in low densities, with a low reproduction rate and a limited opportunity for dispersal and recruitment to existing populations due to the patchy nature of its occurrence, even where it is quite common, as in parts of southern Africa. Where it is threatened, it is primarily a result of human activity in the form of persecution by beekeepers and poultry farmers, killing for bushmeat or body parts, killing as a by-product of trapping and poisoning of other carnivores or loss of habitat – being a species that is not often seen is treated with suspicion or ignored by local communities, it could disappear from some localities without anyone really noticing. Unlike jackals, which breed rapidly and can disperse more widely, the honey badger is hampered by the limitations of its reproductive system and breeding potential, and the progressive fragmentation of populations.

Notes

1 E. Do Linh San et al. (2016) *Mellivora capensis. The IUCN Red List of Threatened Species*, Mellivora capensis (Honey Badger) (iucnredlist.org) accessed 7 April 2022.
2 Ibid.
3 Ibid., Mellivora capensis (Honey Badger) (iucnredlist.org) accessed 7 April.
4 Endangered Wildlife Trust (2012) The red data book of the mammals of South Africa: A conservation assessment, Johannesburg: Endangered Wildlife Trust, https://www.nationalredlist.org/files/2012/11/red-data-book-mammals-south-africa-conservation-assessment.pdf accessed 7 April 2022.
5 Keith & Colleen Begg (no date) *Conservation Status, The Honey Badger*, The Honey Badger - Conservation accessed 7 April 2022.
6 Endangered Wildlife Trust (2016) *Mammal Red List*, Mammal Red List - The Endangered Wildlife Trust (ewt.org.za) accessed 7 April 2022.
7 CITES (no date) *How CITES works*, https://cites.org/eng/disc/how.php#:~-:text=Appendix%20III&text=A%20specimen%20of%20a%20CITES,port%20of%20entry%20or%20exit accessed 21 March 2022.
8 CITES (no date) *Mellivora Capensis*, https://speciesplus.net/species#/taxon_concepts/9613/legal accessed 21 March 2022.
9 CITES (2002) *Eighteenth meeting of the Animals Committee San José (Costa Rica), 8–12 April 2002 Implementation of Decision 11.165 on trade in traditional medicines: List of species traded for medicinal purposes*, p. 8.
10 San Diego Zoo (no date) Ratel/Honey Badger (Mellivora capensis) Fact Sheet: Population & Conservation Status, https://ielc.libguides.com/sdzg/factsheets/ratel/population accessed 19 October 2021.
11 Keith & Colleen Begg (no date).
12 A. Mourad, F. Seddiki & N. Haissoun (2020) Recent records of the Honey Badger Mellivora capensis (Schreber, 1776) in Algeria, *Small Carnivore Conservation*, 2020, 58, p. 1.
13 Ibid.
14 Ibid., p. 3.
15 S. Imad Cherkaoui & A. Bouajaj (2017) Recent records of the elusive Ratel Mellivora capensis (Schreber, 1776) in Morocco and case of human persecution, *Small Carnivore Conservation*, 55, 64–8, p. 64.
16 Ibid., p. 66.
17 Ibid.
18 Ibid., p. 68.
19 J.M. Padial & C. Ibáñez (2005) New records and comments for the Mauritanian mammal fauna. *Mammalia*, 69(2), 239–43, p. 239.
20 J.C. Brito et al. (2010) Data on the distribution of mammals from Mauritania, West Africa, *Mammalia*, 74, 449–55, p. 452.
21 Personal communications, 2021.
22 G.M. Carpanetto & F.P. Germi (1989) The mammals in the zoological culture of the Mbuti pygmies in north-eastern Zaire, *Hystrix*, 1–83, p. 22.
23 Team Africa Geographic (2020) Black honey badgers spotted in Gabon, *Africa Geographic*, 12 February 2020, https://africageographic.com/stories/black-honey-

badgers-spotted-in-gabon/?mc_cid=d44e8a0350&mc_eid=a920cc2294&fbclid=I-wAR0OXqqbz-QZ4BCW0SC2J6KezR_I99Y7S4y3sT5xM7Huc08oGLndyIUP-mAE accessed 30 June 2021.

24 Personal communication, 2021.

25 D. R. Rosevear (1974) *The Carnivores of West Africa*, London: Trustees of the British Museum (Natural History), p. 111.

26 Ibid., p. 126.

27 Ibid., p. 128.

28 A. Cole Burton et al. (2011) Evaluating persistence and its predictors in a West African carnivore community, *Biological Conservation*, 144, 2344–53, p. 2347.

29 D.A. Soewu (2013) Zootherapy and biodiversity conservation in Nigeria, in R.R.N. Alves & I.L. Rosa (eds) *Animals in Traditional Folk Medicine*, Berlin/Heidelberg: Springer-Verlag, 347–65, p. 354.

30 D.R. Rosevear (1953) *Checklist and Atlas of Nigerian mammals*, Lagos: Nigerian Government printer, p. 112 and Plate 174.

31 N. D'Cruze et al. (2020) Snake oil and pangolin scales: insights into wild animal use at "Marché des Fétiches" traditional medicine market, Togo, *Nature Conservation*, 39, 45–71, p. 52.

32 T. Abáigar, M. Cano & C. Ensenyat (2013) Habitat preference of reintroduced dorcas gazelles (Gazella dorcas neglecta) in North Ferlo, Senegal, *Journal of Arid Environments*, 97, 176–81, p. 177.

33 T. Künzel, H.A. Rayaleh & S. Künzel (2000) Status Assessment Survey on Wildlife in Djibouti Final Report December 2000. Munich: Zoological Society for the Conservation of Species and Populations (ZSCSP), p. 42.

34 D.W. Yalden, M.J. Largen & D. Kock (1980) Catalogue of the mammals of Ethiopia, *Monitore Zoologico Italiano. Supplemento*, 13, 1, 169–272, p. 174.

35 M.B. Nimir (1983) *Wildlife values and management in northern Sudan*, Dissertation submitted to for the Degree of Doctor of Philosophy Colorado State University Fort Collins, Colorado Summer 1983, p. 20.

36 S. Mansoiur & A. Osman Eljack (2003) *RAMSAR Information Sheet (RIS) for Dinder National Park, Sudan*, Khartoum: RMSAR, pp. 17–8.

37 Ibid., p. 18.

38 J. Kingdon (1977) *East African Mammals an Atlas of Evolution in Africa, Vol IIIA*, London: Academic Press, pp. 87–8.

39 Ibid., p. 89.

40 Saving the Wild (no date) A Very Hungry Badger, https://www.savingthewild.com/kimana/ accessed 7 April 2022.

41 Ibid.

42 Ibid.

43 Kingdon, 1977, p. 91.

44 K.W. Homewood & W.A. Rodgers (1991) *Maasailand Ecology Pastoralist Development and Wildlife Conservation in Ngorongoro*, Tanzania, Cambridge: Cambridge University Press, p. 66.

45 A. E. Laltaika, personal communication 7 June 2021.

46 N. Pettorelli et al. (2010) Carnivore biodiversity in Tanzania: revealing the distribution patterns of secretive mammals using camera traps, *Animal Conservation*, 13, 131–9, pp. 132 and 134.

47 D. de Luca & N.E. Mpunga (2013) Small carnivores of the Mt Rungwe–Kitulo landscape, southwest Tanzania: presence, distributions and threats, *Small Carnivore Conservation*, 48, 67–82, p. 67.

48 Ibid., pp. 74 and 76.

49 Ibid., pp. 76–7.

50 C. Fischer, R. Tagand & Y. Hausser (2013) Diversity and distribution of small carnivores in a miombo woodland within the Katavi region, Western Tanzania, *Small Carnivore Conservation*, 48, 60–66, p. 63.

51 T. Holmern & E. Røskaft (2013) The poultry thief: subsistence farmers' percep-tions of depredation outside the Serengeti National Park, Tanzania, *African Journal of Ecology*, https://onlinelibrary.wiley.com/doi/abs/10.1111/aje.12124 accessed 10 May 2021, 1–9, p. 2.

52 Ibid., p. 5.

53 K. Begg (2001) *Report on the Conflict between Beekeepers and Honey Badgers Mellivora Capensis, with Reference to their Conservation Status and Distribution in South Africa*, Johannesburg: Endangered Wildlife Trust, p. 8.

54 Ibid.

55 F. Braga-Pereira (2017) First record of Angola's medicinal animals: A case study on the use of mammals in local medicine in Quiçama National Park, *Indian Journal of Traditional Knowledge*, 16, 4, 588–92, p. 588.

56 R. Groom et al. (2018) *Quiçama National Park Angola. A Large and Medium Sized Mammals Survey*, https://erc.undp.org/evaluation/managementresponses/keyaction/documents/download/1002 accessed 11 April 2022, pp. 16–7.

57 Ibid., p. 26.

58 Ibid., pp. 50–1.

59 Ibid., p. 120.

60 Braga-Pereira, 2017, p. 589.

61 M. Wallgren (2009) Influence of land use on the abundance of wildlife and livestock in the Kalahari, Botswana, *Journal of Arid Environments*, 73, 314–21, p. 315.

62 L.N. Rich et al. (2017) Carnivore distributions in Botswana are shaped by resource availability and intraguild species, *Journal of Zoology*, 303, 90–8, p. 92.

63 M.S. Rogan, P. Lindsey & J.W. McNutt (2015) *Illegal Bushmeat Hunting in the Oka-vango Delta, Botswana: Drivers, Impacts and Potential Solutions*, Harare: FAO/Panthera/Botswana Predator Conservation Trust, p. 30.

64 P. Apps (2017) *Botswana Predator Conservation Trust Camera Catches World Record Group of Honey Badgers*, Botswana Predator Conservation Trust, https://www.youtube.com/watch?v=ndaLDg69we8. 27 June 2017 accessed 20 November 2020.

65 Ibid.

66 Botswana Predator Conservation Programme Facebook - www.facebook.com/10014285681767/videos/1051814921971354 - accessed 23 March 2022.

67 Personal communication, 23 March 2022.

68 T. Easter, P. Bouley & N. Carter (2019) Opportunities for biodiversity conservation outside of Gorongosa National Park, Mozambique: A multispecies approach, *Biolog-ical Conservation*, 232, 217–27, p. 220.

69 Biassa Carnivore Project (no date) Alternative Livelihoods, Alternative Livelihoods - Niassa (niassalion.org) accessed 11 April 2022.

70 Captain G.C. Shortridge (1934) *The Mammals of South West Africa*, London: William Heinemann, p. 194.

71 On Mammals Collected in 1923 by Capt. G.C. Shortridge during the Percy Sladen and Kaffrarian Museum Expedition to South-West Africa, as reported to ZSL by Oldfield Thomas, *Proceedings*, 1925, 95, 1, p. 231.

72 S. Edwards, A.C. Gange & I. Wiesel (2015) Spatiotemporal resource partitioning of water sources by African carnivores on Namibian commercial farmlands, *Journal of Zoology*, 297, 22–31, p. 23.

73 Ibid., p. 24.

74 J. Suzman (2017) *Abundance without Affluence the Disappearing World of the Bushmen*, New York: Bloomsbury, p. 148.

75 C.M. Begg et al. (2016) A conservation assessment of Mellivora capensis, in M.F. Child et al. (eds) *The Red List of Mammals of South Africa, Swaziland and Lesotho*. Johannesburg: South African National Biodiversity Institute and Endangered Wild-life Trust, 27.-Honey-Badger-Mellivora-capensis_LC.pdf (ewt.org.za) accessed 7 April 2022, pp. 2–3.

76 Ibid., p. 3.

77 Ibid.
78 Ibid.
79 Ibid.
80 Ibid., citing Begg's PhD thesis of 2001.
81 Begg, 2001, p. 13.
82 Ibid.
83 Begg, 2001, p. 14.
84 Ibid.
85 Cited by beg, 2001, p. 14.
86 Ibid., p. 15.
87 Begg et al., 2016, pp. 5–6.
88 Ibid., p. 6.
89 Keith & C. Begg (2002) The conflict between beekeepers and honey badgers in South Africa: a western cape perspective, *Open Country*, 4, 25–37, p. 28.
90 Ibid., p. 6.
91 Begg, 2001, p. 30.
92 Ibid.
93 Ibid., p. 4.
94 Begg et al., 2016, p. 6.
95 Beg, 2001, p. 4.
96 Ibid.
97 J. Isham, K.S. Begg & C.M. Begg (2005) *Honey Badger and Beekeeper Extension Programme. Final Report.* Midrand, Gauteng: Carnivore Conservation Group, Endangered Wildlife Trust.
98 Begg et al., 2016, p. 7.
99 Ibid.
100 Ibid., p. 8.
101 S. Carter et al. (2017) The honey badger in South Africa: biology and conservation, *International Journal of Avian and Wildlife Biology*, 2, 2, 55–8, p. 55.
102 V.L. Williams & M.J. Whiting (2016) A picture of health? Animal use and the Faraday traditional medicine market, South Africa, *Journal of Ethnopharmacology*, 179, 265–73, p. 265.
103 G. Proulx et al. (2016) World distribution and status of badgers — a review, in G. Proulx & E. Do Linh San (eds) *Badgers: Systematics, Biology, Conservation and Research Techniques*, Sherwood Park, Alberta, Canada: Alpha Wildlife Publications, 31–116, pp. 64–5.
104 W.A. Nieman et al. (2019) Traditional medicinal animal use by Xhosa and Sotho communities in the Western Cape Province, South Africa, *Journal of Ethnobiology and Ethnomedicine*, 34, 1–4, p. 2; and, M.J. Whiting, V.L. Williams & T.J. Hibbitts (2011) Animals traded for traditional medicine at the Faraday market in South Africa: species diversity and conservation implications, *Journal of Zoology*, 284, 84–96, p. 84.
105 Nieman et al., 2019, p. 2.
106 Whiting, Williams & Hibbitts, 2011, p. 88.
107 M. Masseti (2009) Carnivores of Syria, in E. Neubert et al. (eds) *Animal Biodiversity in the Middle East*, Proceedings of the First Middle Eastern Biodiversity Congress, Aqaba, Jordan, 20–23 October 2008. ZooKeys 31, 229–52, p. 246.
108 I. Khorozyan et al. (2014) Patterns of co-existence between humans and mammals in Yemen: some species thrive while others are nearly extinct, *Biodiversity Conservation*, 23, 1995–2013, p. 1996.
109 Ibid., pp. 2004–5.
110 M. Zafar-ul Islam et al. (2011) An attack by Ratel Mellivora capensis on pre-release Asian Houbara Bustards Chlamydotis macqueenii in central Saudi Arabia, *Small Carnivore Conservation*, 44, 35–7, p. 35.
111 Ibid., p. 36.
112 Ibid.

113 C.A. Long & C.A. Killingley (1983) *The Badgers of the World*, Springfield, IL: Charles C. Thomas Publisher, pp. 358–9.

114 V.G. Heptner & N.P. Naumov (1967) *Mammals of the Soviet Union, Volume II, Part 1b*, Moscow; Vysshaya Shkola Publishers, Mammals of the Soviet Union (archive.org) accessed 28 June 2021, pp. 11218–221.

115 Ibid.

116 G. Baryshnikov (2000) A new subspecies of the honey badger Mellivora capensis from Central Asia, *Acta Theriologica*, 45, 1, 45–55, p. 45.

117 A.T. Qashqaei, P. Joslin & P. Dibadj (2015) Distribution and conservation status of Honey Badgers Mellivora capensis in Iran, *Small Carnivore Conservation*, 52/52, 101–7, p. 104.

118 Ibid.

119 H. Sharifi, M. Malekian & G. Shahnaseri (2020) Habitat selection of honey badgers: are they at the risk of an ecological trap? *Hystrix, the Italian Journal of Mammalogy*, 7, 31, 131–6, p. 132.

120 V. Menon (2014) *Indian Mammals a Field Guide*, Gurugram, India: Hachette India, p. 296.

121 Keith & Colleen Begg, no date.

122 B.P. Pati, S. Vijayan & B.S. Mehra (2001) Observations on the food habits and distribution of ratel (Mellivora capensis indica) in Gir, India, *Indian Forester*, 127, 1143–47, pp. 1145–6.

123 Ibid.

124 Ibid., p. 1145.

125 R. Jnawali et al. (2011) *The Status of Nepal's Mammals: The National Red List Series*, Kathmandu, Nepal: Department of National Parks and Wildlife Conservation, p. 68.

INDEX